Springer Series in *Materials Science*

Advisors: M. S. Dresselhaus · K. A. Müller
Editors: U. Gonser · A. Mooradian · R. M. Osgood · M. B. Panish · H. Sakaki
Managing Editor: H. K. V. Lotsch

B. Raveau   C. Michel
M. Hervieu   D. Groult

Crystal Chemistry
of High-T$_c$ Superconducting
Copper Oxides

With 320 Figures

Springer-Verlag

Berlin Heidelberg New York
London Paris Tokyo
Hong Kong Barcelona
Budapest

Professor Dr. Bernard Raveau
Professor Dr. Claude Michel
Professor Dr. Maryvonne Hervieu
Professor Dr. Daniel Groult
Lab. de Cristallographie et Sciences des Matériaux
Boulevard du Maréchal Juin, F-14050 Caen Cedex, France

Series Editors:

Prof. *R. M. Osgood*

Microelectronics Science Laboratory
Department of Electrical Engineering
Columbia University
Seeley W. Mudd Building
New York, NY 10027, USA

Prof. Dr. *U. Gonser*

Fachbereich 12/1
Werkstoffwissenschaften
Universität des Saarlandes
W-6600 Saarbrücken, Fed. Rep. of Germany

M. B. Panish, Ph. D.

AT&T Bell Laboratories
600 Mountain Avenue
Murray Hill, NJ 07974, USA

Prof. *H. Sakaki*

A. Mooradian, Ph. D.

Leader of the Quantum Electronics Group, MIT
Lincoln Laboratory, P. O. Box 73
Lexington, MA 02173, USA

Institute of Industrial Science
University of Tokyo
7-22-1 Roppongi Minato-ku
Tokyo 106, Japan

Managing Editor: Dr. Helmut K. V. Lotsch

Springer-Verlag, Tiergartenstrasse 17
W-6900 Heidelberg, Fed. Rep. of Germany

ISBN-13:978-3-642-83894-1   e-ISBN-13:978-3-642-83892-7
DOI: 10.1007/978-3-642-83892-7

Library of Congress Cataloging-in-Publication Data. Crystal chemistry of high T_c superconducting copper oxides / B. Raveau ... [et al.]. p. cm. – (Springer series in materials science; v. 15) Includes bibliographical references and index. ISBN-13:978-3-642-83894-1   1. High temperature superconductors. 2. Copper oxide superconductors. I. Raveau, B. (Bernard), 1940 –. II. Series. QC611.98.H54C79   1991   537.6'231-dc20   91-15165

© Springer-Verlag Berlin Heidelberg 1991
Softcover reprint of the hardcover 1st edition 1991

The use of registered names, rademark , etc. in this publication does not imply, even in the absence of a specific statement, that such names are exempt from the relevant protective laws and regulations and therefore free for general use.

This text was prepared using the PS™ Technical Word Processor and printed by a Hewlett-Packard Laser Jet III

54/3140-543210 – Printed on acid-free paper

Preface

The recent discovery of high-temperature superconductivity in copper-based oxides is an event of major importance not only with respect to the physical phenomenon itself but also because it definitely shows that solid state chemistry, and especially the crystal chemistry of oxides, has a crucial place in the synthesis and understanding of new materials for future applications. The numerous papers published in the field of high T_c superconductors in the last five years demonstrate that the great complexity of these materials necessitates a close collaboration between physicists and solid state chemists.

This book is based to a large extent on our experience of the crystal chemistry of copper oxides, which we have been studying in the laboratory for more than twelve years, but it also summarizes the main results which have been obtained for these compounds in the last five years relating to their spectacular superconducting properties. We have focused on the structure, chemical bonding and nonstoichiometry of these materials, bearing in mind that redox reactions are the key to the optimization of their superconducting properties, owing to the importance of the mixed valence of copper and its Jahn-Teller effect. We have also drawn on studies of extended defects by high-resolution electron microscopy and on their creation by irradiation effects.

Our objective is to reach a very broad audience: students and teachers, physicists and solid state chemists, whether directly or only indirectly involved in the field of superconductivity. Nonspecialists may skip over many details but will find here the main structural types and the main features which govern the nonstoichiometry of copper oxides, and in this way this book will be useful to students and also to "pure" physicists who are more concerned with the physics of superconductivity. Although far from comprehensive, the lists of references will allow solid state chemists working in this field to find basic tools for their investigations.

Such a book could be written not only for copper oxides but also for many other complex oxides, which have already been synthesized and for which one day suprising physical properties will probably be discovered. For these reasons, we would like to recognize here the tremendous pioneering work which has been done before us by solid state chemists and crystallographers in the field of nonstoichiometric oxides.

The first draft of this book was written during August 1988. We are grateful to our families for the sacrifices they made during those holidays and also at other times before and after the discovery of high-temperature superconductivity.

B. Raveau
C. Michel
M. Hervieu
D. Groult

Caen
June 1991

Contents

1. Introduction:
Superconductivity in Oxides Before 1986

First discovered at the beginning of this century, superconductivity was observed initially in metals and binary alloys, then the focus shifted to the study of chalcogenides and, more recently, to organic compounds. Up until 1986, superconductivity was limited to low temperatures and, in particular, it was linked to the use of liquid helium: the highest critical temperature that had been observed was 23.3 K for Nb_3Ge. In spite of this handicap, superconducting materials played a role in electrotechnical applications, especially in the construction of magnets for high fields. These possibilities were then reinforced with the discovery in 1970 of superconducting ternary chalcogenides, which were studied particularly because of their high critical fields [1.1]. Important progress had been made in recent years in fundamental understanding of solid-state physics, not only dealing with the interactions between superconductivity and paramagnetism but also in the field of organic superconductors [1.2-5]. However, an increase in the critical temperature of about 10 K, or even of only a few degrees, appeared unlikely, in spite of the observations made by *Little* in organic superconductors [1.6]. The discovery in 1986 by *Bednorz* and *Müller* [1.7] of superconductivity in copper oxides close to 35 K necessitated drastic revision of previously accepted ideas. Two years after this event, in 1988, critical temperatures as high as 125 K had been reached. The possibility of using liquid nitrogen as the coolant has now been demonstrated. A huge domain has been opened up for investigation by physicists and solid-state chemists, which should produce important technological developments in coming years in both electrotechnology and electronics. Nevertheless, this spectacular phenomenon is so far not understood, owing to the great complexity of the materials. Thus, the aim of this reference book is to shed some light on the crystal chemistry of these copper oxides, which were, in fact, noted for their metal-like properties as early as 1979 [1.8].

Mixed-valence copper oxides belong to a large family of metal-like oxides called oxygen bronzes by crystal chemists. Famous examples are the oxygen tungsten bronzes Na_xWO_3, isolated for the first time by *Magnéli* and *Hägg* [1.9, 10] in 1935, which opened the door to a promising class of materials. Curiously, the physicists working in the area of superconductivity did not seem to be attracted by the investigation of such metal-like materials, perhaps owing to the lack of interaction between physics and solid-state chemistry.

Among the different oxides that are characterized by a metallic or a semi-metallic behavior, three classes had been found by 1986 to exhibit su-

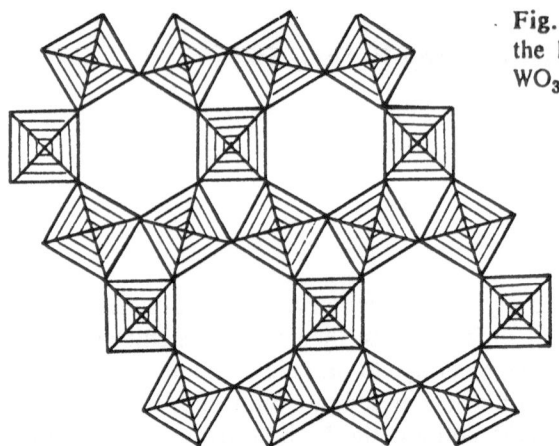

Fig.1.1. Projection of the structure of the hexagonal tungsten bronzes $Rb_x \cdot WO_3$ along the c axis

perconductivity: the hexagonal tungsten bronzes $A_x WO_3$ (A = Rb, Cs), the perovskites $BaPb_{1-x}Bi_xO_3$ and the spinels $Li_{1+x}Ti_{2-x}O_4$.

The hexagonal tungsten bronzes $A_x WO_3$ (A = Rb, Cs) were discovered by *Magnéli* and *Blomberg* [1.11] in the case of the rubidium oxides $Rb_x \cdot WO_3$.[1] The WO_3 host lattice of this bronze is built up of corner-sharing WO_6 octahedra (Fig.1.1), forming wide hexagonal tunnels where the univalent cations are located. The hexagonal cell ($a \simeq 7.4$Å, $c \simeq 7.6$Å) of this oxide contains six WO_3 units and two rubidium ions, leading to the limiting formula $Rb_2W_6O_{18}$. The homogeneity range of this oxide, $0.20 \le x \le 0.33$, which corresponds to a random distribution of the rubidium ions and cationic vacancies in the hexagonal tunnels, induces a variation of the mean valence of tungsten. The mixed valence of tungsten in these oxides, intermediate between V (d^1 configuration) and VI (d^0), is of course formal, as shown by the electrical and magnetic properties, which clearly show the absence of W^{5+} and W^{6+} ions. These oxides do indeed exhibit a metallic conductivity and a paramagnetism independent of the temperature, characteristic of a metal. This metallic conductivity has been explained by *Goodenough* [1.12] in terms of a band model (Fig.1.2) resulting from the overlapping of the d orbitals of tungsten and the p orbitals of oxygen, leading to a π_p band belonging to the valence band and a narrow π^* conduction band. In this model the electrons are delocalized over the WO_3 framework whereas the univalent ions are completely ionized.

The superconducting properties of these bronzes have been demonstrated by *Stanley* et al. [1.13] for $Rb_x WO_3$ and by *Skokan* et al. [1.14] for $Cs_x WO_3$. The critical temperatures of these phases range from less than 1 to 6.7 K. It is remarkable that the critical temperature of $Cs_x WO_3$ decreases as x increases, and tends to disappear for the limiting bronze $Cs_{0.33} WO_3$ (Fig. 1.3), i.e. as the W(V) content increases. A rather similar phenomenon is also observed in the case of the rubidium bronzes. However, in that case an

---

[1] A multiplication dot is used in the text where it has proved necessary to split a relatively long chemical formula for typographical reasons.

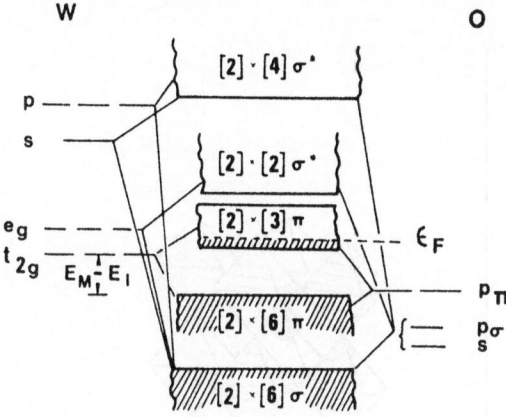

W   O   Fig.1.2. Energy diagram for $A_x \cdot$ WO_3 bronzes according to *Goodenough* [1.12]

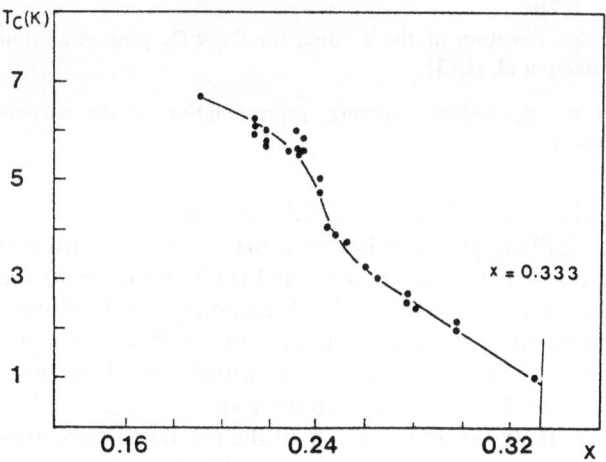

Fig.1.3. Critical temperature versus x value for $0.19 \leq x \leq 0.33$ $Cs_x WO_3$ samples according to *Skokan* et al. [1.14]

anomaly is observed for the composition corresponding to x = 0.24, which exhibits a very low critical temperature ($T_c \simeq 1 K$). The magnitudes of the Seebeck coefficients observed for these oxides (Fig.1.4) clearly show that the carriers are electrons and that their evolution is typical of metals. The ac susceptibility studies, which lead to a fractional volume ranging from 70% to 90% for $Cs_x WO_3$, confirm the existence of a bulk superconductivity. The superconducting properties of these bronzes were interpreted in terms of the electron-phonon interaction. Nevertheless, it does not appear that a detailed study of the band structure for these oxides has been carried out.

The superconductor $(BaPb_{1-x}Bi_x)O_3$ was discovered by *Sleight* et al. [1.15] in 1975. This oxide exhibits the perovskite structure (Fig.1.5): its host lattice is built up of corner-sharing BiO_6 and PbO_6 octahedra forming cages bounded by twelve oxygen atoms where the barium ions are located. However, it differs from the classical cubic perovskite by its orthorhombic

3

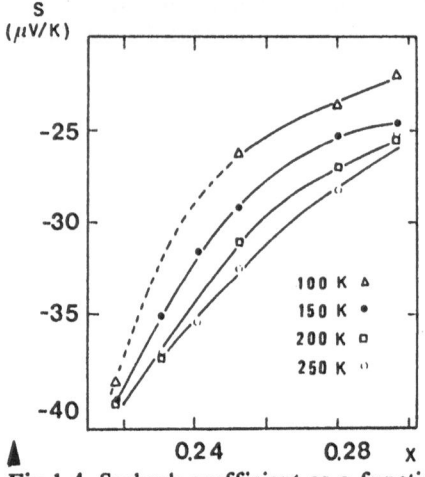

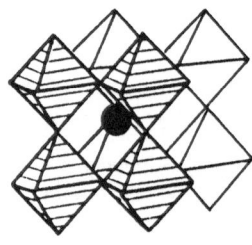

Fig.1.4. Seebeck coefficient as a function of the x value for $Cs_x WO_3$ plotted at four temperatures according to *Skokan* et al. [1.13]

Fig.1.5. Perspective view of the perovskite structure: representation of the corner-sharing BiO_6 and PbO_6 octahedra

symmetry: an orthorhombic cell with a = 6.024Å, b = 6.065Å and c = 8.056Å was observed for $BaPbO_3$ [1.16], whereas a primitive cell with a = 4.343 Å, b = 4.358 Å and c = 4.333 Å was determined for $BaBiO_3$, with the possibility of the existence of a larger true cell. A complete solid solution $Ba(Pb_{1-x}Bi_x)O_3$ of these two oxides can be synthesized for $0 \le x \le 1$, by heating appropriate mixtures of oxides, carbonates or nitrates in air at temperatures ranging from 800° to 1000°C. Superconductivity was observed in this system from x = 0.05 to 0.30, while for x > 0.30 the electrical measurement indicated a semiconducting behavior. Meissner-effect measurements showed that the critical temperature increases from 9 K at x = 0.05 to 13 K at x = 0.3. The sharp resistive transition observed at 11 K for the $BaPb_{0.8} \cdot Bi_{0.2}O_3$ crystal (Fig.1.6) confirms the magnetic measurements and shows that the temperature dependence is metal-like in the range 11-298 K. The origin of superconductivity in this oxide has not been clearly established. A model involving the interaction of the lead 6s band and the oxygen 2p band was proposed by *Sleight* on the basis of the highly covalent Pb-O bonding. Nevertheless, he drew attention to the fact that the complete mixing of the Bi 6s state into the Pb 6s band is not at all certain. Thus the situation in these compounds appears to be very different from that in the tungsten bronzes, since the cubic perovskite-type bronzes $Na_x WO_3$, which are metal-like, do not exhibit any superconducting properties. However, it must be pointed out that a common feature of the bronzes is the mixed valence of bismuth Bi(V)/Bi(III), and that a possibly dynamic exchange, corresponding to the equilibrium Bi(III)+Pb(IV) $\leftrightarrows$ Bi(V)+Pb(II) should perhaps be taken into consideration for the explanation of the superconductivity of these oxides.

4

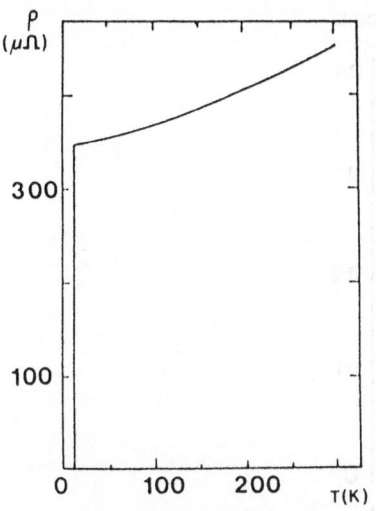

Fig.1.6. Electrical resistivity versus temperature for a crystal of $BaPb_{0.8}Bi_{0.2}O_3$ according to *Sleight* et al. [1.15]

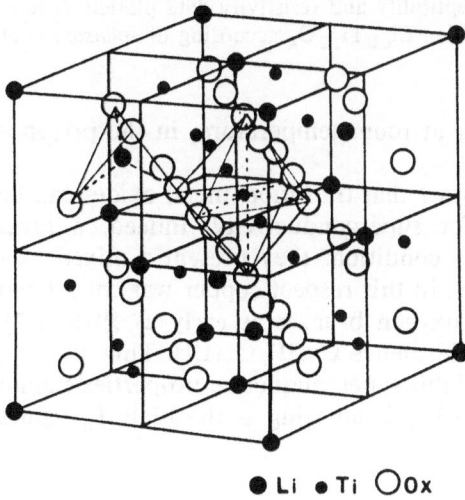

● Li ● Ti ○ ox

Fig.1.7. Perspective view of the spinel $LiTi_2O_4$

Superconductivity at high temperatures in the ternary Li-Ti-O system was detected by *Johnson* et al. [1.17] in 1973. The phase responsible for superconductivity is the spinel $Li_{1+x}Ti_{2-x}O_4$, which was prepared and whose structure was established for the first time by *Deschanvres* et al. [1.18] in 1971. This oxide exhibits the normal spinel configuration with lithium in tetrahedral coordination while titanium occupies the octahedral sites (Fig. 1.7). The metal-like behavior of this compound results from the electronic delocalization over the Ti-O polyhedra due to the mixed valence Ti(III)/Ti(IV). The superconductivity data of this phase (Fig.1.8) show that the critical temperature is close to 12 K; the researchers concluded that there is a filamentary superconductivity due to the fact that the resistive transition is slightly above the magnetic-susceptibility transition temperature. Curiously, an extensive study of these materials has not been carried out, per-

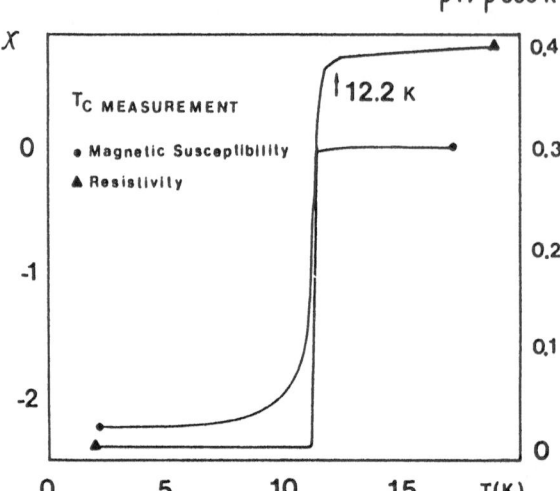

Fig.1.8. Low-temperature magnetic susceptibility and resistivity data plotted vs temperature for a sample of nominal composition $Li_{0.8}Ti_{2.2}O_4$ according to *Johnston* et al. [1.17]

haps owing to their instability in air at room temperature, in the presence of carbon dioxide and humidity.

From this brief overview it appears that transition-metal oxides can be considered as potential materials for superconductivity. Indeed, a great number of them offer one necessary condition: they present a mixed valence, leading to metal-like properties. In this respect copper was considered as a potential element for forming oxygen bronzes as early as 1979 [1.7], owing to the possibility of the mixed-valence Cu(II)/Cu(III). Thus the fact that mixed valence copper oxides exhibit superconducting properties cannot be considered very surprising, but what is amazing is the high T_c values that can be reached in these oxides.

2. Phases of the Systems A-La-Ca-Cu-O and A-Y-Ca-Cu-O (A = Ca, Sr, Ba): Structural Aspects

Consideration of the relationships between the crystal chemistry of the oxides and their electron transport properties suggests that the metal atoms M that contribute to the framework of MO_n polyhedra should exhibit a mixed valence in order to obtain metallic or semimetallic conductivity. This concept is of central importance in the search for new materials with such physical properties. It leads to the conclusion that the metal atom M should present at least two oxidation states simultaneously in the same compound, and thus should be a transition metal. This is indeed the case for oxides involving Ti(III)/Ti(IV) or W(VI)/W(V) or V(V)/V(IV), which are called oxygen bronzes. The case of copper is exceptional due to the fact that the two oxidation states present are Cu(II) and Cu(I), and that Cu(I) implies a localization of the electron on the atom, which would not lead to any metallic behavior. The possibility of generating copper oxides with metallic properties was considered in Caen as early as 1979 on the basis of a mixed valence Cu(II)/Cu(III). Nevertheless, such an idea appeared at that time difficult to implement easily, since the only Cu(III) oxide which was known was the perovskite $LaCuO_3$, which had been synthesized only under a high oxygen pressure of about 60 kbar in Bordeaux. Here we describe first the main features governing the chemistry of copper in the oxides that determined the choice of systems in the search for new phases with anisotropic metallic properties. Then the different structures which are known in those systems will be described, with particular attention being paid to the oxides that are related to the perovskite phase.

2.1 Copper Chemistry in Oxides: Oxidation States and Coordination

As previously stated, the generation of metallic properties in a MO_n framework requires an overlapping of the d orbitals of the metallic M atoms with the p orbitals of the oxygen atoms in order to form bands favourable to the delocalization of the electrons on the metal-oxygen framework. A second condition, which should allow the electron transport properties to be modified in a flexible way, concerns the possibility of oxygen nonstoichiometry that does not greatly change the structure. Such a characteristic, which allows a significant variation of the number of carriers, implies that the metallic element M is able to present several coordinations in the same structure. The transition elements of the first period of the periodic table are potential candidates: several oxidation states and several coordinations are possible for each of them. However, most of them are not favorable for the generation of new metallic or semimetallic conductors. For example, elements like iron or manganese, which exhibit several oxidation states and several coordinations, are in fact magnetic ions, so their electrons will not be delocalized over the metal-oxygen framework but will be confined close to the metallic center, leading to poor semiconducting or insulating properties. In this respect nickel and copper are interesting elements. The metallic properties of the perovskites $LaCuO_3$ and $LaNiO_3$ were interpreted by *Goodenough* et al. [2.1] in terms of a degenerate gas of spin polarons. Copper is very attractive owing to its great number of coordinations - octahedral, pyramidal, tetrahedral, square planar - and to its Jahn-Teller effect: $3d^9 Cu(II)$ is a classical example of distortion of a CuO_6 octahedron given by *Orgel* [2.2]. In an octahedral field two configurations can be distinguished concerning the splitting of the e_g orbitals of Cu(II). The first one, which is characterized by an elongated octahedron, corresponds to a stabilization of the d_{z^2} level with respect to $d_{x^2-y^2}$, i.e. it corresponds to the electronic configuration $(t_{2g})^6 (d_{z^2})^2 (d_{x^2-y^2})^1$ (Fig.2.1a), whereas the second one, which is given for a compressed octahedron, corresponds to a stabilization of the $d_{x^2-y^2}$ level, i.e. to the configuration $(t_{2g})^6 (d_{x^2-y^2})^2 \times (d_{z^2})^1$ (Fig.2.1b). Band models explaining electron transport properties of copper oxides can be established on the basis of similar considerations.

In order to build a copper-oxygen framework capable of high conductivity, the CuO_n polyhedra must share at least their corners. Taking into account the octahedral coordination for copper, which is the highest that can be obtained, it appears that the most oxidized framework that can be synthesized should be built up from corner-sharing CuO_6 octahedra. This

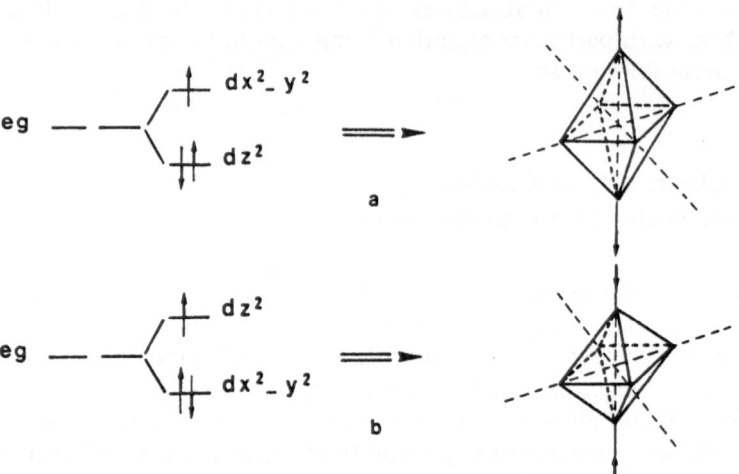

Fig.2.1. Jahn-Teller effect of $3d^9$-Cu(II) leading to (a) a stabilization of the d_{z^2} level, i.e. to an elongated octahedron, and (b) a stabilization of the $d_{x^2-y^2}$ level, i.e. to a compressed octahedron

would correspond to the formulation $[CuO_3]_\infty$. In order to realize such a host lattice, one should avoid introducing small metallic ions, i.e. transition and post-transition elements, as well as magnesium or lithium, which exhibit a size close to that of copper. Such elements would have a possibility of being located on the copper sites and would break the interactions Cu-O-Cu. Consequently, one has to introduce "invited" ions that are as large as possible and highly charged in order to obtain the $[CuO_3]_\infty$ framework. The rare-earth elements, and especially lanthanum, are the most suitable, owing to their size (ionic radii greater than 1 Å) and to their oxidation state (III). The synthesis of the metal-like perovskite $LaCuO_3$ [2.3] is in agreement with this point of view, but unfortunately necessitates high oxygen pressure.

It is well known from solid-state chemistry that trivalent copper is usually not stable at high temperatures under normal pressure conditions, and that even Cu(II) tends to be destabilized into Cu(I), as shown, for instance, for the decomposition of CuO into Cu_2O and oxygen in air above 1050°C. However, unusual oxidation states of elements can sometimes be stabilized in oxides by combinations involving cations that exhibit basic properties. This is the case for antimony and bismuth: Sb(V) can be stabilized by alkaline earths in $CaSb_2O_6$ and $SrSb_2O_6$, and similarly Bi(V) in $KBiO_3$. Thus one positive point, which was at the origin of our investigation in 1979, was the introduction of alkaline-earth ions besides lanthanum as invited elements in order to stabilize Cu(III). The success of the synthesis stems from the great flexibility of the perovskite structure [2.4]. In this ReO_3-type framework it is possible to eliminate rows of oxygen atoms parallel to the [001] direction of the cubic cell (Fig.2.2). This leads to the formation of hexagonal tunnels, and implies for the metal atoms the formation of MO_5 pyramids. Such a behavior, which was observed in strontium and calcium manganese oxides [2.5], could be transferred with success to copper because it too can present this coordination. In the same way, the elimination of rows of oxygen atoms along the [110] direction of the cubic cell (Fig.2.3) was considered in order to generate layer structures involving layers of corner-sharing CuO_5 pyramids. Another way of creating anisotropic electron transport properties was based on the possible intergrowths between the rock-salt-type and perovskite-type structures. The two-dimensional agreement between the sodium-chloride-type structure and the perovskite structure in the (001) plane of both cubic cells has allowed a large family of titanates $SrO(SrTiO_3)_m$ to be synthesized [2.6] that can be

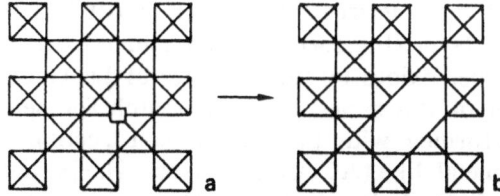

Fig.2.2. Elimination of rows of oxygen atoms along [001] in the perovskite framework (a) leading to the formation of hexagonal tunnels (b)

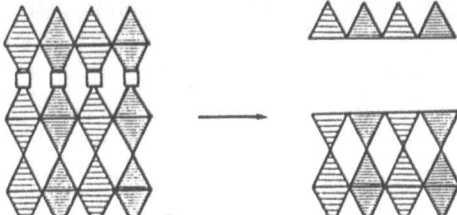

Fig.2.3. Elimination of rows of oxygen atoms along [110] in the perovskite framework (a) leading to the formation of a layer structure (b)

Fig.2.4a–c. The structure of different members of the family of the titanates $SrO(SrTiO_3)_m$. (a) m = 1, (b) m = 2, (c) m = 3. They correspond to the intergrowth of single SrO rock-salt-type layers with multiple perovskite layers $(SrTiO_3)_m$

described as the intergrowth of single SrO rock-salt-type layers and multiple perovskite layers, in which m characterizes the number of octahedral sheets forming the perovskite-type layer (Fig.2.4). This latter investigation was encouraged in the case of copper oxides by the existence of La_2CuO_4 [2.7, 8], which is the first member of this series. Thus the search for new phases $(AO)(ACuO_{3-x})_m$ was performed with the idea of constructing a stack of insulating AO layers and metal-like $(ACuO_{3-x})_m$ layers. The partial replacement of lanthanum ions in $LaCuO_3$ by alkaline-earth ions was performed to create anionic vacancies, and paradoxically to favor the introduction of additional oxygen on those sites, i.e. to stabilize the mixed valence Cu(II)/Cu(III). The anisotropic character of the structures was induced either by ordering of the anionic vacancies in the perovskite structure or by creating intergrowths between the oxygen-deficient perovskite structure and the rock-salt-type structure.

2.2 The Ternary Systems La-Cu-O and A-Cu-O (A = Ca,Sr,Ba)

Few phases have been isolated in the ternary systems La-Cu-O and A-Cu-O. The reaction of oxides La_2O_3 and CuO, or carbonates A_2CO_3 (A = Ca, Sr) and CuO carried out in air or in an oxygen flow leads to Cu(II) oxides, whatever the temperature may be. The only Cu(III) oxide that seems to be synthesized at normal pressure is $BaCuO_{2+x}$. However, its composition is not established with certainty.

Only two phases were isolated in the system La-Cu-O, namely $LaCuO_3$ and La_2CuO_4. The Cu(III) perovskite $LaCuO_3$ was prepared by *Demazeau* et al. [2.3] by heating mixtures of La_2O_3 and CuO at 900°C under an oxygen pressure of 65 kbar. The rhombohedral perovskite (a= 5.43 Å, α=60° 51′) exhibits a metal-like behavior. The oxide La_2CuO_4 [2.7, 8] exhibits the K_2NiF_4-type structure. It can be described as an intergrowth of $[LaO]_\infty$ single rock-salt-type layers and of single $[LaCuO_3]_\infty$ perovskite-type layers (Fig.2.5), similar to Sr_2TiO_4. However, unlike Sr_2TiO_4, the cell is not tetragonal but orthorhombic and related to the perovskite-type cell in the following way: a = 5.366 Å $\simeq \sqrt{2}a_p$, b = 5.402 Å $\simeq \sqrt{2}a_p$, c = 13.149 Å $\simeq 2a_p + a_{SrO}$.

This multiple orthorhombic cell results from a monoclinic distortion of the tetragonal cell in the basal plane (001) (Fig.2.6). Thus this structure differs from the tetragonal K_2NiF_4-type oxides by a puckering of the octahedral layers. Moreover, examination of the Cu-O distances clearly shows the Jahn-Teller effect of Cu(II) characterized by an elongated octahedron with two long Cu-O distances along the c axis (2.46 Å) and four shorter distances in the basal plane (001), 1.91 Å. No deviation from stoichiometry could be detected for La_2CuO_4 by X-ray diffraction or by neutron diffraction. However, a recent study of the electron transport properties of this oxide has shown that drastic variations of conductivity can be observed either by changing the La/Cu ratio in a very narrow range close to 1, or by applying an oxygen pressure. These very small deviations from stoichiometry, which allow a wide range of properties from semiconducting to semimetallic and even superconducting will be studied in the next chapter, taking into account the possibility of the mixed valence Cu(II)/Cu(III) in this structure.

It should be pointed out that, among the lanthanide copper oxides with the formulation Ln_2CuO_4, La_2CuO_4 is the only one that exhibits the K_2NiF_4-type structure. The other Ln_2CuO_4 oxides [2.9], although characterized by a very similar tetragonal cell (a $\simeq a_p \simeq$ 3.9 Å and c $\simeq$ 12 Å), have

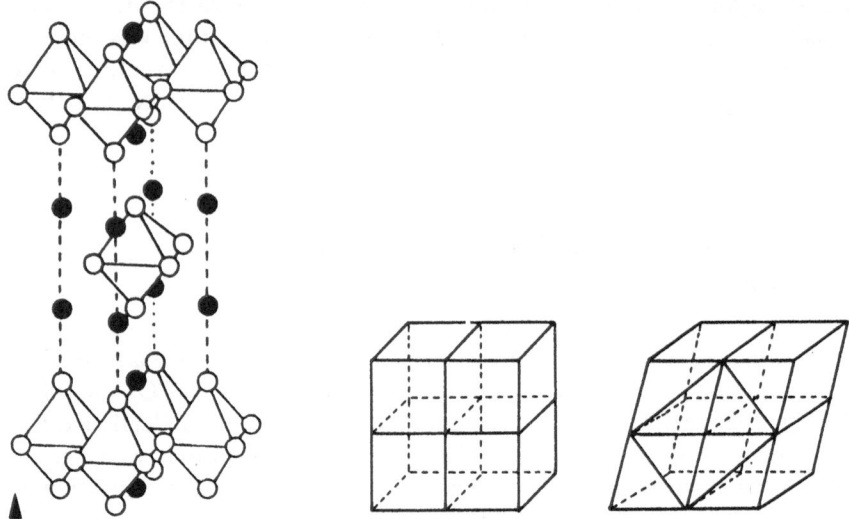

Fig.2.5. Schematized structure of La_2CuO_4: (•) La, (○) O

Fig.2.6. Relationships between the orthorhombic cell of La_2CuO_4 and the cubic perovskite cell

a very different structure, which will be denoted as the Nd_2CuO_4 type. This latter structure (Fig.2.7), although related to the La_2CuO_4-type structure, differs in the positions of the oxygen atoms leading to a square coordination of copper. Both structures exhibit similar positions of the Cu^{2+} and Ln^{3+} cations, as well as of the oxygen atoms forming the basal planes of the CuO_6 octahedra in La_2CuO_4 (Fig.2.7a), i.e. the $[CuO_2]_\infty$ layers are parallel to (001) in both cells, so forming a $LnCuO_2$ lattice common to the two structural types. The remaining two oxygen atoms are located in very different positions: the oxygen atoms of the $[LaO]_\infty$ layers in La_2CuO_4 (Fig. 2.7a) have migrated between two neodymium layers in Nd_2CuO_4 in such a

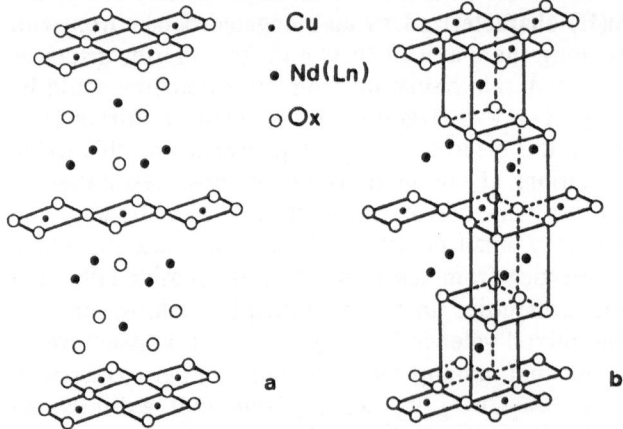

• Cu

• Nd(Ln)

○ Ox

Fig.2.7. Structure of Nd_2CuO_4 showing its relationships with (a) La_2CuO_4 and (b) the fluorite structure

12

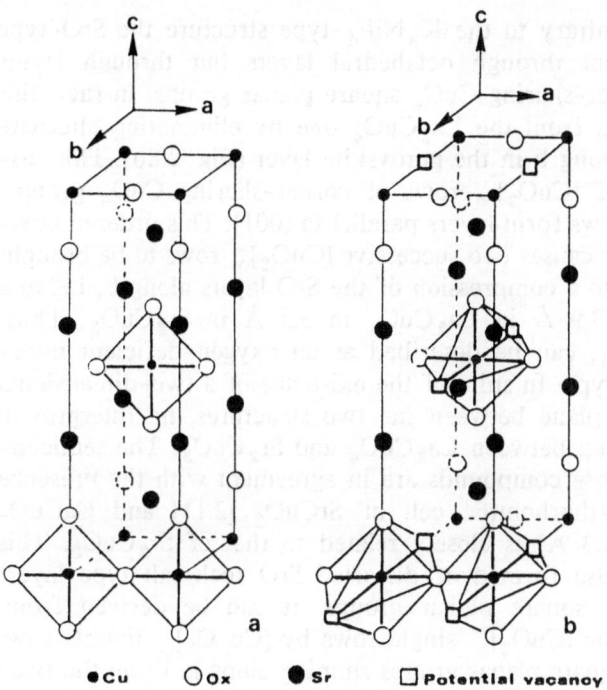

● Cu   ○ Ox   ● Sr   □ Potential vacancy

Fig.2.8. The structure of Sr_2CuO_3 (a), which can be deduced from the La_2CuO_4-type structure by eliminating alternate rows of oxygen atoms along **b** (b)

way that they form a fluorite-type lattice with the oxygen atoms of the $[CuO_2]_\infty$ layers (Fig.2.7b). Consequently, the Nd_2CuO_4 structure can be described as being formed of double Nd_2O_4 fluorite-type layers parallel to (001) in which the NdO_8 cubes share their edges (Fig.2.7b). Along **c**, the $[Nd_2O_4]$ fluorite layers share the faces of their NdO_8 cubes. Thus, the $[Nd_2O_4]_\infty$ lattice can be described as a fluorite structure in which every third plane of neodymium ions has been translated by a $\frac{1}{2}a_{Fluorite}$; the copper ions play the role of doping elements in the square planar groups formed by the oxygen atoms at the junction of two double $[Nd_2O_4]_\infty$ fluorite-type layers. Nd_2CuO_4 structure observed for numerous Ln_2CuO_4 oxides appears to be intermediate between the fluorite structure and the K_2NiF_4 type. Most of these oxides are poor semiconductors, whatever their method of synthesis. However recently three superconductors have been synthesized by replacing partially in this structure neodymium by cerium or thorium [2.10-12]. Although the critical temperatures of these oxides remain rather low (smaller than 24K), they are of the highest interest since they are the only n-type superconducting cuprates known so far.

The Sr-Cu-O and Ca-Cu-O systems have common features. Indeed, one observes two families of isostructural oxides A_2CuO_3 and $ACuO_2$ for A = Ca, Sr. The oxides A_2CuO_3 [2.13] with an orthorhombic symmetry (a $\simeq3.9$Å, b$\simeq3.5$Å, c$\simeq12.7$Å) are closely related to the La_2CuO_4 structure. Their structure (Fig.2.8a), like that of La_2CuO_4, is formed of SrO rock-salt-type layers, but contrary to the K_2NiF_4-type structure the SrO-type layers are connected not through octahedral layers but through layers formed of rows of corner-sharing CuO_4 square planar groups. In fact, this structure can be derived from the La_2CuO_4 one by eliminating alternate rows of oxygen atoms along b in the perovskite layer (Fig. 2.8b). This results in the formation of $[CuO_2]_\infty$ rows of corner-sharing CuO_4 groups, running along **a**; these rows form layers parallel to (001). This ordered creation of anionic vacancies causes two successive $[CuO_2]_\infty$ rows to be brought closer together, leading to a compression of the SrO layers along **b**, i.e. to a contraction of b from 3.9 Å in La_2CuO_4 to 3.5 Å in Sr_2CuO_3. Thus, Sr_2CuO_3, and Ca_2CuO_3, can be described as an oxygen-deficient intergrowth of the K_2NiF_4 type. In spite of the existence of a two-dimensional agreement in the (010) plane between the two structures, no intergrowth along **c** has been observed between La_2CuO_4 and Sr_2CuO_3. The semiconducting properties of those compounds are in agreement with the presence of only Cu(II). The orthorhombic cell of $SrCuO_2$ [2.11] and $CaCuO_2$ (a~3.9Å, b~3.5Å, c~16.3 Å) is closely related to that of Sr_2CuO_3. This structure (Fig.2.9a) is also formed of distorted SrO rock-salt-type layers and of layers of CuO_4 square planar groups. It can be derived from Sr_2CuO_3 by replacing the $[CuO_2]_\infty$ single rows by $[Cu_2O_3]_\infty$ double rows of edge-sharing CuO_4 square planar groups running along **a**. From the two-dimensional agreement between Sr_2CuO_3 and $SrCuO_2$ in (001), structural relationships between the two structures can be established. One can indeed consider, in a first step, a series of hypothetical structure of formulation $Sr_2CuO_3(CuO_2)_n$ in which multiple rows of corner-sharing CuO_4 groups replace the single row $[CuO_2]_\infty$. In this series "$Sr_2Cu_2O_5$" represents the member n = 2 (Fig.2.9b), and $SrCuO_2$ can be deduced from this hypothetical member by a shearing along **a** of the $[CuO_2]_\infty$ rows to form $[Cu_2O_3]_\infty$ rows. Thus, the $SrCuO_2$-type structure can be described as the result of a double operation, creation of anionic vacancies in a ReO_3-type layer, followed by shearing similar to that observed in the tungsten suboxides.

The oxide $CaCu_2O_3$ [2.15] of orthorhombic symmetry (a~ 4.1Å, b~ 3.5Å and c~9.8Å) exhibits a two-dimensional agreement in the (001) plane with Sr_2CuO_3 and $SrCuO_2$ as can be seen from the a and b values. The structure of this phase (Fig.2.10a) differs from $SrCuO_2$ and Sr_2CuO_3 by the absence of $[CaO]_\infty$ rock-salt-type layers. On the other hand, one observes $[Cu_2O_3]_\infty$ of distorted CuO_4 square planar groups parallel to (001), sometimes sharing their corners as in Sr_2CuO_3 and sometimes their edges as in $SrCuO_2$. In fact, this structure is closely related to the shear structures W_nO_{3n-1} of the [001] type. It can be derived from the structure of the hypothetical member "W_2O_5" (n = 2) (Fig.2.10b) by a simple elimination of oxygen atoms located on each side of the basal plane of the CuO_6 octahedra. Consequently the $[Cu_2O_3]$ layers are brought closer together (b ~ 3.5Å) with respect to the distances between the basal planes of the octahedral layers of the shear structures (3.9Å). Like $SrCuO_2$ and $CaCuO_2$, $CaCu_2O_3$ exhibits a low conductivity, in agreement with the presence in the structure of only Cu(II).

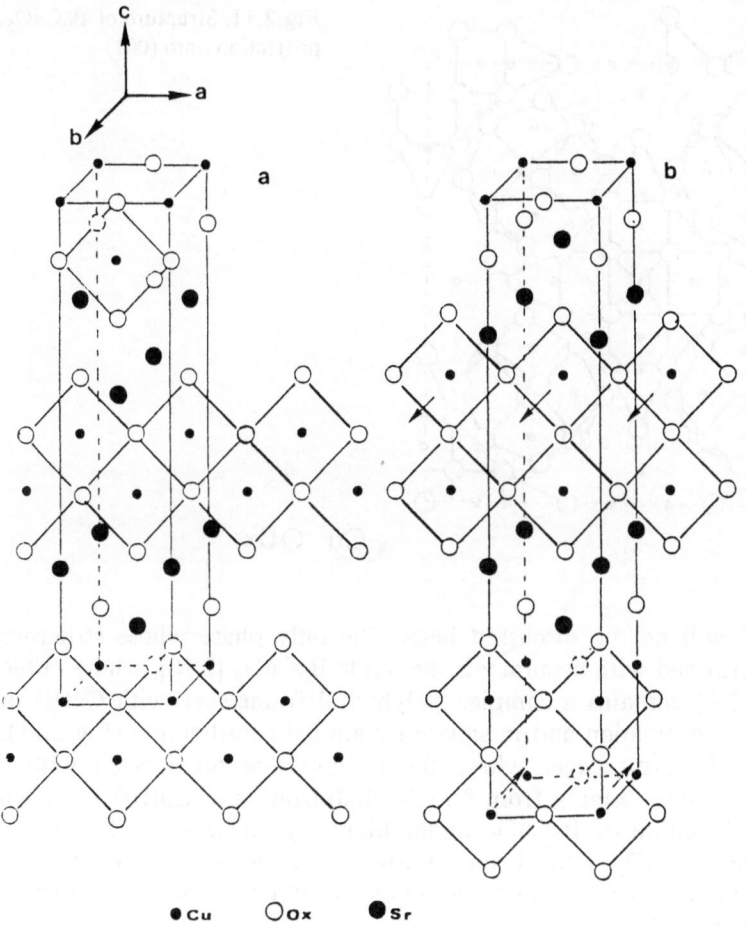

● Cu   ○ ox   ● Sr

Fig.2.9. The structure of SrCuO$_2$ (a) can be obtained from the hypothetical oxide "Sr$_2$Cu$_2$O$_5$" (b) by a shearing mechanism

Fig.2.10. The structure of CaCu$_2$O$_3$ (a) results from the hypothetical n = 2 member of the series W$_n$O$_{3n-1}$ (W$_2$O$_5$) (b) by elimination of rows of oxygen atoms

The system Ba-Cu-O is very different from those of strontium and calcium. Several phases related to the perovskite structure have been detected recently for the compositions Ba$_2$CuO$_{3+\delta}$. There is an orthorhombic phase [2.16] (a≃4.09Å, b≃3.90Å, c≃12.94Å) and a tetragonal one [2.17] (a≃3.9Å, c≃7.8 Å). The structure of these phases is at present not well es-

15

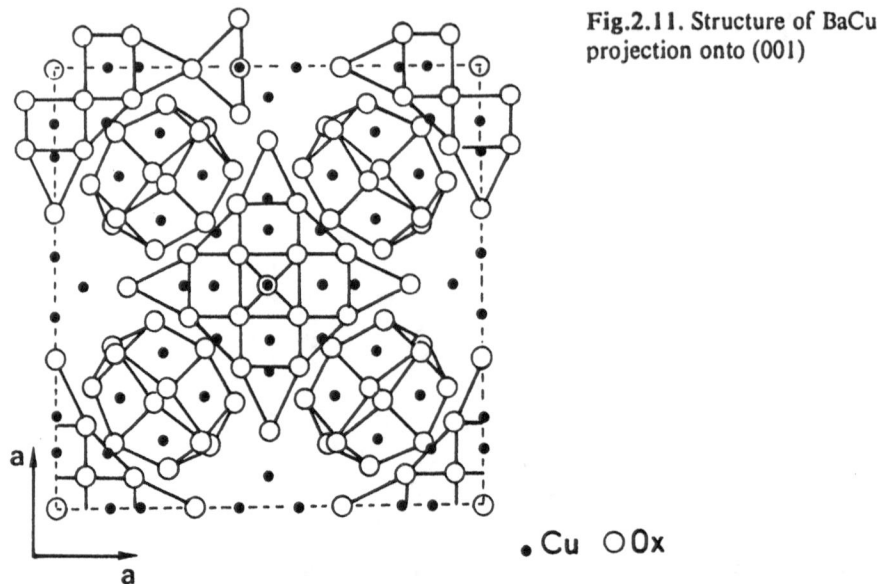

Fig.2.11. Structure of $BaCuO_2$: projection onto (001)

• Cu   ○ Ox

tablished and will not be discussed here. The only phase whose structure has been determined with accuracy is the oxide $BaCuO_2$ [2.18], whose cubic cell (a = 18.27Å) contains a complex polyhedral framework with Cu(II) in square planar coordination and in square pyramidal coordination (Fig.2.11). This framework forms cages where the barium ions are located with a coordination number ranging from 8 to 24. Stabilization of Cu(III) and even Cu(IV) by oxidization of $BaCuO_2$ under high oxygen pressure, leading to $BaCuO_{2.50}$ and $BaCuO_{2.63}$ has been reported [2.19]. No characterization of the structure of these latter oxides and no confirmation by other researchers have been given, so far.

Although the Cu(I) oxides are not of great interest for an understanding of superconductivity, it is worth noting that $BaCu_2O_2$ and $SrCu_2O_2$ are isostructural [2.20]. Their structure is formed of zigzag CuO chains with Cu(I) in twofold coordination and barium or strontium in octahedral coordination.

2.3 The Pseudoternary Systems A-La-Cu-O
(A = Ca,Sr,Ba)

No systematic study of the systems A-La-Cu-O (A = Ca, Sr, Ba) has been carried out. All attempts at synthesis were made with the aim of realizing perovskite structures or phases related to the perovskite involving the mixed valence Cu(II)/Cu(III), i.e. characterized by a metallic or semi-metallic conductivity. Two main classes of materials can be distinguished: the ordered oxygen-deficient perovskite structures and the intergrowths of the rocksalt-type and perovskite structures. During these investigations an insulating

16

phase $La_{4-2x}Ba_{2+2x}Cu_{2-x}O_{10-2x}$ was isolated, which will be described here. It should be noted that all these phases are prepared in a simple way by heating mixtures of oxides and carbonates in air, and that annealing at low temperatures in air or in an oxygen flow sometimes allows an increase of the oxygen content, i.e. of the Cu(III) content.

2.3.1 Phases with Oxygen-Deficient Perovskite Structures

Four phases can be described as oxygen-deficient perovskite structures in this system: $BaLa_4Cu_5O_{13+\delta}$, $La_{8-x}Sr_xCu_8O_{20}$, $Ba_3La_3Cu_6O_{14}$ and $LaBa_2\cdot Cu_3O_{7-\delta}$.

a) $BaLa_4Cu_5O_{13+\delta}$

$BaLa_4Cu_5O_{13+\delta}$ [2.21] was, at an early stage, considered by *Bednorz* and *Müller* [2.22] as potential material for superconductivity. This oxide which is a good metallic conductor, is in fact not superconducting owing to its three-dimensional framework. The structure of this oxide, which has been resolved by powder neutron diffraction, is closely related to that of perovskite, with a tetragonal cell ($a=8.647\text{Å}\simeq\sqrt{5}a_p$ and $c=3.859\text{Å}\simeq a_p$). The $[Cu_5O_{13}]$ host lattice (Fig.2.12) is built up from corner-sharing CuO_5 pyramids and CuO_6 octahedra forming hexagonal tunnels and perovskite cages where the lanthanum and barium ions are located in an ordered manner. The Ba^{2+} ions are located in the perovskite tunnels whereas the La^{3+} ions occupy the hexagonal tunnels. One typical feature of the $[Cu_5O_{13}]_\infty$ network is the geometry of the hexagonal tunnels, which is rather different from the ideal model derived from the stoichiometric perovskite structure. The O-O-O angles are close to 70° instead of 90°, and some oxygens are displaced towards the center of the tunnels. With four Cu-O distances of 2.039 Å and two of 1.93 Å, the CuO_6 octahedra of this mixed valence copper oxide are slightly flattened, contrary to those of the Cu(II) oxides described above, which are strongly elongated. On the other hand, one observes for the CuO_5 pyramids a long Cu-O distance (2.27Å) compared to the four distances of the basal plane (1.875-1.935Å), which means that copper tends to present a square coordination.

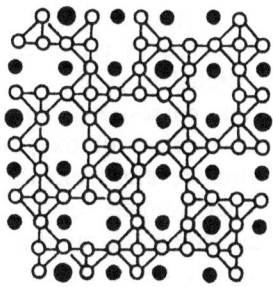

$La_4 Ba Cu_5 O_{13+\delta}$ Fig.2.12. Structure of $La_4BaCu_5O_{13+\delta}$: projection onto (001)

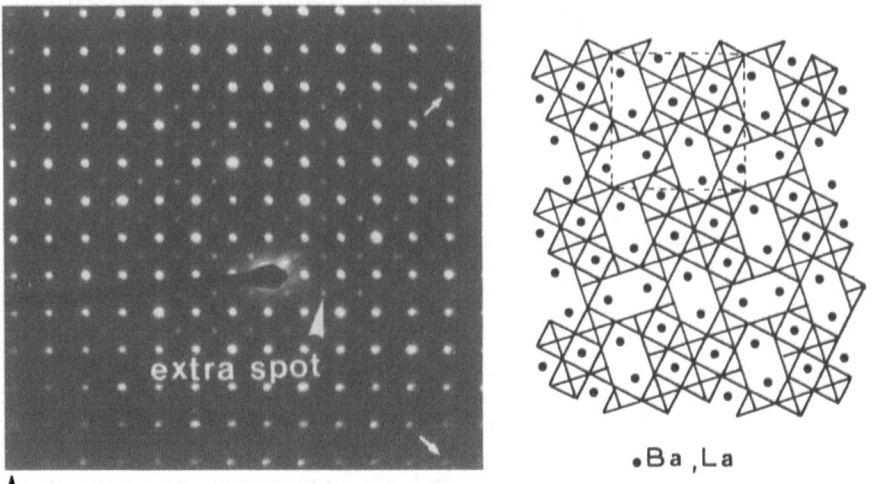

Fig.2.13. $La_4BaCu_5O_{13}$: electron diffraction pattern with [001] exhibiting extra spots corresponding to a double cell: $\sqrt{10}a_p \times \sqrt{10}a_p$

Fig.2.14. Idealized drawing of a hypothetical model corresponding to three oxygen vacancies per cell in the $La_4BaCu_5O_{13}$ $\sqrt{10}a_p \times \sqrt{10}a_p \times a_p$ cell

An interesting feature deals with the additional oxygen ($\delta \simeq 0.12$), which can be deduced from both chemical analysis and neutron data that show that anionic positions in the hexagonal tunnels are partially occupied. A systematic high-resolution electron microscopy study shows that the large majority of crystals exhibit the structure described above with the ideal composition $La_4BaCu_5O_{13}$, but the electron diffraction investigation reveals that besides these crystals, there exist crystals characterized by a double $BaLa_4Cu_5O_{13}$-type cell $\sqrt{10}a_p \times \sqrt{10}a_p \times a_p$. Such a pattern is depicted in Fig.2.13; the intensity of the superstructure reflections varies from one crystal to the other. Two hypotheses can be considered to explain this phenomenon; both are related to an excess of oxygen with regard to the stoichiometric $BaLa_4Cu_5O_{13}$ oxide. The first one corresponds to a formula $BaLa_4Cu_5O_{13.5}$, with three oxygen vacancies per double cell; an idealized drawing of such a model is shown in Fig.2.14. It is built up from four octahedra and six pyramids. As in the $BaLa_4Cu_5O_{13}$ structure, hexagonal tunnels exhibit alternate orientations; it could be considered as an intergrowth of slabs of $BaLa_4Cu_5O_{14}$ and $BaLa_4Cu_5O_{13}$ parallel to $[130]_p$, i.e. parallel to the a axis of the double cell. The second one corresponds to a simple distortion of the $\sqrt{5}a_p \times \sqrt{5}a_p$ cell, due to a small excess of oxygen randomly located in the hexagonal tunnels; this distortion results from the introduction of additional oxygen into the hexagonal tunnels of the $BaLa_4Cu_5O_{13}$ structure, involving displacement of the atoms surrounding the 2(c) sites and especially of the O(5) atoms. It should also be pointed out that such superstructure spots generally induce a monoclinic distortion of the tetragonal lattice (Fig.2.13). Unfortunately, no high-resolution image could be obtained from these crystals, but a modification of the electron

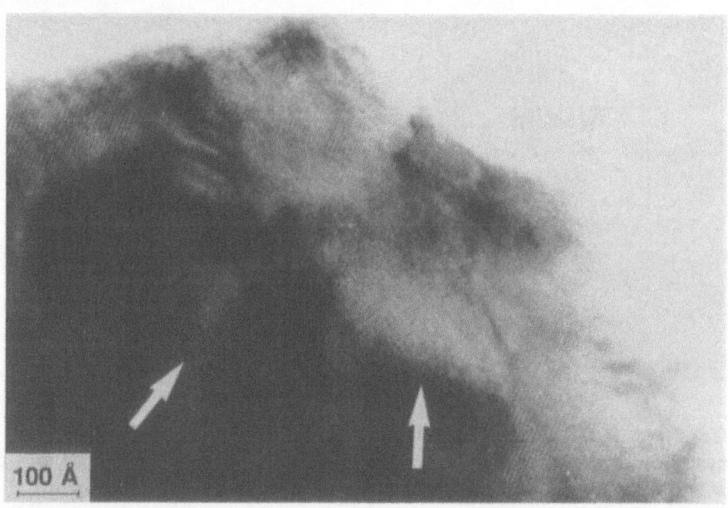

Fig.2.15. $BaLa_4Cu_5O_{13+\delta}$ matrix. HREM image of a crystal characterized by the presence of microdomains corresponding to modulations of the intensity and loss of the periodicity (*arrows*)

diffraction pattern was indeed observed after exposure to the electron beam. All the extra spots disappeared and the contrast was then similar to that observed for the $\sqrt{5}a_p \times \sqrt{5}a_p$ cell. This phenomenon suggests a loss of oxygen and is consistent with both hypotheses but does not allow a choice between a new ordering scheme (with $\delta = 0.5$) and a random distribution (with unknown δ) of the additional oxygen. However, the uneven contrast observed in small areas of some microcrystals, such as modulations of the intensity and loss of periodicity in the form of microdomains (Fig. 2.15), could be interpreted in terms of local variations of the arrangements of the anionic vacancies.

A small number of crystals exhibit an electron diffraction pattern more closely related to the parent perovskite subcell but characterized by a monoclinic distortion. Micrographs corresponding to such a crystal are exhibited in Fig.2.16. The regions labelled 1 and 3 of the crystals are "pseudocubic" (a_p) and region 2 is tetragonal ($\sqrt{5}a_p \times \sqrt{5}a_p$). The different parts are slightly misaligned (Fig.2.16). Such pseudocubic domains have been previously observed in the oxygen-deficient oxides $Sr_2Mn_2O_5$ built up from MnO_5 square pyramids. They have grown coherently in the matrix in the form of smaller or large microdomains. It was shown that they could correspond of either to the formulae AMO_3 and $AMO_{2.5}$ (with a random arrangement of the deficient $AMO_{2.5}$ layers) or to an intermediate oxygen content.

In conclusion, the occurrence of such periodicity appearing in the form of single microcrystals, or of misaligned parts of one crystal, suggests that the introduction of oxygen to the $BaLa_4Cu_5O_{13}$ structure induces a distortion of the framework and explains the oxygen nonstoichiometry in the oxide $BaLa_4Cu_5O_{13+\delta}$.

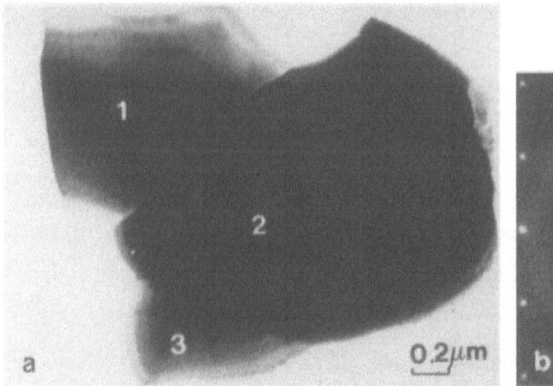 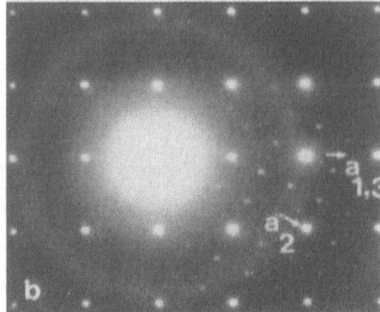

Fig.2.16. $La_4BaCu_5O_{13+\delta}$. (a) Low resolution image and (b) electron diffraction pattern of a microcrystal exhibiting areas with a pseudo-cubic cell (*areas 1* and *3*) and a tetragonal one (*area 2*). It should be noted that the different areas are slightly misaligned

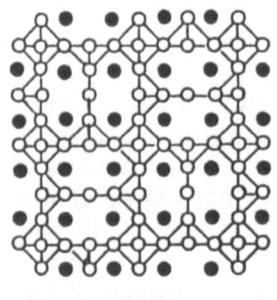

$La_{8-x}Sr_xCu_8O_{20-\epsilon}$

Fig.2.17. Structure of the oxygen-deficient perovskite-type phase $La_{8-x}Sr_xCu_8O_{20}$

b) $La_{8-x}Sr_xCu_8O_{20}$

The oxygen-deficient perovskite-type phase $La_{8-x}Sr_xCu_8O_{20}$ [2.23] has been isolated for $1.28 \leq x \leq 1.92$. Its structure has been determined by neutron diffraction (Fig.2.17). It crystallizes in the tetragonal system (a = $10.868-10.875$Å ~ $2\sqrt{2}a_p$, c=$3.856-3.863$Å ~ a_p). The $[Cu_8O_{20}]_\infty$ framework shows a great similarity with that of $BaLa_4Cu_5O_{13}$. In both structures one can recognize $[Cu_5O_{22}]$ groups of five corner-sharing polyhedra built up from one CuO_6 octahedron and four CuO_5 pyramids, forming hexagonal tunnels running along c. However, the $[Cu_5O_{22}]$ groups are arranged in different ways in the two structures: they share the corners of their pyramids in $BaLa_4Cu_5O_{13}$ whereas they are linked through CuO_4 square planar goups in the strontium oxide. It is also worth pointing out that the CuO_4 groups exhibit two perpendicular orientations, parallel to [110] and [1$\bar{1}$0], respectively. It results that each CuO_5 pyramid is linked to one CuO_4 group, two CuO_6 octahedra and two CuO_5 pyramids in $La_{6.4} \cdot Sr_{1.6}Cu_8O_{20}$, whereas it is surrounded by four other pyramids and one octahedron in $BaLa_4Cu_5O_{13}$. Consequently the hexagonal tunnels are differently oriented in the two structures: they share only their corners in $La_4BaCu_5O_{13}$, whereas they form pairs by sharing one CuO_4 group in $La_{6.4}Sr_{1.6}Cu_8O_{20}$. The fact that La^{3+} and Sr^{2+} ions are distributed in a statistical way in the tunnels is dictated by the space group, and is quite in agreement with the similarity of their size.

The geometry of the copper-oxygen polyhedron is close to that observed for $La_4BaCu_5O_{13+\delta}$. The CuO_6 octahedra, with two short Cu-O distances (1.93 Å) and four longer ones (2.03 Å), are compressed along the c axis as in $La_4BaCu_5O_{13+\delta}$. The CuO_5 pyramids are also characterized by four short Cu-O distances (1.88-1.93 Å) and a longer one (2.39 Å). The CuO_4 groups are approximately square planar, with Cu-O distances ranging from 1.91 to 1.93 Å.

An oxygen deficiency is observed by chemical analysis, leading to the formula $La_{8-x}Sr_xCu_8O_{20-\epsilon}$ (with $\epsilon \simeq 0.1-0.3$). This deviation from stoichiometry is however too low to be refined even by neutron diffraction, with the occupancy factor of all the oxygen sites tending always towards 1.0. These additional oxygen vacancies are probably distributed over the O(1) sites: the creation of oxygen vacancies on those sites leads to the formation of CuO_5 pyramids, whereas the creation of oxygen vacancies on the other sites would involve an unusual coordination of copper. The doubling of the c parameter of a minority of crystals suggests that these additional vacancies are ordered along the c axis.

The mixed valence Cu(II)/Cu(III) induces a metallic conductivity as for $La_4BaCu_5O_{13}$, but no superconducting properties were detected for this oxide, in agreement with its three-dimensional character.

c) $La_3Ba_3Cu_6O_{14}$ and $LaBa_2Cu_3O_{7-\delta}$

$La_3Ba_3Cu_6O_{14+\delta}$, which was synthesized for the first time in 1981 [2.24], was found from the X-ray powder diffraction study to exhibit a layer structure formed of corner-sharing CuO_6 octahedra, CuO_5 pyramids and CuO_4 square planar groups (Fig.2.18a). A recent reinvestigation by neutron diffraction [2.25] of this compound confirmed the relative position of all the metal atoms - Ba,La,Cu - and of 12 oxygen atoms out of 14 (Fig. 2.18b). However, a divergence was observed for the two remaining oxygen atoms, labelled a, which was attributed to the insufficient information available from the X-ray powder spectra. Thus, in a first step the tetragonal cell (a=3.90 Å$\simeq a_p$; c=11.71 Å$\simeq 3a_p$) can be described as formed of triple layers of corner-sharing CuO_5 pyramids and CuO_4 square planar groups. The cohesion between these layers is ensured by La^{3+} cations, while the barium ions and remaining lanthanum ions are distributed inside the layers. This oxide can intercalate oxygen when annealed in an oxygen flow at 450°C, leading to a progressive transition from a semiconducting to a semimetallic state.

The orthorhombic form of $LaBa_2Cu_3O_{6.7}$ [2.26], superconducting below 75 K, is closely related to $La_3Ba_3Cu_6O_{14}$. This oxide, which is isostructural with the 92 K superconductor $YBa_2Cu_3O_7$, exhibits the following parameters: a = 3.895 Å $\simeq a_p$, b = 3.944 Å $\simeq a_p$ and c = 11.829 Å $\simeq$

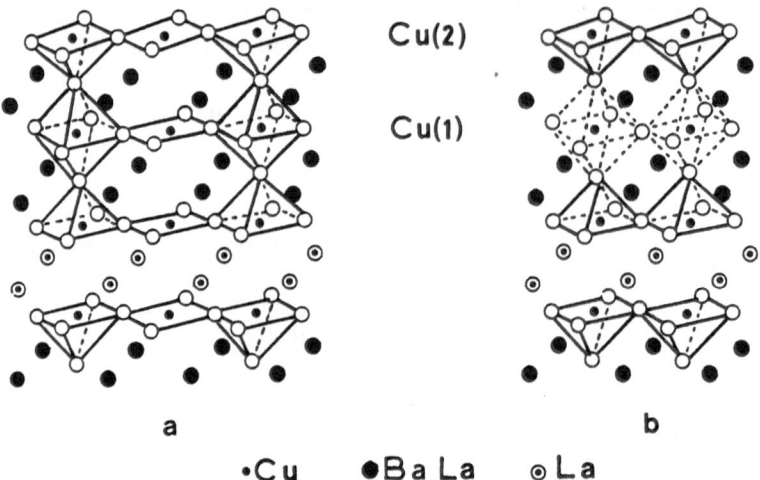

$Cu(2)$

$Cu(1)$

a   b

• Cu   ● Ba,La   ◎ La

Fig.2.18. Structure of $La_3Ba_3Cu_6O_{14}$ deduced from (a) powder X-ray diffraction data and (b) neutron diffraction data

$3a_p$. This structural type will be analyzed in detail for $YBa_2Cu_3O_{7-\delta}$ (Sect. 2.4.1). It should however be pointed out that the orthorhombic form can transform into the tetragonal form on being heated to temperatures higher than 900°C in air. This latter oxide, $LaBa_2Cu_3O_{7-\delta}$, which corresponds to an oxygen loss ($\delta \sim 0.7$), is closely related to the orthorhombic form with a = 3.91 Å and c = 11.74 Å. It can also be considered as belonging to the same structure as $La_{3-x}Ba_{3+x}Cu_6O_{14+\delta}$, with x equal to 0 and 1 for $La_3Ba_3Cu_6O_{14}$ and $LaBa_2Cu_3O_7$ respectively. Thus the tetragonal $LaBa_2\cdot Cu_3O_{7-\delta}$ corresponds to a strict arrangement of La^{3+} and Ba^{2+} ions, which are located between and inside the $[Cu_3O_7]_\infty$ layers, respectively. In fact, a careful examination of these structures by electron microscopy (Chap. 6) shows that the nonstoichiometry of these structures is much more complex than described here, so that $La_3Ba_3Cu_6O_{14-\delta}$ cannot be considered as a true single phase but is formed of various microdomains.

2.3.2 The Intergrowths Between Perovskite and Rock–Salt–Type Structures

Two families of mixed valence copper oxides corresponding to the intergrowth $(AO)_n(ACuO_{3-x})_m$ have been isolated for A = La+Ba or Ca or Sr (n = 1) and A = La+Sr or Ca (n = 2).

a) The Oxides $La_{2-x}A_xCuO_{4-x/2+\delta}$

The oxides $La_{2-x}A_xCuO_{4-x/2+\delta}$ exhibit an oxygen-deficient K_2NiF_4-type structure (Fig.2.19): one observes single SrO-type layers intergrown with

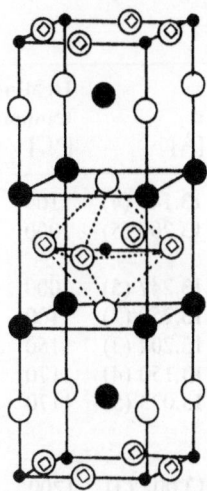

Fig.2.19. Structure of the oxides $La_{2-x}A_xCuO_{4-x/2+\delta}$. The oxygen vacancies appear in the basal planes of the CuO_6 octahedra

single oxygen-deficient perovskite-structure layers. The homogeneity range of these phases depends on the nature of the alkaline-earth ion : $0 \leq x < 0.20$ for A = Ca, Ba and $0 \leq x \leq 1.33$ for A = Sr. The symmetry of the cell and the parameters depend on the substitution rate x and also on the oxygen content, i.e. on the oxygen pressure and temperature used during the synthesis.

The barium and calcium oxides prepared in air at 1100°C [2.27] and quenched to room temperature are characterized by a monoclinic distortion of the tetragonal cell, i.e. by an orthorhombic symmetry ($a \sim \sqrt{2}a_p$, $b \sim \sqrt{2}a_p$, $c \simeq 13$ Å) whatever the x value may be ($0 \leq x \leq 0.20$) (Table 2.1). Slow cooling of such samples in an oxygen flow or annealing at a low temperature shows a transition from the orthorhombic symmetry for x > 0.05 to tetrago-

Table 2.1. Crystallographic data and analytical results (quenched materials) for the oxides $La_{2-x}A_xCuO_{4-x/2+\delta}$ (A = Ca, Ba) at a heating temperature of 1100°C

| A | x | δ | Cell parameters [Å] | | |
			a	b	c
	0		5.366	5.402	13.149
Ba	0.05	0.001	5.361	5.380	13.201
	0.1	0.02	5.359	5.364	13.245
	0.2	0.05	5.356		13.320
Ca	0.05	0.01	5.351	5.387	13.150
	0.1	0.02	5.356	5.385	13.174
	0.1	0.05	5.357	5.380	13.210

Table 2.2. Crystallographic data of $La_{2-x}Sr_xCuO_{4-x/2+\delta}$ compounds

Range	x	δ	Composition	a [Å]	b [Å]	c [Å]	Heating temperature [°C]
I	0	0	La_2CuO_4	5.366 (2)	5.402 (2)	13.149 (4)	1100
	0.08	0.030 (1)	$La_{1.92}Sr_{0.08}CuO_{3.99}$	5.368 (1)	5.368 (2)	13.200 (5)	1000
II	0.25	0.060 (4)	$La_{1.75}Sr_{0.25}CuO_{3.935}$	3.775 (2)		13.247 (5)	1000
	0.33_3	0.119 (4)	$La_{1.66_6}Sr_{0.33_3}CuO_{3.95}$	3.776 (1)		13.250 (2)	1100
	0.50	0.100 (4)	$La_{1.50}Sr_{0.50}CuO_{3.85}$	3.773 (1)		13.204 (3)	1160
	0.66_6	0.092 (4)	$La_{1.33}Sr_{0.66_6}CuO_{3.75_8}$	3.775 (1)		13.150 (4)	1170
	0.88_0	0.088 (4)	$La_{1.12}Sr_{0.88}CuO_{3.64_8}$	3.773 (1)		13.073 (5)	1170
III	1.00	0.0	$LaSrCuO_{3.50}$	3.767ª (1)		13.002 (3)	1200
	1.28	0.0	$La_{0.72}Sr_{1.28}CuO_{3.36}$	3.761ª (2)		12.922 (9)	1200
	1.34	0.0	$La_{0.66}Sr_{1.34}CuO_{3.33}$	3.759ª (3)		12.907 (9)	1200

The number in parentheses expresses the standard deviation
ª The a parameter of the range-III compounds are those of the tetragonal subcell

Table 2.3. Values of n observed by electron diffraction for compounds of range III

Composition	x	n
$LaSrCuO_{3.5}$	1.00	1, 4.5
$La_{0.88}Sr_{1.12}CuO_{3.44}$	1.12	1, 4.5, 5
$La_{0.80}Sr_{1.20}CuO_{3.40}$	1.20	5
$La_{0.72}Sr_{1.28}CuO_{3.36}$	1.28	1, 4.6, 5, 5.3, 5.4
$La_{0.66}Sr_{1.34}CuO_{3.33}$	1.34	1, 4, 5, 5.6, 6

nal symmetry ($a \sim a_p$, $c \sim 13 Å$) [2.28]. It is worth pointing out that only those samples close to x = 0.10–0.20 and annealed in oxygen are superconducting, with T_c ranging from 20 to 38 K.

Inspection of the X-ray diffraction patterns of the strontium compounds [2.29] allowed three composition ranges to be distinguished (Table 2.2), which were then studied by electron diffraction. The a parameters of these compounds (range III) are those of the tetragonal subcell; n values for every composition are given in Table 2.3.

(I) $0 \leq x < 0.10$. The X-ray patterns very similar to that of La_2CuO_4 were indexed in an orthorhombic cell with $a_I \simeq \sqrt{2}a_p \simeq a_{La_2CuO_4}$, $b_I \simeq \sqrt{2}a_p \simeq b_{La_2CuO_4}$ and $c \simeq c_{La_2CuO_4}$.

(II) $0.10 \leq x < 1$. The symmetry is tetragonal; the cell parameters are related to those of $LaSrCuO_4$ and to I in the following manner: $a_{II} \simeq a_I/\sqrt{2} \simeq a_p \simeq a_{LaSrCuO_4}$, $c_{II} \sim c_I \sim c_{LaSrCuO_4}$.

(III) $1 \le x \le 1.34$. The X-ray diffractograms are characterized by the existence of a system of strong peaks, which were already observed for the compounds (II), involving at least the existence of a tetragonal subcell of the same type. However, for all these patterns, weak peaks were always observed that could not be indexed in this cell. An electron diffraction study was thus undertaken: about 50 crystals were examined for each value of x given in Table 2.3. Two main types of crystals were isolated:

- A small number of crystals, about 10%, were characterized by a tetragonal cell similar to that of $LaSrCuO_4$:

$$a_{III} \sim a_{II} \sim a_p \sim a_{LaSrCuO_4} ,$$

$$c_{III} \simeq c_{II} \simeq c_I \sim c_{LaSrCuO_4} .$$

- Most of the crystals i.e. about 90%, presented, in addition to the fundamental reflections previously described, superstructure reflections with a variable intensity. The electron diffraction patterns allowed the following relations for the actual tetragonal cell for a composition x to be found:

$$a_{III}^x = na_{III} \sim na_{II} ,$$

$$c_{III}^x = c_{III} \ne c_{II} \simeq c_I .$$

For a same composition x, several sorts of superstructures were generally observed, characterized either by integral n values (n = 4, 5) or by nonintegral values of n (n ranging from 4.5 to 5.6), as shown for several compositions in Table 2.3. Figure 2.20 shows, as an example, the electron diffraction patterns of the (001) planes for $La_{2/3} Sr_{4/3}CuO_{3.33}$. From Table 2.3 it can be seen that a pure member, characterized by a superstructure with an integral value of n (n = 5), is obtained only for x = 1.20. Calculations of intensities for the X-ray diffractogram of the oxide $La_{0.8}Sr_{1.2}CuO_4$, performed in the actual cell (a=3.76Å, c≃12.14Å), have shown that copper atoms are not significantly displaced from their ideal position in the stoichiometric perovskite structure, while bigger cations like La, Sr and the oxygen atoms are only slightly displaced from their ideal positions, but enough to produce the superstructure reflections. These small displacements can be assumed to be the consequence of an ordering of the oxygen vacancies, whose contribution to the intensities is too small to be detected by X-ray powder diffraction.

The stabilization, in this system, of Cu(III) just by heating the oxides in air is worth noting, but the most important characteristic of this system concerns the existence of a Cu(III) composition range (0 < x < 1) that lies between two Cu(II) regions (x=0 and x≥1), for structures closely related to each other. This can be explained by two opposite effects which are competitive: the trend to preserve a stoichiometric K_2NiF_4 structure as for

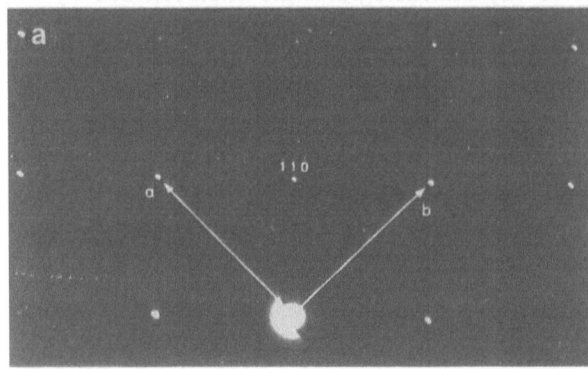

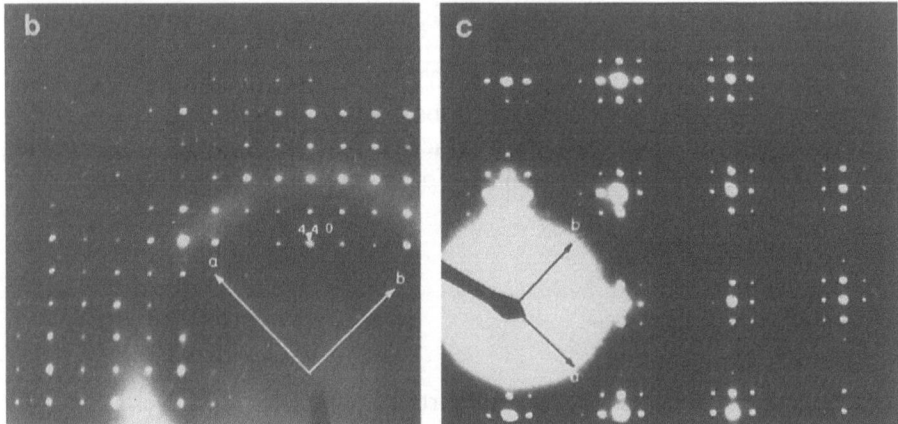

Fig.2.20. Electron diffraction patterns of the (001) planes for $La_{2/3}Sr_{4/3}CuO_{3.33}$: (a) n = 1, (b) n = 4, (c) n = 5.6

La_2CuO_4 and the trend to form a related defect structure but with an ordering of the oxygen vacancies. Thus, rather close to the stoichiometric compound La_2CuO_4 (x < 1), the trend to stoichiometry is favored and the vacancies formed from the nominal compositions involving only Cu(II) are partly balanced by the oxidation of Cu(II) into Cu(III). For x ≥ 1, i.e. far from stoichiometry, the La_2CuO_4 or "LaSrCuO$_4$" stoichiometric compounds cannot be stabilized any more and orderings of the oxygen vacancies appear, leading to different microphases, as observed from the electron diffraction study, favoring Cu(II) with smaller coordinations.

Structure is not, of course, the only factor governing the relative stability of Cu(II) and Cu(III) in these oxides. Kinetics plays an important part in determining the ratio Cu(III)/Cu(II) in the oxides richer in Cu(III). For 0 < x < 1, it was noticed that the pure compounds could only be synthesized by heating for at least 12 h at the formation temperature (Table 2.2) in order to ensure a good crystallization. Annealing the same samples at the same temperature for longer periods (24h) in air allows pure phases to be prepared with the same structure, but with a greater amount of Cu(III). The

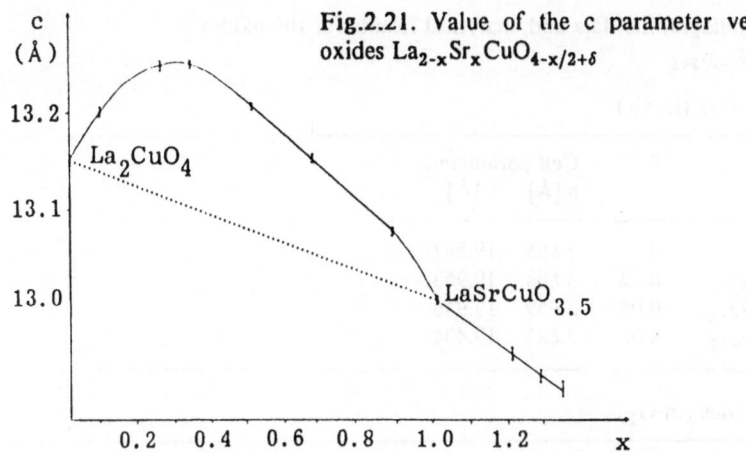

Fig.2.21. Value of the c parameter versus x for the oxides $La_{2-x}Sr_xCuO_{4-x/2+\delta}$

oxygen pressure will also influence the Cu(III)/Cu(II) ratios. Heating, for example, some Cu(III) samples at a low temperature under vacuum produces a decrease of Cu(III) without destroying the structure. In the same way, reaction in an oxygen flow allows the Cu(III) to be increased. Finally we should point out that $LaSrCuO_4$ characterized by only Cu(III) was prepared under high oxygen pressure by *Goodenough* et al. [2.30].

The influence of the amount of Cu(III) can also be detected by considering the structural evolution, especially the c parameter, of these compounds as a function of composition (Fig.2.21). This evolution is rather complex and quite different from that usually observed for single solid solutions. The substitution of strontium for lanthanum should not affect this evolution, due to the similar size of these cations. It seems interesting to take Cu(II) compounds as a reference (dotted line). It can be seen that from $La_2Cu^{II}O_4$ to $La_{0.7}Sr_{1.3}Cu^{II}O_{3.35}$ a continuous decrease of a and c parameters could be predicted for all Cu(II) compounds, as x increases, in agreement with the increase of oxygen vacancies. The evolution is not linear, probably due to ordering of the vacancies observed for different compositions. What is worth noting is the large deviation from this law observed only for the compounds containing Cu(III) (continuous line): the c parameter is greater than that obtained from the "reference line" corresponding to the presence of Cu(II) only, while the a parameter is smaller. Moreover, the largest deviations are observed for x = 0.33, which corresponds to the maximum value of δ (δ = 0.119), i.e., for the greatest amount of Cu(III). It can thus be observed that the c/a ratio increases with the Cu(III)/Cu(II) ratio. There is no doubt that the oxygen content in this phase can be modified by changing the thermal treatment, especially in the ranges I and II. The ability of this phase to intercalate and disintercalate oxygen will greatly influence electron transport properties, especially in the range $0 \leq x \leq 0.25$. This is in agreement with the fact that for this range one observes a transition from semiconducting to semimetallic properties for specimens prepared in air and quenched to room temperature, whereas superconductivity below 40 K is obtained for samples corresponding to x = 0.10-0.20 annealed at

Table 2.4. Crystallographic data and analytical results for the oxides $La_{2-x}A_{1+x}Cu_2O_{6-x/2+\delta}$

(a) Quenched oxides (in air)

Composition	δ	Cell parameters a [Å]	c [Å]
$La_2SrCu_2O_6$	0	3.865	19.887
$La_{1.9}Sr_{1.1}Cu_2O_{5.97}$	0.02	3.863	19.963
$La_{1.86}Sr_{1.14}Cu_2O_{5.97}$	0.04	3.959	19.956
$La_{1.9}Ca_{1.1}Cu_2O_{5.97}$	0.02	3.825	19.404

(b) Annealed oxides (in O_2)

Composition	δ	Cell parameters a [Å]	c [Å]
$La_2SrCu_2O_{6.2}$	0.20	3.865	20.065
$La_{1.86}Sr_{1.14}Cu_2O_{6.22}$	0.29	3.868	20.051
$La_{1.9}Ca_{1.1}Cu_2O_{6.02}$	0.08	3.825	19.404

400°C in an oxygen flow. The low conductivity of the oxides of domain III is in agreement with the existence of only Cu(II) in these materials.

b) The Oxides $La_{2-x}A_{1+x}Cu_2O_{6-x/2+\delta}$

As for the $La_{2-x}A_xCuO_{4-x/2+\delta}$ oxides, the oxygen content of $La_{2-x}A_{1+x} \cdot Cu_2O_{6-x/2+\delta}$ depends on the method of synthesis. We can distinguish first the oxides $La_{2-x}A_{1+x}Cu_2O_{6-x/2+\delta}$ prepared in air. These compounds are synthesized at 900°C, and at 1100°C for A = Sr and then quenched to room temperature. No barium compound could be synthesized in these experimental conditions. Their homogeneity ranges are narrow: $0 \leq x \leq 0.14$ for strontium and $x = 0.10$ for calcium.

These oxides [2.31] exhibit a tetragonal symmetry (Table 2.4) closely related to that of the perovskite and SrO-type structures: $a \simeq a_p$, $c \simeq 2(2a_p + \frac{1}{2}a_{SrO})$.

Their tetragonal structure is closely related to that of $Sr_3Ti_2O_7$, as shown for $La_2SrCu_2O_6$, in which copper has only the oxidation state II (Fig. 2.22). The metal atoms are located on the same positions and the oxygen atoms are close to those observed for the titanate. However, oxygen vacancies are localized inside the perovskite layers at the levels z = 0 and 1/2. This structure can thus be described as an intergrowth of stoichiometric SrO-type layers and of double oxygen-deficient perovskite layers. This ordered distribution of the oxygen vacancies leads to a pyramidal coordination of copper. As in other oxides, copper is located near the basal plane of the pyramid, forming four short Cu-O distances (1.91–1.94Å) and one long Cu-O distance (2.26–2.37Å).

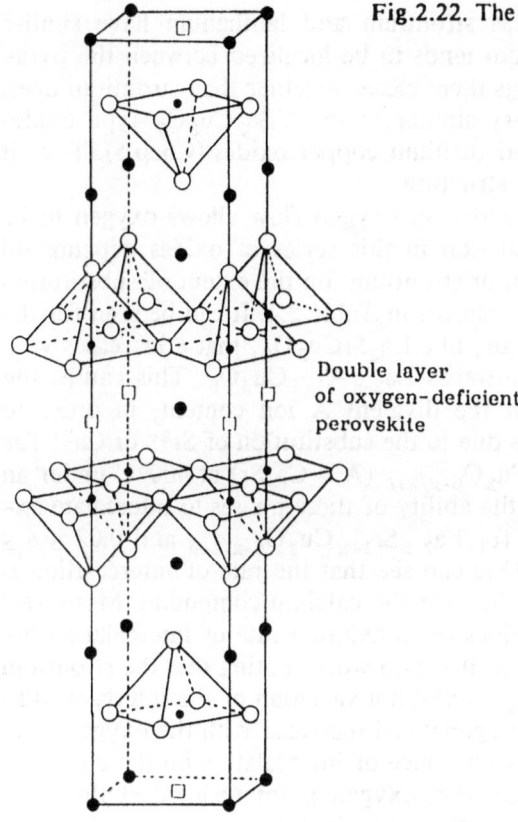

Fig.2.22. The $La_2SrCu_2O_6$-type structure

Double layer
of oxygen-deficient
perovskite

Two sorts of sites are available to lanthanum, strontium or calcium, labelled A_1 and A_2. The A_1 sites are located inside the perovskite layers and ensure the cohesion between two layers of corner-sharing CuO_5 pyramids: in those sites the cations present an eightfold coordination, derived from the twelvefold coordination of the perovskite structure by a simple elimination of four oxygen atoms at the level $z = \frac{1}{2}$ or 0. Thus the oxygen atoms which surround the A_1 sites are located at the corners of a square prism with A_1-O distances ranging from 2.46 Å to 2.56 Å The A_2 sites correspond to the metal atoms located in the SrO-type layers. The A_2 coordination is similar to that observed in the intergrowths $(SrTiO_3)_n$ SrO, i.e. one observes eight long A_2-O distances (2.66- 2.79Å) and a shorter one along c (2.26-2.33Å). This latter geometry of the A_2 sites is very similar to that observed for lanthanum in La_2CuO_4, which exhibits one very short La-O distance (2.30Å) and eight long distances (2.65-2.90Å). The distribution of the La^{3+} and Sr^{2+} or Ca^{2+} cations over the A_1 and A_2 sites is remarkable. A preferential occupancy of the A_2 sites of SrO-type layers by lanthanum is observed: 70%-90% of these sites contain La^{3+}. Moreover, the occupancy factor of these sites by La^{3+} is much greater for calcium oxides (90%) than for strontium compounds (70%-75%). This greater tendency to ordering in the case of calcium is easily understood because calcium is smaller than lanthanum, whereas strontium and lanthanum have similar sizes. Thus it appears that calcium tends to be localized between the pyramidal $[CuO_{2.5}]_\infty$ layers and brings them closer together than strontium does. In this respect, this oxide is very similar to the $YBa_2Cu_3O_7$-type oxides (Sect.2.4) and to the bismuth and thallium copper oxides (Chap.5). Thus it can be considered as a true layer structure.

Annealing of these oxides under an oxygen flow allows oxygen to be intercalated. The oxygen intercalation in this series of oxides depends on the nature of the A ions, calcium or strontium, on the extent of substitution x, and on the oxygen pressure, as shown in Table 2.4. It can be seen for the strontium oxides synthesized in air, like $La_2SrCu_2O_6$, that δ increases with the strontium content, tending towards $La_{2-x}A_{1+x}Cu_2O_6$. This causes the Cu(III) content to increase with the divalent A ion content, in order to compensate the oxygen vacancies due to the substitution of Sr^{2+} or Ca^{2+} for La^{3+}. Annealing of $La_{2-x}A_{1+x}Cu_2O_{6-x/2+\delta}$ (A = Ca, Sr) at 400°C under an oxygen pressure of 1 bar shows the ability of these phases to intercalate oxygen, δ ranging from 0 to 0.29 for $La_{2-x}Sr_{1+x}Cu_2O_{6-x/2+\delta}$ and $0.02 \leq \delta \leq 0.08$ for $La_{1.9}Ca_{1.1}Cu_2O_{5.95+\delta}$. One can see that the rate of intercalation is higher for the strontium oxides than for the calcium compound. Moreover, it seems that in the strontium oxides the maximum rate of intercalation increases with the strontium content. It is also worth noting that the strontium compounds $La_{2-x}Sr_{1+x}Cu_2O_{6-x/2+\delta}$ exhibit a variation of the interlayer distances: the c parameter of the tetragonal cell increases with the oxygen content δ, for the same x value. This influence of intercalation on the c parameter can be explained by the fact that oxygen is intercalated at the same level as the strontium ions between the pyramidal copper layers, which are then displaced. On the other hand, no variation of c was observed for calcium: one can assume that the oxygen content does not exceed O_6 in the limit of accuracy, so that all the excess oxygen has been intercalated inside the copper pyramidal layers. This can easily be explained by the smaller size of calcium, which does not allow oxygen to be intercalated between the layers. A similar situation will be observed for yttrium, which has about the same size, in the 92 K superconductor $YBa_2Cu_3O_7$ (Sect.2.4). The reversible character of this phenomenon is similar to that observed for $La_2Cu_4 \cdot O_4$-type oxides. For instance, oxygen disintercalation can be implemented by heating the most-oxidized compound $La_2SrCu_2O_{6.20}$ at 400°C under low oxygen pressures ($\sim 10^{-3}$ bar), which leads progressively to the reduced phase $La_2SrCu_2O_6$.

It will be seen further that the Cu(II) oxides exhibit a poor semiconducting behavior, whereas the oxygen intercalation allows semimetallic properties to be obtained. No superconductivity has been observed for these oxides when prepared at normal oxygen pressure, but a critical temperature of 60 K can be reached with oxygen pressures of about 20 bars [2.32].

Another oxide, $La_{1.1}Sr_{1.9}Cu_2O_{5.83}$, with a structure closely related to $La_2SrCu_2O_6$ was recently synthesized (Sect.2.5).

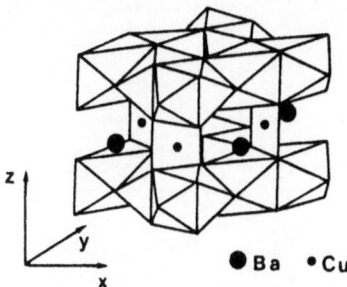

Fig.2.23. Structure of the $La_{4-2x}Ba_{2+2x}Cu_{2-x}O_{10-2x}$ oxide

● Ba • Cu

c) $La_{4-2x}Ba_{2+2x}Cu_{2-x}O_{10-2x}$

The oxide $La_{4-2x}Ba_{2+2x}Cu_{2-x}O_{10-2x}$ [2.33] appears to be the only Cu(II) compound of the system La-Ba-Cu-O whose structure is known. It has been synthesized by heating intimate mixtures of La_2O_3, CuO and $BaCO_3$ at 900° - 1000°C. The X-ray powder diffraction and electron diffraction study of this phase allowed the structure to be determined. It crystallizes in the tetragonal system with a = 6.86 Å, c = 5.87 Å The idealized formula of this oxide is in fact $A_6Cu_2O_{10}$. Copper (II) exhibits a square planar coordination with short Cu-O distances (1.852Å). The CuO_4 groups are isolated and oriented along the (110) and (1̄10) planes alternately, as shown in the perspective view of the structure (Fig.2.23). The barium and lanthanum ions ensure that the CuO_4 groups join. These cations exhibit different sorts of coordination according to whether they are located at the level z = 1/2 like copper [A(1) sites] or at z = 0 [A(2) sites]. The cations in A(1) are in tenfold coordination: the $A(1)O_{10}$ polyhedron (Fig.2.24a) can be described as a bicapped square-based prism: the A(1)-O distances (2.935-2.175Å) suggest that these sites are occupied by barium ions. The A(2) sites form a bicapped trigonal prism $A(2)O_8$ (Fig.2.24b); they are assumed to be mainly occupied by the lanthanum ions in agreement with the shorter A(2)-O distances (2.47-2.68Å).

The structure of this oxide can be described as a stacking along c of two sorts of layers with the ideal compositions $A(1)_2Cu_2O_5$ and $A(2)_4O_5$, respectively. The $A(1)_2Cu_2O_5$ layers have their polyhedra located at the level z = 1/2 (Fig.2.24c). In these sheets the $A(1)O_{10}$ polyhedra share their edges parallel to c, forming rows running along [110] and [1̄10]. The CuO_4 groups also share their edges parallel to c with the $A(1)O_{10}$ polyhedra. Clearly, the cohesion of these layers is principally ensured by the $A(1)O_{10}$ polyhedra, which form a host lattice for the copper ions. The $A(2)_4O_5$ layers (Fig.2.24d) can be described as being formed of groups of face-sharing trigonal prisms. These groups share their edges along c. The $A(1)Cu_2O_5$ and $A(2)_4O_5$ layers share the edges of their three sorts of polyhedra in the (001) plane. This phase is a very poor semiconductor, in agreement with the existence of Cu(II) in the form of isolated CuO_4 groups.

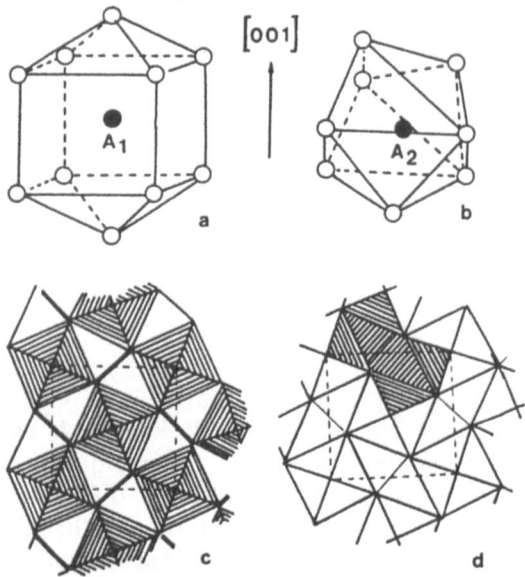

Fig.2.24. $La_{4-2x}Ba_{2+2x}Cu_{2-x}O_{10-2x}$. (a) The $A(1)O_{10}$ polyhedra occupied by Ba^{2+}. (b) $A(2)O_8$ polyhedra mainly occupied by La^{3+}. (c) The $A(1)_2Cu_2O_5$ layers. (d) The $A(2)_4O_5$ layers

2.4 The Pseudoternary System Y-Ba-Cu-O

The equilibrium phase diagram of the Y-Cu-O system has recently been reported by *Roth* et al. [2.34]. The only oxide whose structure has been determined [2.35] in this system is $Y_2Cu_2O_5$. It is isostructural with $Ho_2Cu_2O_5$ [2.36]. The orthorhombic cell (a=10.83Å, b=12.49Å, c=3.49Å) contains eight distorted YO_6 octahedra and eight distorted CuO_4 square planar groups. The arrangement of these polyhedra is shown in Fig.2.25. Each YO_6 octahedron shares an edge with two of its neighbors and a corner with the other six, while the edge- and corner-sharing CuO_4 groups form infinite chains running along **a**. Two other oxides have been synthesized: the Cu(I) oxide $YCuO_2$ [2.37], which was prepared by heating $Y_2Cu_2O_5$ [2.19] in a nitrogen flow at 1100°C, and the Cu(III) oxide $YCuO_3$, which was obtained under a high oxygen pressure ($4 \cdot 10^7$ Pa). The structure of those three compounds has so far not been determined.

2.4.1 The Orthorhombic 92 K Superconductor $YBa_2Cu_3O_{7-\delta}$

After the discovery by *Wu* et al. [2.38] of superconductivity at 92 K in the system Y-Ba-Cu-O, several groups tried to isolate the phase responsible for this phenomenon. The 92 K superconductor $YBa_2Cu_3O_7$ was rapidly isolated by several groups almost simultaneously [2.39-41]. This oxide, which presents the mixed valence Cu(II)/Cu(III), can be prepared by heating mix-

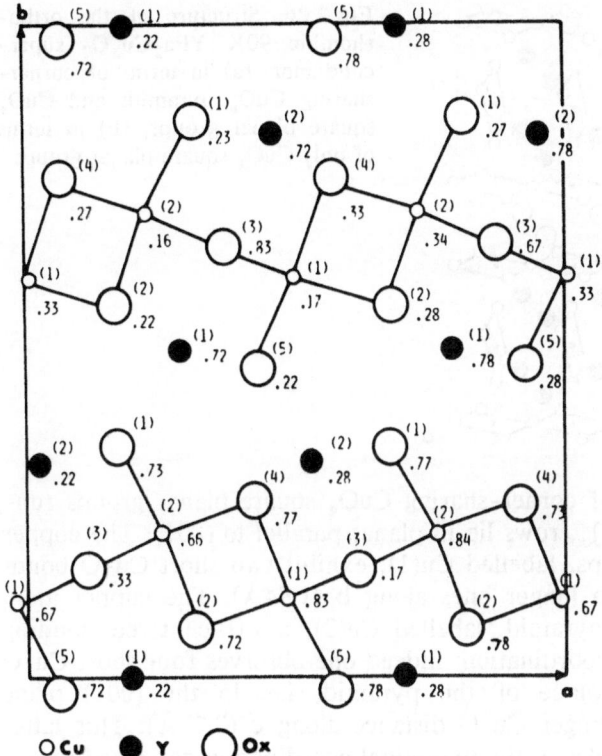

O Cu ● Y ◯ Ox

Fig.2.25. Projection of the structure of $Y_2Cu_2O_5$ onto (001). The z values are indicated and the number of each atom is noted in parentheses

tures of Y_2O_3, $BaCO_3$ or BaO_2 and CuO at 900°C in air, followed by sintering at 1000°C for some hours in an oxygen flow and finally cooling slowly or annealing under oxygen at a low temperature (400°- 500°C). The crystal structure of this phase was studied by X-ray diffraction [2.39-42] and neutron powder diffraction [2.43, 44]. The X-ray diffraction study of "single crystals" did not allow the correct structure to be determined owing to the systematic twinning of the crystals. This latter phenomenon will be described in Sect.2.4.3.

The 92 K superconductor $YBa_2Cu_3O_7$ is orthorhombic (a= 3.81 Å $\simeq a_p$, b=3.88Å$\simeq a_p$, c=11.63Å$\simeq 3a_p$). The structure of this superconductor (Fig. 2.26a), already pointed out above, can be considered as an oxygen-deficient perovskite type: it is derived from the stoichiometric perovskite structure by elimination of rows of oxygen atoms parallel to the [010] direction at the levels z = 0 and z = 1/2 in the orthorhombic cell and by an ordered distribution of the Ba^{2+} and Y^{3+} ions in the A sites of the ABO_3 perovskite in such a way that two barium planes alternate with one yttrium plane along c. This structure is built up from triple layers $[Cu_3O_7]_\infty$ of corner-sharing polyhedra, parallel to (001), whose cohesion is ensured by planes of yttrium cations. Each $[Cu_3O_7]_\infty$ layer consists itself of two $[CuO_{2.5}]_\infty$ sheets of corner-sharing CuO_5 pyramids. These sheets are connected to each other

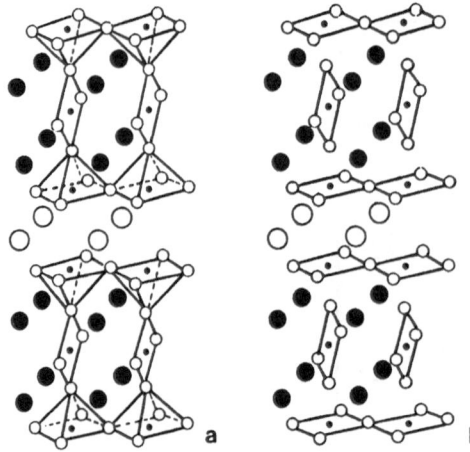

Fig. 2.26. Structure of the ortho-rhombic 90 K $YBa_2Cu_3O_7$ super-conductor: (a) in terms of corner-sharing CuO_5 pyramids and CuO_4 square planar groups, (b) in terms of only CuO_4 square planar groups

through $[CuO_2]_\infty$ rows of corner-sharing CuO_4 square planar groups running along **b**. The $[CuO_2]_\infty$ rows lie in planes parallel to (001). The copper atoms of the CuO_4 groups, labelled Cu(1), exhibit two short Cu-O bonds along **c** (1.85 Å) and two longer ones along **b** (1.94 Å). The copper atom belonging to the CuO_5 pyramid, labelled Cu(2), is off-centered, tending towards a square planar coordination: indeed one observes four short Cu-O distances in the basal plane of the pyramid, i.e. In the (001) plane (1.92-1.96 Å), and one longer Cu-O distance along **c** (2.27 Å). This latter configuration is very similar to the pyramidal coordination described previously for the other oxides. Nevertheless, it tends to increase the two-dimensional character of this structure. Assuming the Cu(2)-O bond parallel to **c** to be rather weak, owing to its length, one is led to consider that the $[Cu_3O_7]_\infty$ are formed of $[CuO_2]_\infty$ planes of corner-sharing CuO_4 square planar groups [corresponding to the Cu(2) atoms] separated by $[CuO_2]_\infty$ rows of corner-sharing CuO_4 groups [corresponding to the Cu(1) atoms]. Such a description (Fig. 2.26b) may also be of interest for the interpretation of the superconductivity of this phase, and especially of its high T_c. One important feature concerns the close relationships of this oxide with the two nonsuperconducting oxides $La_{2-x}Sr_{1+x}Cu_2O_6$ and Sr_2CuO_3 (see above). Sr_2CuO_3 exhibits $[CuO_2]_\infty$ rows of corner-sharing CuO_4 groups similar to those observed for $YBa_2Cu_3O_7$. In the same way, the double layers of corner-sharing CuO_5 pyramids interleaved with a plane of yttrium ions are very similar to those observed in $La_2SrCu_2O_6$ or $La_{1.9}Ca_{1.1}Cu_2O_6$, yttrium being replaced by calcium or strontium.

2.4.2 The Tetragonal Phases $YBa_2Cu_3O_{7-\delta}$

All researchers have observed that the orthorhombic superconducting form transforms into a tetragonal nonsuperconducting form when heated at temperatures ranging from 750°C to 900°C in air. This phase can also be prepared by heating a mixture of CuO, Y_2O_3 and $BaCO_3$ in air at 900°-1000°C and by quenching it to room temperature. The problem concerning

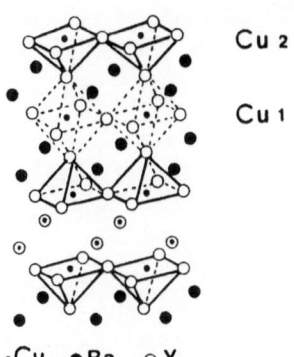

Cu 2

Cu 1

Fig.2.27. Structure of the tetragonal nonsuperconducting oxide $YBa_2Cu_3O_{7-\delta}$ ($\delta \simeq 0.7$)

•Cu ●Ba ◉Y

this phase transformation can be stated as follows: is the orthorhombic-tetragonal transformation a true transition, i.e. taking place without variation of composition, or does it result from an oxygen loss? Several neutron diffraction and electron microscopy studies of this phase [2.45-48] answer the question. This tetragonal form denoted T1 (a=3.838Å, c=11.792Å) differs (Fig.2.27) from the orthorhombic one only in the positions of some oxygen atoms, and by the oxygen content. One observes again triple layers of corner-sharing CuO_n polyhedra, whose cohesion is ensured by a plane of yttrium cations. The ordering of the yttrium and barium planes of the type "1-2", along c remains unchanged. In the triple layers of polyhedra one recognizes also the $[CuO_{2.5}]_\infty$ sheets of corner-sharing CuO_5 pyramids with similar Cu(2)-O distances (four 1.9Å and one 2.2Å). The CuO_4 groups of the orthorhombic phase corresponding to the Cu(1) atoms at z = 0 are replaced by "CuO_6 octahedra". The refinement by neutron diffraction of the oxygen occupancy of the different sites shows that the oxygen sites corresponding to the corner of the basal planes of the "CuO_6 octahedra" are less than half occupied, leading to a composition ranging from $YBa_2Cu_3O_{6.20}$ to $YBa_2Cu_3O_{6.40}$, according to the authors. Clearly, the orthorhombic-tetragonal transformation corresponds to an oxygen loss, and the composition of this tetragonal form seems to be rather well defined in these experimental conditions of synthesis in air. It is worth pointing out that the composition of the studied sample deduced from chemical analysis generally exhibits a higher oxygen content than that deduced from the neutron diffraction refinement. This difference is easily explained by the presence of an amorphous phase besides the crystallized matrix $YBa_2Cu_3O_{6.30}$. Such an assumption was supported first by the observation of high background of the neutron diffraction patterns. It was definitely confirmed by the electron diffraction study. The electron diffraction patterns give evidence of a crystallization quality of the particles ranging from very poor to quite good, and, moreover, low resolution images show than even crystallized particles often exhibit complex microstructures, which need to be specified.

The first example (Fig.2.28) is an Electron Diffraction (ED) pattern of a highly disturbed crystal: the broad intense reflections are those of the perovskite subcell, the superstructure spots characteristic of the ordering of the ions in triple layers are very weak, sometimes absent, and the presence

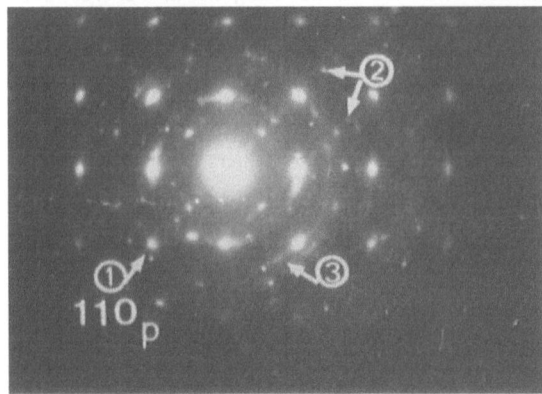

Fig.2.28. ED pattern of a highly distorted crystal of $YBa_2 \cdot Cu_3O_{7-\delta}$, 1: intense perovskite subcell; 2: weak superstructure $c \simeq 3a_p$; 3: polycrystalline rings

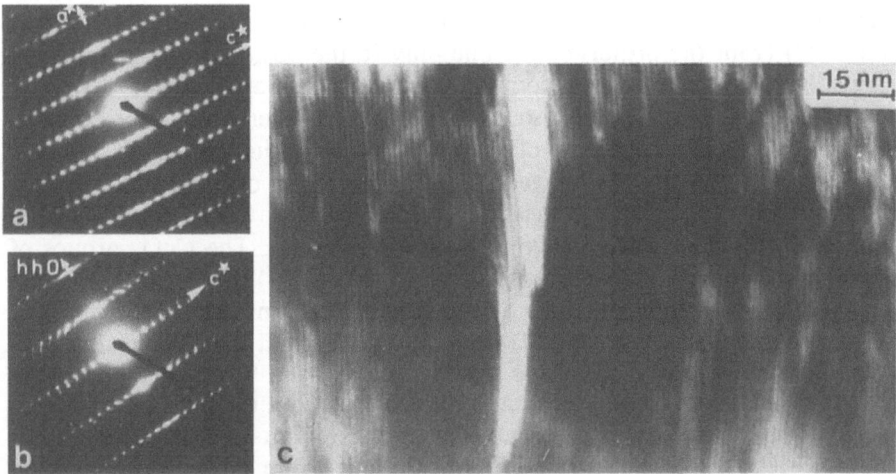

Fig.2.29. $YBa_2Cu_3O_{7-\delta}$. (a) h0l ED pattern with diffuse streaks along **c**. (b) hhl ED pattern showing arcs. (c) Corresponding low-resolution image

of rings shows a partly polycrystalline state. This pattern is typical of particles made up of misoriented crystallized areas embedded in a very poorly crystallized matrix, in agreement with the corresponding images.

A large majority of the crystals are characterized by diffuse streaks along the **c** axis (Fig.2.29a) and/or spots h0l or hhl arranged in quite continuous arcs (Fig.2.29b). These features can be correlated to the existence of numerous defects such as variations in the spacing of the layers and misorientations between different crystal areas, as shown from the corresponding low resolution image (Fig.2.29c). We note especially the presence of an amorphous layer at the edge of the crystal, fractures significantly more important than those encountered in the superconductor matrix, and undulation of the fringe systems, which are no longer resolved in some areas. In the same way, numerous crystals characterized by well-resolved

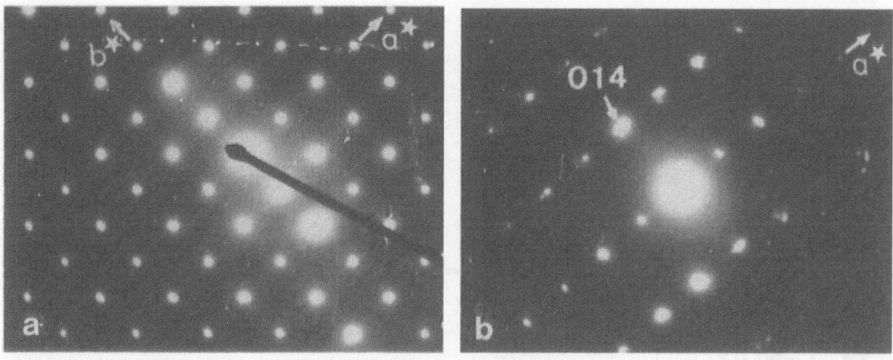

Fig.2.30. $YBa_2Cu_3O_{7-\delta}$. (a) hk0 ED diffraction pattern showing sharp dots. (b) hkl ED pattern of the crystal after tilting around [100]; hkl dots are split and diffuse

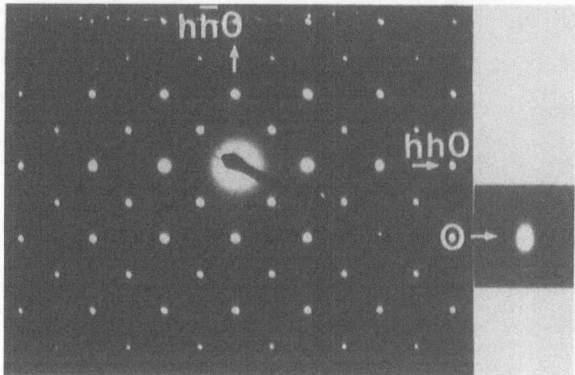

Fig.2.31. $YBa_2Cu_3O_{7-\delta}$. hk0 ED pattern with a slight splitting of dots, evidence of twinning

spots on [001] patterns (Fig.2.30a) exhibit split and diffuse spots 0kl, (Fig. 2.30b), when the crystal is tilted around [100].

Careful examination of the particles along [001] showed that the matrix is composed of both true tetragonal crystals and orthorhombic ones, as shown from the splitting of hh0 reflections (Fig.2.31). The latter corresponds to an orthorhombic distortion, usually much smaller than that observed in the superconducting oxide $YBa_2Cu_3O_7$. Such phenomena do indeed suggest variations of the oxygen content from one crystal to another.

The variation of oxygen stoichiometry in the crystals, without destroying the structure, can be carried further, leading to the limiting oxide $YBa_2Cu_3O_6$ [2.49-51]. This oxide, denoted T2, can be synthesized by heating the orthorhombic superconductor to 600°C in vacuum. The samples obtained by this treatment are tetragonal ($a=3.86\,\text{Å}\simeq a_p$ and $c=11.85\,\text{Å}$) and exhibit no superconductivity. Exactly as for the tetragonal $YBa_2Cu_3O_{6.3}$ crystals, these crystals do not show any twinning, contrary to the orthorhombic superconducting crystals, so that their structure could be determined from single-crystal X-ray diffraction data. The perspective view of this crystal structure (Fig.2.32) shows that it is very closely related to that

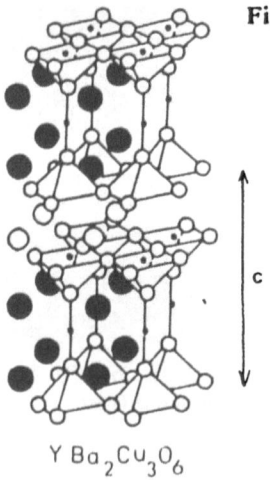

Fig.2.32. Structure of the limiting tetragonal oxide $YBa_2Cu_3O_6$

$Y Ba_2 Cu_3 O_6$

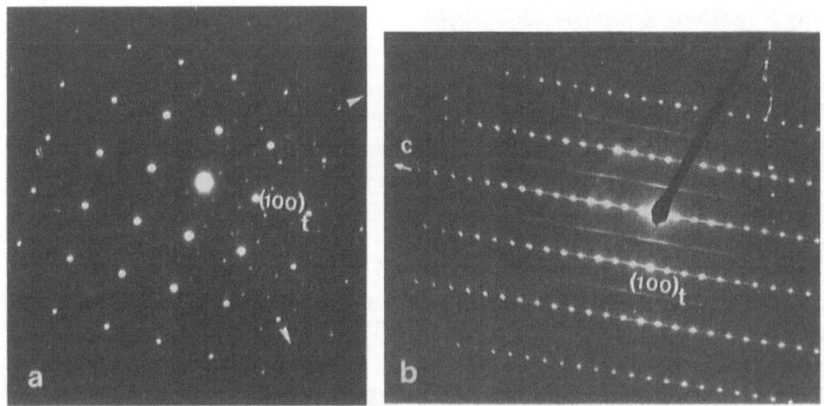

Fig.2.33a,b. ED pattern of $YBa_2Cu_3O_{6+\delta}$ crystals showing a $2\sqrt{2}a \times 2\sqrt{2}a$ superstructure. (a) a^*a^* plane (sharp spots). (b) a^*c^* plane (diffuse streaks along c^*)

of the orthorhombic superconductor $YBa_2Cu_3O_7$. It is indeed derived from this latter structure by simply removing the oxygen atoms of the $[CuO_2]_\infty$ rows located at the same level as the Cu(1) (z = 0). In this structure the pyramidal layers $[CuO_{2.5}]_\infty$ remain unchanged with four short Cu(2)-O distances (1.94Å) and one long Cu(2)-O bond (2.47Å) and can be assumed to be occupied by Cu(II), while univalent copper in twofold coordination [Cu(1) site] ensures the connection between these layers, with rather short Cu-O distances (1.80Å). Besides those numerous crystals which can be formulated $YBa_2Cu_2^{II}Cu^IO_6$, some crystals were found to exhibit superstructure reflections (Fig.2.33a) corresponding to a $2\sqrt{2}a \times 2\sqrt{2}a \times c$ cell. These crystals are also tetragonal within the resolution of the electron diffraction method. The superstructure reflections are weak but quite sharp in the a^*a^* plane (Fig.2.33a) but form diffuse streaks along c^* with an almost uniform intensity distribution (Fig.2.33b). This means that there is long-range order

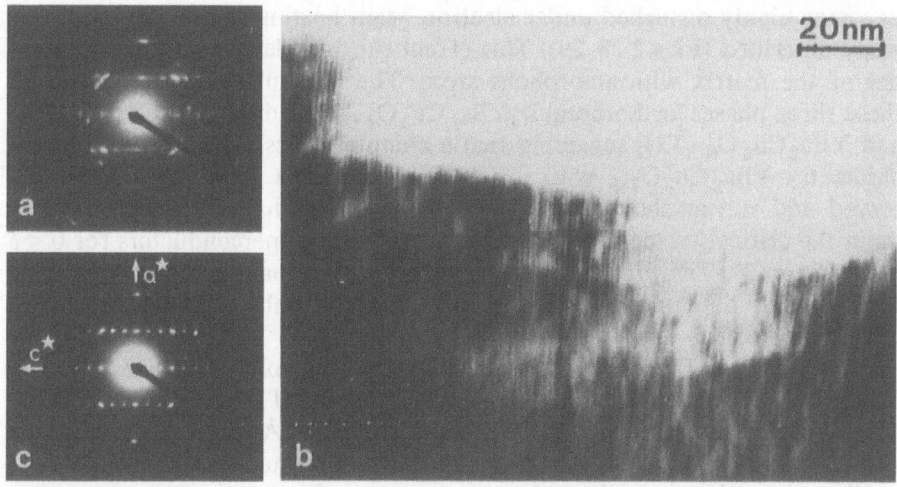

Fig.2.34. $YBa_2Cu_3O_{6+\delta}$ h0l ED pattern (**a**) and low resolution image (**b**) of a poorly crystallized crystal and h0l ED pattern (**c**) of the same crystal after electron irradiation

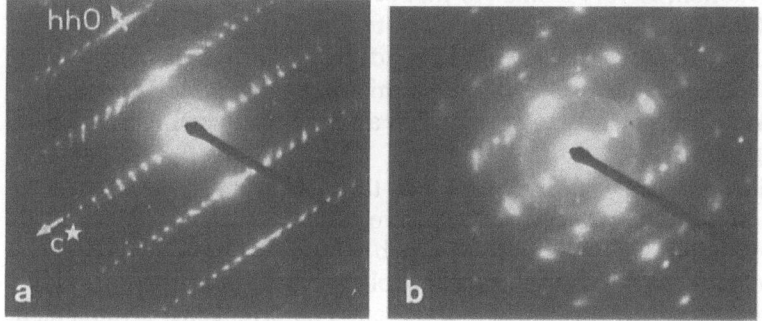

Fig.2.35. $YBa_2Cu_3O_{6+\delta}$. hhl ED pattern (**a**) before and (**b**) after electron irradiation. The superstructure $c \simeq 3a_p$ dots gave way to rings due to polycrystalline structure

of the structural feature causing the superstructure in the **aa** plane but almost no correlation along the **c**-axis. The superstructure is very sensitive to heating by the electron beam. The $2\sqrt{2}a \times 2\sqrt{2}a$ pattern converts irreversibly into the ordinary tetragonal pattern, suggesting that the phase exhibiting this superstructure might be thermodynamically unstable or might lose oxygen on heating by the beam. Such a behavior under the electron beam was also observed for the tetragonal phase $YBa_2Cu_3O_{6.3}$ [2.48]. Two examples illustrate this feature. The first crystal can be classified as poorly crystallized from its [010] ED pattern (Fig.2.34a) and the corresponding low-resolution image (Fig.2.34c) exhibits quite sharp spots with no more streaks or arcs. This behavior can be related to a loss of oxygen, corresponding to a transformation of a crystal $YBa_2Cu_3O_{6.3}$ into $YBa_2Cu_3O_6$. This feature suggests that, in such disturbed crystals, disorder is only related to local variations in the distribution of the oxygen in the copper planes, between the Ba layers. In contrast, the second crystal (Fig.2.35a) becomes highly disturbed under electron beam heating (Fig.2.35b) as previously described (Figs.2.28, 29). This effect can be related to a greater disorder of the matrix with amorphous areas. The structural characterization of these three phases [orthorhombic $YBa_2Cu_3O_7$, tetragonal $YBa_2Cu_3O_{6.3}$ (T1) and $YBa_2Cu_3O_6$ (T2)] suggested that a complete possible range of nonstoichiometry $YBa_2Cu_3O_{7-\delta}$ with $0 \leq \delta \leq 1$ could exist. This was indeed observed and it was shown that the oxygen content has a drastic influence upon the critical temperature of the orthorhombic superconductors for $0 \leq \delta \leq 0.6$ (Chap.3). The problems of nonstoichiometry for those oxides that exhibit various defects are very complex, and can only be solved by high-resolution electron microscopy (Chap.6).

Besides the T1 and T2 tetragonal phases, an oxygen rich 123 phase, $YBa_2Cu_3O_{6.8}$, with tetragonal symmetry, denoted T3, was isolated [2.52]. This T3 form is characterized by $c/a = 3$ (a=3.866Å, c=11.599Å). Such a metastable oxide can only be prepared by ceramic techniques using a wet medium for grinding and heating in an oxygen flow at temperatures ranging between 850° and 930°C. As for the other two tetragonal forms, no superconductivity was detected. Its mean structure, determined by powder neutron diffraction, is very similar to that of T1, showing an increase in the occupancy factor of the oxygen sites located at the same level as Cu(1) (Fig.2.36) along c. In fact the HREM study of this phase shows that its structure is more complex. One observes microdomains oriented at 90° resulting from the lack of mismatch between a and c (c is rigorously equal to 3a). Such domains can be easily imaged; some examples are shown in Fig.2.37 to illustrate the way they set up in the bulk, varying in size as well as in shape. For instance, the small [001] area shown in the bulk of a [100] oriented crystal (Fig.2.37a) exhibits a and b parameters parallel to a and c parameters of the crystal matrix, respectively. Note that in the [001] area, twin boundaries (curved arrows) are observed. The orthogonal arrangement of the c axis is illustrated in Fig.2.37b. In these two examples domains are adjacent and well shaped, but in some crystals (Fig.2.37c) the three orientations are observed without any sharp boundary between the domains, suggesting their superposition in the different slices. Some striped domains cross over the whole crystal. An example of six adjacent domains, numbered 1-6, are shown in Fig.2.37d; the widths of these domains are $w_1 = 29$

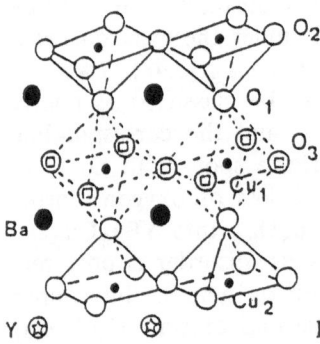

Fig.2.36. Structure of $YBa_2Cu_3O_{6.8}$, T_3 form

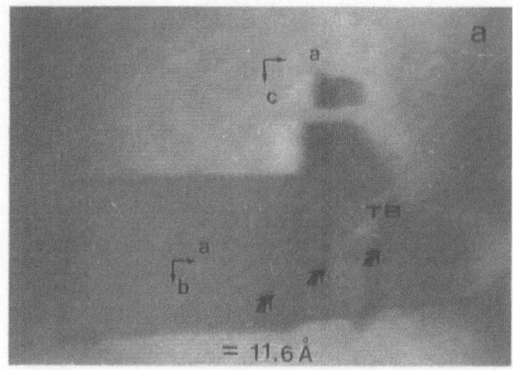

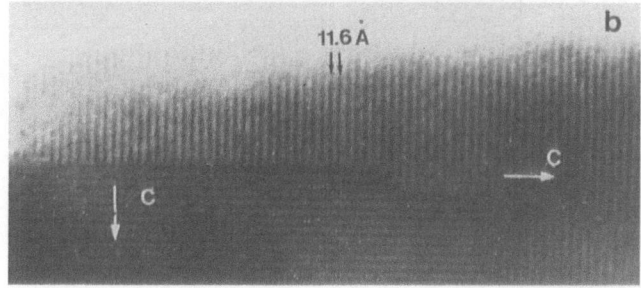

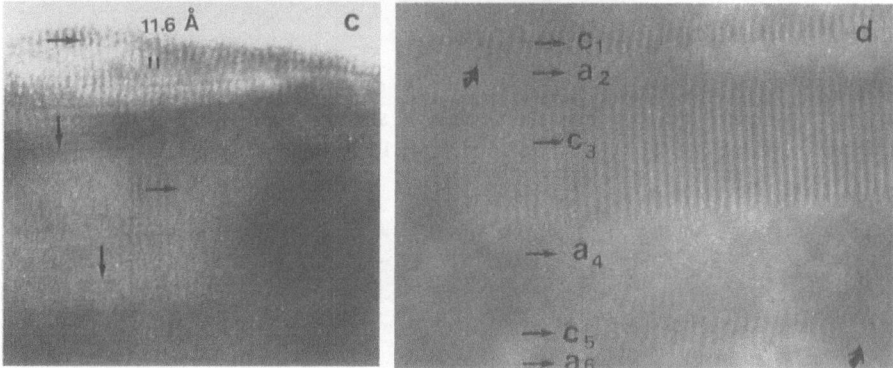

Fig.2.37a-d. Examples of oriented domains. (a) Small [001] areas can be observed in [100] oriented crystals. (b) Orthogonal arrangement of two c-axes: Note that the boundaries are straight and sharp. (c) Another orthogonal arrangement of c-axes: Note that the orientations are observed without any boundary between the domains. (d) Examples of striped domains

Å, w_2 = 35 Å, w_3 = 230 Å, w_4 = 100 Å and w_5 = 35 Å. It should be noted that in the arrowed areas (domains *1* and *5*) a misorientation is observed. Models corresponding to the arrangement of such domains are shown in Fig.2.38, but these are not the only ones and many other types of junctions between 123 domains can be proposed.

The second important feature deals with microtwinning, as shown, for instance, in Fig.2.39a, where is can be seen that the twinning domains are spaced at distances ranging from 22 Å (i.e., $4a_p\sqrt{2}$) to 65 Å (i.e., $12a_p\sqrt{2}$).

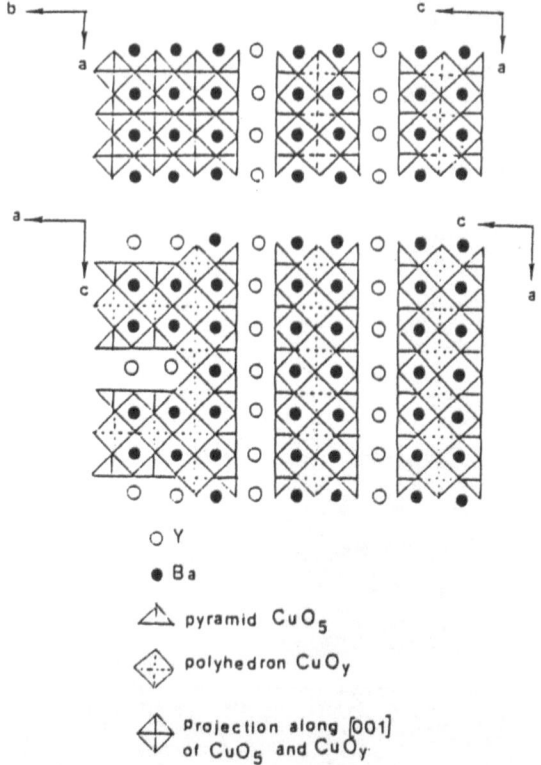

○ Y

● Ba

△ pyramid CuO$_5$

◇ polyhedron CuO$_y$

◇ Projection along [001]
  of CuO$_5$ and CuO$_y$

Fig.2.38. Two examples of idealized models of straight junctions. The nature of the polyhedron is not specified, except for the CuO$_5$ pyramid screwed along [100] or [010]

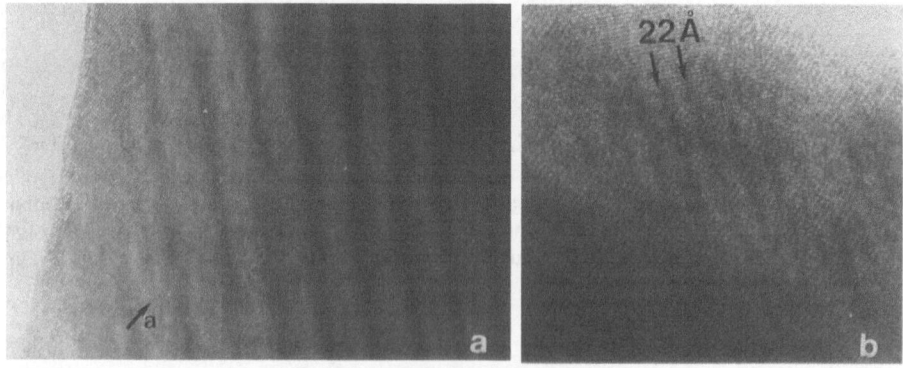

Fig.2.39. (a) [001] high resolution image showing the small width of the twinning domains. (b) Example of twinning domains spaced at 22 Å

An example of the smallest domains is depicted in Fig.2.39b with a regular 22 Å spacing. The mean width of the domains is often regular in quite a large area of the crystals. The twin boundaries are rather broad. These observations are of fundamental importance, since they definitely show that

42

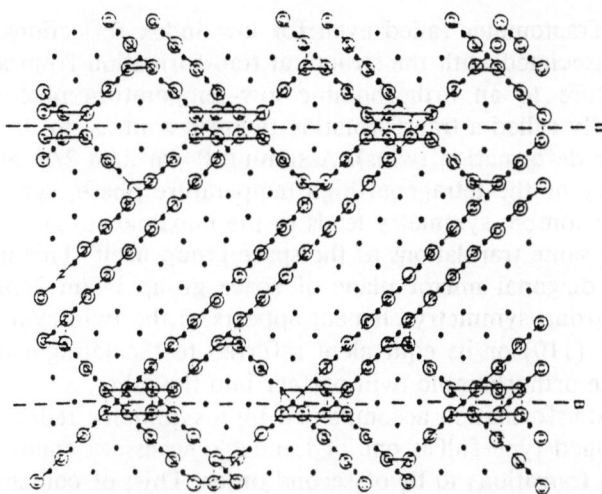

Fig.2.40. Structural model showing twin boundaries containing alternately two univalent copper atoms in twofold coordination and two mixed valence copper atoms Cu(II)-Cu(III) in octahedral coordination

the true symmetry of this phase is not tetragonal, but may be orthorhombic. They strongly support the formation of CuO_5 and CuO_6 polyhedra around Cu(1) due to an extra copper disproportionation. One can indeed propose a structural model (Fig.2.40) in which the twinning boundaries contain alternately two univalent copper atoms in twofold coordination and two mixed valence copper atoms [Cu(II)-Cu(III)] in octahedral coordination. In this model one observes in each twinning domain double rows mainly built up of CuO_4 square planar groups parallel to **a**, which alternate with double rows formed principally of Cu^IO_2 rods. These rows are connected, from time to time, in the orthogonal direction through CuO_4 groups and CuO_5 pyramids, corresponding to a spacing of $6a_p \sqrt{2}$ of the twinning boundaries. Of course, very similar models can be proposed for the different spacings of the boundaries just by changing the number of copper atoms in the chains. In the same way, the ordering of the "O_6" and "O_7" chains along $\langle 110 \rangle$ can also be changed. The fact that this oxide does not exhibit superconducting properties can easily be explained by its microstructure. The microtwinning effects involve a three-dimensional connection between the copper chains within the triple copper layers. Moreover, the existence of oriented domains shows that the triple copper layers are connected to each other, forming a three-dimensional framework, which is detrimental to superconductivity.

2.4.3 Problems of Microtwinning in the Orthorhombic $YBa_2Cu_3O_{7-\delta}$

Concerning the crystallographic quality of the orthorhombic crystals for the X-ray study, we pointed out above that not a single one turned out be an untwinned crystal. In these conditions attempts to measure integrated intensities with a 4-circle diffractometer failed even for low-index reflections. This twinning [2.42] is associated with the structural transformation from a tetragonal high-temperature to an orthorhombic low-temperature phase. This kind of twin is usually called a transformation twin (in contrast to, for instance, growth twins or deformation twins). Assuming P4/m 2/m 2/m as the space group symmetry of the tetragonal high temperature phase, symmetry reduction to orthorhombic symmetry leads to the maximal subgroup P2/m 2/m 2/m, with the same translations as the space group itself. This is equivalent to losing the diagonal mirror plane of space group P4/m 2/m 2/m and this lost space group symmetry element appears as the twin symmetry element [the plane (110) or its equivalent referred to the tetragonal cell] which transforms one orthorhombic twin partner into the other.

A structural phase transformation accompanied by a symmetry reduction of the above-mentioned type fulfils one of Landau's necessary conditions for structural phase transitions to be of second order. This, of course, does not mean that the transition actually is of second order.

Starting from a tetragonal cell (Fig.2.41a) there are two ways of distorting this cell into an orthorhombic one of given a/b while retaining a particular diagonal plane. Figure 2.41 depicts a projection of two cells of two orthorhombic twins (I and II) onto the ab plane. They are related to each other by reflection at the tetragonal (110) plane. The axes a_I and b_{II} (or a_{II} and b_I) include a characteristic angle Δ, which depends on the distortion of the cell from the tetragonal metric and amounts to 0.45° for the orthorhombic cell given above.

The corrresponding projection in reciprocal space is depicted in Fig. 2.41b. In reciprocal space the angle Δ appears between directions [100]I and [010]II together with angle 2Δ between directions [110]I and [1$\bar{1}$0]II. This schematic picture exactly reveals the features of the TEM-SAD pattern of $YBa_2Cu_3O_{7-\delta}$ seen in Fig.2.42a: splitting of reflections hh0 (and $\bar{h}h0$) into two peaks at equal distances from the origin and relatively large separation (angle 2Δ), small splitting of reflections h00 and 0k0 (angle Δ) at different distances from the origin and no splitting of reflections of type $\bar{h}h0$ (and hh0).

Superposition of the contributions of many twins is a general feature always observed in X-ray diffraction experiments on bulk samples. Figure 2.43 shows a precession photograph, hh0 plane, taken with filtered Cu Kα radiation, of a bulk crystal of $YBa_2Cu_3O_{7-\delta}$; Fig.2.41c gives the corresponding sketch. This pattern is easily obtained by rotating Fig.2.41b by 90°C around c* and superimposing it onto the original pattern. This leads to a 4-fold splitting along a* (referred to the tetragonal cell) and a 3-fold splitting along the diagonal with the central peak twice as strong as the "satellites" (indicated by a large black spot and two smaller ones in Fig. 2.41c).

Figures 2.44 and 45 show two examples of diffractometer scans recorded with a small counter aperture of about 2mm × 2mm to get optimum angular resolution. The positions of the scans are indicated in Fig. 2.41c. The first one is a ω scan through the position of reflection 200 (tetragonal).

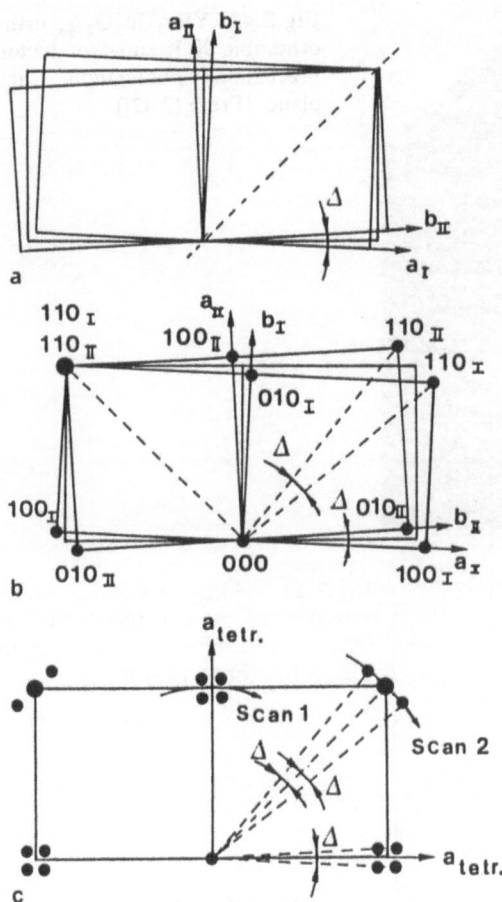

Fig.2.41a-c. YBa$_2$Cu$_3$O$_{7-\delta}$, ortho-rhombic 90 K superconductor. (a) Schematic representation of twin orientations I and II in direct space (projection along **c**). (b) Diffraction pattern constructed from the ED results. (c) Diffraction pattern for bulk samples as seen in X-ray diffraction experiments. (From [2.42])

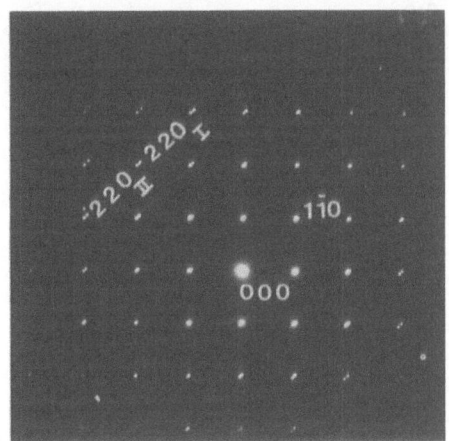

Fig.2.42. YBa$_2$Cu$_3$O$_{7-\delta}$, orthorhombic 90 K superconductor. ED patterns of very thin samples

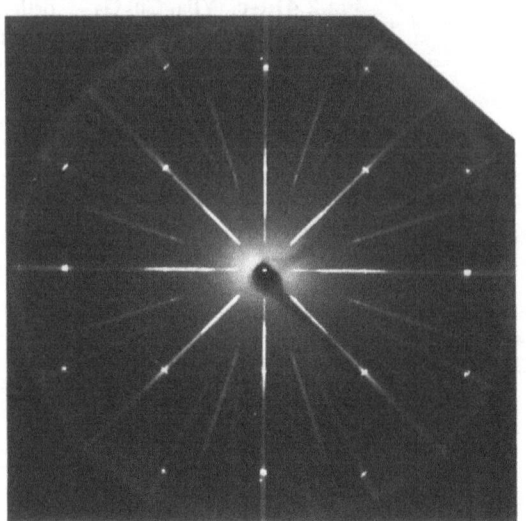

Fig.2.43. $YBa_2Cu_3O_{7-\delta}$, orthorhombic 90 K superconductor. Precession photograph, hk0 plane. (From [2.42])

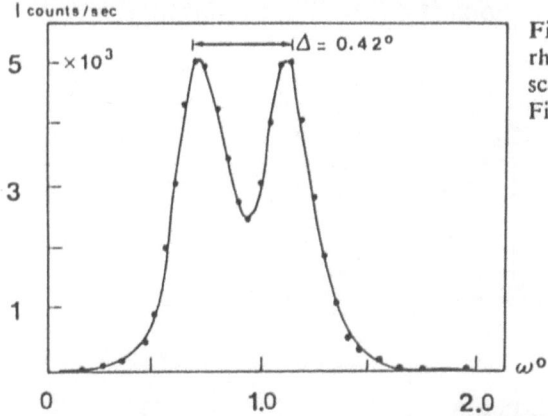

Fig.2.44. $YBa_2Cu_3O_{7-\delta}$, orthorhombic 90 K superconductor. ω scan of reflection 200 (*scan 1* in Fig.2.41c). (From [2.42])

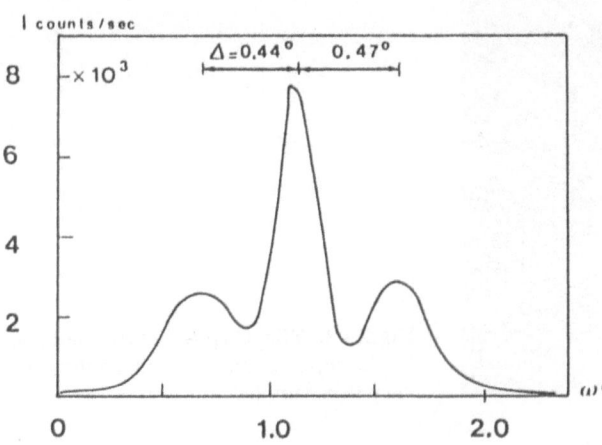

Fig.2.45. $YBa_2Cu_3O_{7-\delta}$, orthorhombic 90 K superconductor. ω scan of reflection 110 (*scan 2* in Fig.2.41c). (From [2.42])

The fourfold splitting in Fig.2.41c appears in these scans as a pair of partially overlapping peaks of equal intensity and with an angular separation of $\Delta = 0.42°$.

Scan 2 for reflection 110 shows three peaks with an integral intensity ratio of about 1:2:1 and an angular separation of 0.44° and 0.47°, respectively. This compares very well with the "theoretical" value of 0.45° calculated from the actual orthorhombic distortion from powder X-ray diffraction studies (see above).

The fact that no single crystals of orthorhombic $YBa_2Cu_3O_{7-\delta}$ are available implies that the acquisition of accurate structural data from neutron or X-ray diffraction experiments on bulk twinned samples is greatly influenced by the process of space-averaging that is inherent to these methods. Moreover, as already pointed out, the deviation from oxygen stoichiometry in the orthorhombic superconductor $YBa_2Cu_3O_{7-\delta}$ makes the understanding of the structure of this phase more difficult, especially for $0 < \delta < 0.6$. This structural problem, together with the formation of different defects, and microtwinning form the subject of high-resolution electron microscopy investigations (Chap.6).

2.4.4 The "Green Phase" Y_2BaCuO_5

During recent investigations of the Y-Ba-Cu-O systems for superconductivity, a great number of researchers drew attention to the formation of a green phase with semiconducting properties. This oxide [2.52] has been known since 1982, and belongs to a large family of isostructural compounds A_2BaCuO_5 (A = Y, Sm, Eu, Gd, Dy, Ho, Er, Yb). It crystallizes in the orthorhombic system (a= 7.13Å, b=12.18Å, c=5.66Å). The structure of this compound can be described in terms of a framework based on YO_7 polyhedra. The yttrium ions exhibit a monocapped trigonal prismatic coordination: two YO_7 prisms share one triangular face, forming Y_2O_{11} units (Fig.2.46). In the orthorhombic cell, these structural units are located at z = 1/4 and 3/4. At the same z level they share their edges parallel to c, forming zigzag chains along b (Fig.2.47); at a different z level each unit shares its edges parallel to (001) with three other units of the copper level and three other units of the upper level, forming rings (Fig.2.48). This framework forms cavities where the barium and copper ions are located. The barium ions are surrounded by eleven oxygens whereas copper exhibits a pyramidal coor-

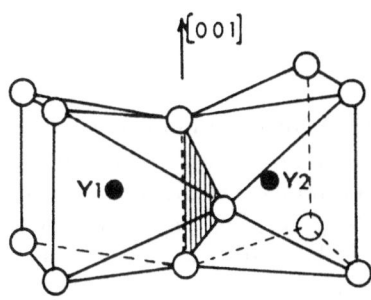

Fig.2.46. Y_2BaCuO_5. Units of Y_2O_{11} resulting from the association of two monocapped YO_7 prisms, sharing one triangular face (hatched)

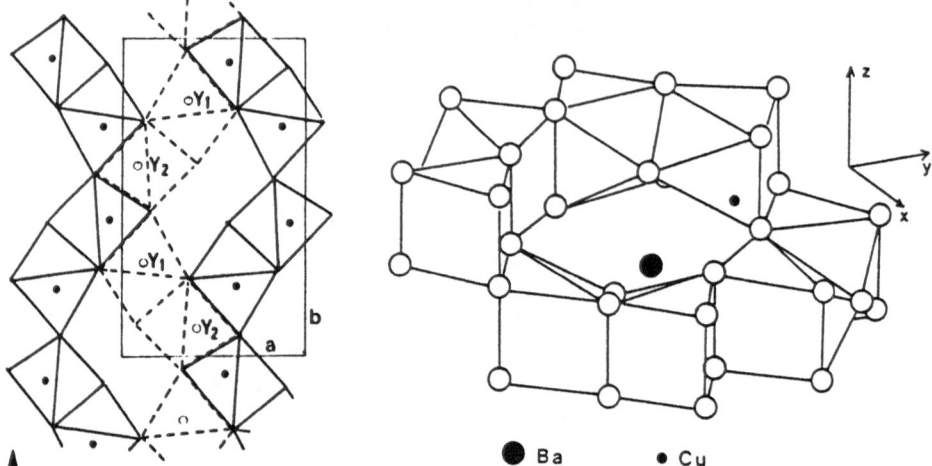

● Ba   • Cu

Fig.2.47. Y_2BaCuO_5. Projection on (001) of the strings of Y_2O_{11} units. Solid lines correspond to the units at z = 3/4, while dashed lines correspond to the units at z = 1/4

Fig.2.48. Y_2BaCuO_5: Schematized perspective view of the rings formed by the Y_2O_{11} units

dination. The CuO_5 pyramids are isolated and rather strongly distorted: one observes two short Cu-O distances (2.01 Å), two medium distances (2.18 Å) and one long distance (2.29 Å). The low conductivity of this oxide is in agreement with the presence of only divalent copper.

2.4.5 Other Phases of the System Y–Ba–Cu–O

Several other phases closely related to the perovskite phase have been observed recently in this system. *De Leeuw* et al. [2.16] report three phases: $YBa_4Cu_3O_{9-x}$, tetragonal with a = 8.069 Å, and c = 4.034 Å; $Y_3Ba_8Cu_5 \cdot O_{18}$, tetragonal with a = 5.78 Å, c = 8.01 Å; and $YBa_5Cu_2O_9$, orthorhombic with a = 4.03 Å, b = 4.09 Å and c = 21.6 Å, whereas *Chevalier* et al. [2.54] have observed an oxygen-deficient perovskite-type $YBa_4Cu_3O_{9-x}$, tetragonal with a = 5.803 Å and c = 8.03 Å. The structures of these oxides are not yet established with certainty and for this reason will not be discussed here. None of these compounds seem to present superconducting properties.

2.5 The Systems Ln–Sr–Cu–O (Ln≠La)

The system Y–Sr–Cu–O has not been investigated correctly until now, so it is not useful to give a description here of compounds whose compositions and structures have not been determined with certainty. On the other hand, a series of oxides with the general formula $Ln_{2-x}Sr_{1+x}Cu_{6-x/2+\delta}$ with rich strontium contents (0.70 ≤ x ≤ 0.90) have been isolated for Ln = La, Sm, Eu and Gd [2.55, 56]. The crystallographic characteristics of those oxides (Table

Table 2.5 Crystallographic parameters of $Ln_{2-x}Sr_{1+x}Cu_2O_{6-x/2+\delta}$

Ln	Compound	x	a [Å]	b [Å]	c [Å]	V [Å³]
Nd	$NdSr_2Cu_2O_{5.66}$ (or $Sr_6Nd_3Cu_6O_{17}$)	1	3.755	11.488	20.098	8.669
	$NdSr_2Cu_2O_{5.76}$	1	3.770	11.438	20.094	866.5
Sm	$Sm_{1.30}Sr_{1.70}Cu_2O_{5.65}$	0.70	3.746	11.416	20.068	858.2
	$Sm_{1.10}Sr_{1.90}Cu_2O_{5.55}$	0.90	3.747	11.439	20.058	859.7
Eu	$Eu_{1.40}Sr_{1.60}Cu_2O_{5.70}$	0.60	3.742	11.325	20.052	849.8
	$Eu_{1.30}Sr_{1.70}Cu_2O_{5.65}$	0.70	3.744	11.337	20.047	850.9
	$Eu_{1.10}Sr_{1.90}Cu_2O_{5.55}$	0.90	3.742	11.380	20.020	852.5
Gd	$Gd_{1.30}Sr_{1.70}Cu_2O_{5.65}$	0.70	3.738	11.350	20.045	850.4
	$Gd_{1.10}Sr_{1.90}Cu_2O_{5.55}$	0.90	3.738	11.385	20.015	851.8
La	$La_{1+x}Sr_{2-x}Cu_2O_{5.5+\delta}$	0.05–0.15	3.80	11.48	20.23	882.5

2.5) are very similar to those of $La_2SrCu_2O_6$ (Sect.2.3), but differ from the latter by a tripling of one of the parameters leading to an orthorhombic cell with $a \simeq a_p$, $b \simeq 3a_p$, $c \simeq 20$ Å ($a_p \simeq 3.8$ Å is the parameter of the cubic perovskite cell). These compounds can also be distinguished from the lanthanum-rich $La_2SrCu_2O_6$-type oxides by a significant oxygen deficiency with respect to the "O_6" ideal formula given above.

The system "$Nd_2SrCu_2O_6 - NdSr_2Cu_2O_{6-y}$" also exhibits the oxides $Nd_{2-x}Sr_xCu_2O_{6-x/2+\delta}$ (x = 0.14) [2.57] isotypic of $La_2SrCu_2O_6$. A recent investigation [2.58, 59] allowed a new layered cuprate $Sr_6Nd_3Cu_6O_{17}$ to be isolated by solid state reaction under an argon flow, and its structure to be determined by neutron diffraction. This phase, which can also be formulated $Sr_2NdCu_2O_{5.66}$, exhibits similar parameters to those observed for the other strontium-rich compounds. The structure of this oxide is characterized by a similar arrangement of the metallic atoms and a great part of the oxygen atoms with respect to the $La_2SrCu_2O_6$-type structure (Fig.2.49). However, several of the oxygen atoms and anionic vacancies are distributed in a different way. Consequently, it appears that this structure is intermediate between the structure of the "123" phase $YBa_2Cu_3O_7$ [2.44] and the "0212" structure of $La_2SrCu_2O_6$. One can describe it as a regular intergrowth of the "123"-type structure with the rock salt structure. It is indeed built up from "123"-type ribbons $[Sr_{7/6}Nd_{5/6}Cu_2O_{4.66}]_\infty$ running along b and $[Sr_{5/6}Nd_{1/6}O]_\infty$ rock-salt-type layers parallel to (100). Thus $Sr_6Nd_3 \cdot Cu_6O_{17}$ can be considered as the member m = 2, n = 1 of a series of intergrowths of general formula $[ACuO_{2.33}]_m^{"123"}[AO]_n^{Rs}$, in which A corresponds to a suitable content of alkaline earth cations (Sr, Ba) and of lanthanides. The "123"$[Sr_{7/6}Nd_{5/6}Cu_2O_{4.66}]_\infty$ ribbons, which are three polyhedra wide along b ($2CuO_5 + 1CuO_4$), exhibit a significant difference from the classical $YBa_2Cu_3O_7$ structure. The Cu-O interatomic distances show indeed that

49

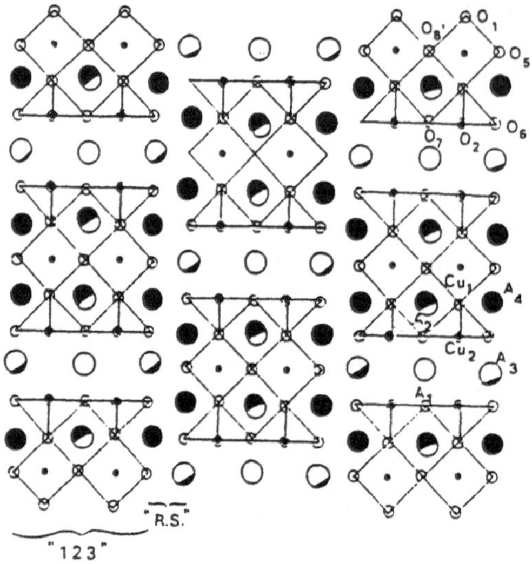

Fig.2.49. Crystal structure of $Sr_6Nd_3Cu_6O_{17}$

the CuO_4 groups, like the CuO_5 pyramids, are strongly distorted, with Cu–O distances ranging from 1.85 Å to 2.22 Å The apical Cu–O distance of the CuO_5 pyramid (2.22 Å) is shorter than that observed for $YBa_2Cu_3O_7$ (2.29 Å), whereas a large distortion is observed in the basal plane (1.88–2.02 Å) compared to $YBa_2Cu_3O_7$ (1.93–1.96 Å). The CuO_4 groups are no longer planar: Cu(1), two O(1) atoms and O(5) remain in approximately the same plane, whereas O(8'), which ensures the junction between two CuO_4 groups, is significantly out of this plane, leading to a Cu–O–Cu angle of 110° instead of 180°. These features are easily understandable if one considers the sequence of "123" ribbons along c (Fig.2.49). Two successive ribbons are shifted by about 3.8 Å along b, so that at the same level along this latter direction a neodymium plane is replaced by a copper plane. Such an arrangement induces strains in the structure which can only be alleviated by a distortion of the CuO_4 and CuO_5 polyhedra and by their tilting.

The distribution of the strontium and neodymium cations is remarkable and is certainly an important factor for the stabilization of such a structure. One observes that in the middle of the "123" type ribbons the smaller cations, neodymium, take the place of yttrium in eightfold coordination [Nd(1) sites], whereas in rock-salt-type layers the sites which exhibit an eightfold coordination [Sr(4) sites] are occupied only by strontium. On the other hand, the cationic sites, which are located at the intersection of the rock salt layer and the basal planes of the CuO_5 pyramids [Nd/Sr(3) sites], are occupied half by strontium and half by neodymium. In the same way, it is worth pointing out that the barium sites, in the "123" ribbons, are occupied by much smaller cations, half neodymium and half strontium [Nd/Sr(2) sites]. This latter occupancy may be at the origin of the distortion of the CuO_4 square groups and may allow an adjustment of the "123" and rock-salt layers.

The structure of the other strontium-rich compounds, with oxygen contents intermediate between "$O_{5.66}$" and "O_6" has not yet been completely

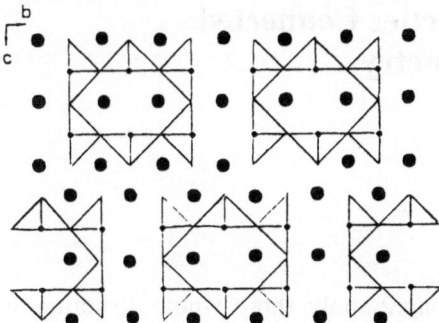

b
c

Fig.2.50. Schematic drawing of the hypothetical compound $Sr_2NdCu_2O_6$

elucidated. Nevertheless, the neutron diffraction studies of the oxides, $La_{1.1}Sr_{1.9}Cu_2O_{5.83}$ [2.60], $NdSr_2Cu_2O_{5.76}$ [2.59] and $Nd_{1.4}Sr_{1.6}Cu_2O_{5.79}$ [2.61] clearly show that their structure is closely related to that of $Sr_6Nd_3 \cdot Cu_6O_{17}$. The oxygen distribution in the structure can indeed be interpreted as the result of the intimate intergrowth of two extreme structures, the first corresponding to $Sr_6Nd_3Cu_6O_{17}$ (Fig.2.49) and the second to the hypothetical compound $Sr_2NdCu_2O_6$, whose ideal structure (Fig.2.50) would consist of six-sided tunnels built up from rings of six corner-sharing CuO_5 pyramids. Such tubes would be arranged so as to form fluorite-type cages and rock-salt-type layers where the Nd and Sr cations would be located.

From all these results it is clear that the oxygen nonstoichiometry in these oxides is not yet completely understood and that it should be possible to synthesize new phases by playing with ordering of oxygen and vacancies, i.e., by changing the thermal treatments (temperature, time) and also the nature of the gaseous atmosphere. Although these oxides do not superconduct, knowledge of their magnetic and electron transport properties will be crucial for our understanding of superconductivity in layered cuprates. Similar layered structures can be expected for other lanthanides in the systems Sr–Ln–Cu–O and Sr–Y–Cu–O.

3. Electron Transport Properties Connected with Oxygen Nonstoichiometry

Of the copper oxides described in Chap.2, only those whose structure is either an oxygen-deficient perovskite, or an intergrowth of the rock-salt-type structure and the perovskite structure, present various electron transport properties leading to semiconducting, metallic or superconducting behavior. These properties are due to copper, which exhibits a mixed valence in these oxides that seems to be enhanced by the localization of copper in the perovskite parts of the structures. In this chapter we focus our attention on the electron transport properties of cuprates whose structures are related to the perovskite structure, whether they are superconductors or not. As will be shown, these properties are greatly influenced by the oxygen content, which governs the mixed valence of copper, and depends on the thermal treatment and on the composition. For superconducting materials, only the properties of the 40 and 90 K superconductors and the influence of cationic substitutions not on copper sites will be discussed here; substitutions for copper will be treated in Chap.4.

3.1 Electron Transport Properties in Cuprates Related to the Perovskite: General Considerations

As is well known by physical chemists, when the transition metal M of a framework MO_x exhibits mixed valence (nonintegral mean oxidation state), charge carriers are generated that can be delocalized in bands built up from the overlap of the nd, (n+1)s and (n+1)p orbitals of the transition metal M with the 2s and 2p orbitals of the oxygen atom or from the direction overlap of the nd orbitals of neighboring M atoms. In the perovskite structure, no direct d-d interactions occur because the M-M distance is greater than the critical distance defined by *Goodenough* [3.1], so the bands mainly result from M-O interactions.

A well-known example is that of the cubic oxygen tungsten bronzes Na_xWO_3 [3.2-5]. The metallic properties of these mixed valence W(V) – W(VI) oxides can be explained from the schematic band diagram established for ReO_3 [3.1], which is also applicable to WO_3. In an isolated MO_6 octahedron, σ and π M-O bonds are formed by interactions between M d,s,p orbitals and O s,p orbitals according to their symmetry (Fig.3.1a), leading to the formation of bonding σ and π levels and antibonding σ^* and π^* levels, which in the solids are broadened into bands (Fig.3.1b). The σ

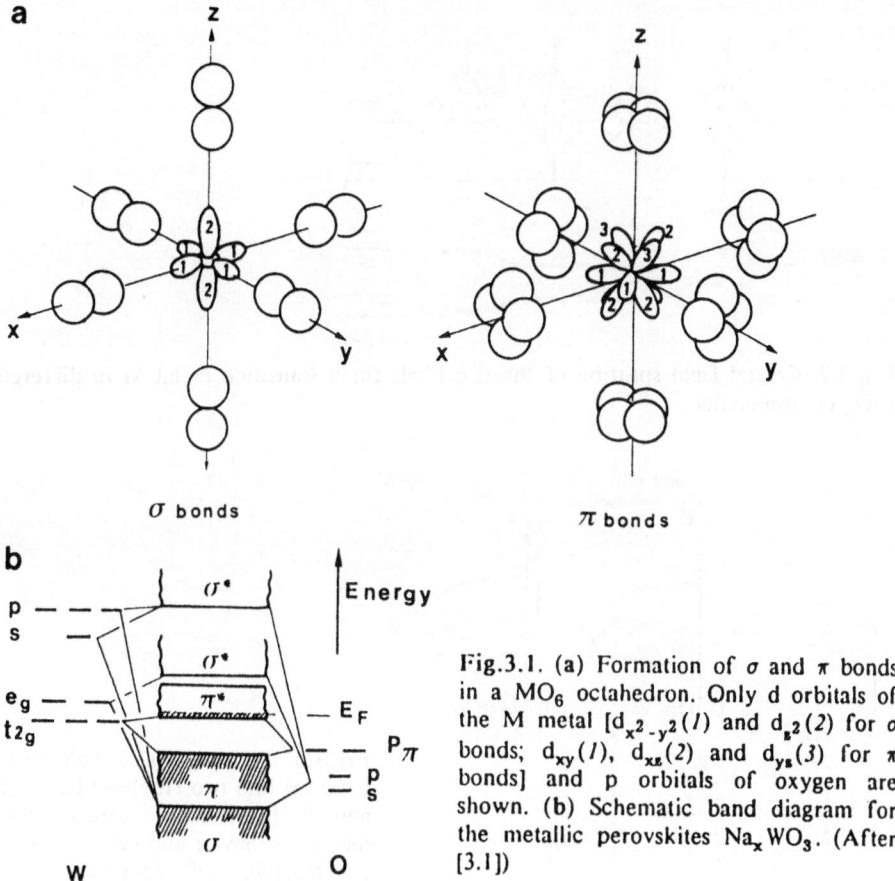

a

σ bonds

π bonds

b

Fig.3.1. (a) Formation of σ and π bonds in a MO_6 octahedron. Only d orbitals of the M metal [$d_{x^2-y^2}(1)$ and $d_{z^2}(2)$ for σ bonds; $d_{xy}(1)$, $d_{xz}(2)$ and $d_{yz}(3)$ for π bonds] and p orbitals of oxygen are shown. (b) Schematic band diagram for the metallic perovskites Na_xWO_3. (After [3.1])

and π bands are filled and the σ^* and π^* bands are empty in WO_3, which is an insulator. In the case of Na_xWO_3 each sodium atom brings one extra electron, which is delocalized in the π^* band (Fig.3.1b), leading to a metallic conductor. The case of the copper oxides is more complex owing to the important Jahn-Teller effect of copper, which involves an axial distortion of the copper polyhedra and favors the formation of coordinations lower than six: fivefold coordination (square pyramidal) and fourfold coordination (square planar). Different crystal field splittings result for the 3d orbitals (Fig.3.2), which will influence the electronic structure of these materials. In particular, the π^* band and the lower σ^* band, which have a prominent d character, will be highly modified. For example, the lower σ^* band is composed of a $\sigma^*_{x^2-y^2}$ band and a $\sigma^*_{z^2}$ band, which lie at the same energy level in a pure cubic crystal field; these two bands will be split like the corresponding 3d levels by the axial distortion. The same phenomenon occurs for the π^* band. In most cases, the copper polyhedra are distorted so that the $3d_{x^2-y^2}$ level is the highest in energy and is occupied by the unpaired electron due to the $3d^9$ configuration of Cu(II). In these conditions, the $\sigma^*_{x^2-y^2}$ band is half-filled and is the conduction band. In a system contain-

53

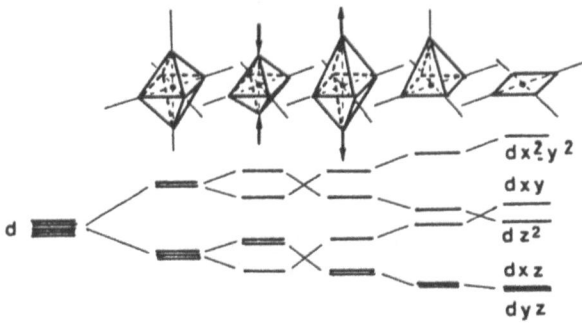

Fig.3.2. Crystal field splitting of the d orbitals for a transition metal M in different MO_n environments

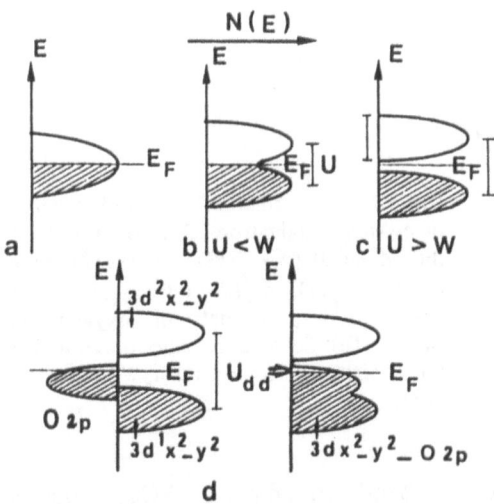

Fig.3.3. Splitting of a half-filled band (a) into two Hubbard bands (b and c). (a) and (b) correspond to metallic behavior and (c) to insulating behavior. (d) Schematic band diagram near the Fermi level for the mixed valence copper oxides

ing more than one outer electron per site, strong electron-electron interactions may occur, splitting each band into two subbands, as expected from Hubbard's theory [3.6]. The two Hubbard bands are separated by an intraatomic Coulomb energy U and have a width W, whose values are dominant for the electron transport properties (Fig.3.3a-c). The interpretation of these properties can be made more complicated by the fact that in the same oxide, copper sometimes exhibits different coordinations, e.g. in $BaLa_4 \cdot Cu_5O_{13}$, where CuO_6 and CuO_5 polyhedra coexist (Chap.2).

The structural anisotropy, and consequently the anisotropy of the electron transport properties, is a major factor for superconductivity, as was pointed out by theorists such as *Labbé* and *Bok* [3.7] and *Friedel* [3.8] in France, and *Mattheis* [3.9] in the USA, a few months after the discovery of the first high-T_c superconductors. These researchers showed, for the first time, that the two-dimensional character of the structures (and of the band diagram) is essential for the appearance of superconductivity with high critical temperature.

54

The second main characteristic is the mixed valence of copper. This mixed valence, Cu(II)-Cu(III), must be considered as formal. It is only deduced from electroneutrality calculations and does not imply that Cu^{3+} ions exist in these compounds, even with very short lifetimes, due to the delocalization of the charge carriers. Many studies by X-ray absorption spectroscopy at the Cu L_3 edge have been carried out on the mixed valence superconductors $La_{1.85}Sr_{0.15}CuO_4$ [3.10] and $LnBa_2Cu_3O_7$ (Ln = Gd, Ho, Y) [3.11, 12] and also on the nonsuperconducting oxides exhibiting only formally trivalent copper, e.g. $NaCuO_2$ [3.12, 13], $La_2Li_{0.5}Cu_{0.5}O_4$ and K· Cu(III)(biuret)$_2$ [3.12]. They showed that the formation of the Cu $3d^8$ configuration as the ground state in formally trivalent Cu ions was excluded. However, a $3d^9\,\underline{L}$ (where $\underline{L}$ is a ligand hole) configuration was mainly observed. This configuration corresponds to a transfer of the hole from copper to the oxygen atoms, leading to holes delocalized in a conduction band mainly built up from the 2p oxygen orbitals. In other words, the O 2p hole is the major carrier in the mixed valence copper oxides. A schematic electronic structure in agreement with the dominant O 2p hole character of the Fermi level is depicted in Fig.3.3d [3.14]. It has been established for copper in the square coplanar configuration, which is often considered in the first approximation when a high axial distortion of the polyhedra occurs. This model supposes a O 2p band between the two pure Cu $3d_{x^2-y^2}$ Hubbard bands. The $3d^9\,\underline{L}$ configuration is also called the $[Cu^{2+}\text{-}O^-]$ or $[Cu\text{-}O]^+$ species by some researchers.

Whatever the exact nature of the mixed valence, for a given oxide the number of formally trivalent copper ions (or $[Cu\text{-}O]^+$ species) and consequently the number of holes will depend greatly on the oxygen stoichiometry, which is itself influenced by the thermal treatment (quenching, annealing in pure oxygen or in various atmospheres).

3.2 Oxides with the K_2NiF_4-Type Structure

Of the oxides with the K_2NiF_4-type structure, the most important for superconductivity are La_2CuO_4 and the substituted compounds $La_{2-x}A_xCu\cdot O_{4-x/2+\delta}$ (A = Ba, Ca, Sr), which are known as 40 K superconductors. Their structure (Fig.2.5) can be considered as being built up by stacking conducting perovskite monolayers alternating with insulating rock salt monolayers along the c axis.

3.2.1 La_2CuO_4

a) "Normal" Properties

In contrast to the other Ln_2CuO_4 oxides such as Pr_2CuO_4, Sm_2CuO_4 and Nd_2CuO_4 [3.15], which are semiconductors and in which copper exhibits a square coplanar configuration, La_2CuO_4 can be considered as essentially semimetallic near room temperature. This particular behavior was reported

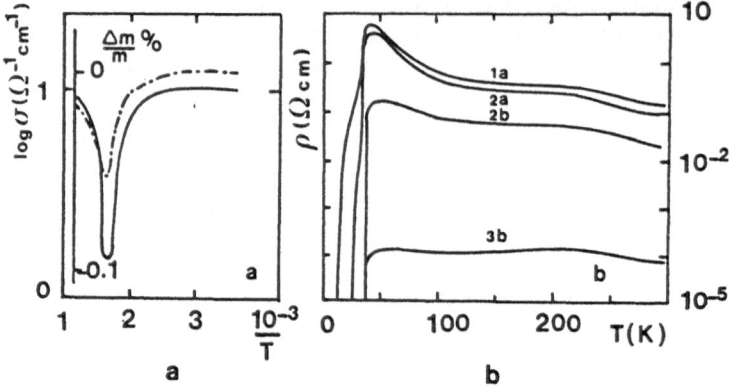

Fig.3.4. (a) Thermal dependence of the conductivity in air for La$_2$CuO$_4$ (*solid line*), and corresponding thermogravimetric curve (*dash-dotted line*) in the same experimental conditions. (b) The logarithm of the resistivity of La$_2$Cu$_x$O$_{4-y}$ vs temperature for x = 0.98 (*1*), x = 1.02 (*2*) and x = 1.04 (*3*). The samples a were annealed at 450°C under a pressure of 1 bar of oxygen. The samples b are annealed at 900°C in oxygen at 500 bar. (After [3.36])

by several researchers: first by *Foex* [3.16], then by *Ganguly* and *Rao* [3.15], and *Kenjo* and *Yajima* [3.17]. For as-synthesized samples, the conductivity measured on ceramics is about 10 Ω^{-1}cm^{-1} at room temperature and nearly temperature independent above 200 K. A small minimum is observed on the curve $\sigma = f(T)$ near 530 K, which is associated with the phase transition orthorhombic $\rightleftarrows$ tetragonal which occurs at this temperature [3.15]. More recently, *Nguyen* et al. [3.18] have shown that the decrease in the σ value can be of about one order of magnitude at the temperature of transition if the sample is heated very slowly. By means of a careful thermogravimetric study, they showed that the evolution of the conductivity could be associated with variations of the oxygen content of the sample (Fig.3.4a). At temperatures lower than 200 K, La$_2$CuO$_4$ exhibits semiconducting behavior [3.19-21] with $\sigma_{10K}/\sigma_{200K} \sim 10^{-4}$ [3.20]. This behavior is considered as a metal–insulator transition, which is assumed to be correlated with a phase transition to a space group of lower symmetry [3.22]. The thermoelectric power is positive, in agreement with hole conduction; it is nearly temperature independent above 200 K and increases as the temperature decreases below 200 K [3.22]. Its value at room temperature is high, indicating a low density of charge carriers near the Fermi level, but differs from one publication to another: $\alpha \sim 100~\mu V \cdot K^{-1}$ [3.21], 250 $\mu V \cdot K^{-1}$ [3.15], 300 $\mu V \cdot K^{-1}$ [3.18]. The concentration of mobile charge carriers per copper atom, deduced from the conductivity value, is c ~ 0.1. This is consistent with a temperature-independent Seebeck coefficient given for a small c value by

$$\alpha = - (k/e)\ln c \sim 200~\mu V \cdot K^{-1} \tag{3.1}$$

and agrees with the magnetic measurements [3.19].

The electrical properties of La_2CuO_4 were interpreted theoretically by *Goodenough* [3.24] and a schematic band structure was proposed which was very close to that of Fig.3.3d except that the contributions of Cu $3d_{x^2-y^2}$ and O 2p were not distinguished. The important axial distortion of the copper octahedra involves a high stabilization of the $3d_{z^2}$ levels compared to the $3d_{x^2-y^2}$, leading to a conduction band built up mainly from the overlap of the Cu $3d_{x^2-y^2}$ and O 2p orbitals, in agreement with the two-dimensional character of the structure. The strong electron correlation on the Cu sites has been deduced from X-ray photoelectron spectroscopy (XPS) and Auger Valence Electron Spectroscopy (AVES) experiments [3.25,26]. Values of about 6 eV for the interatomic Coulomb energy U and 4-5 eV for the width W of the Cu 3d Hubbard bands are mainly obtained [3.26-28]. In pure La_2CuO_4, copper is only bivalent and the Cu $3d_{x^2-y^2}$-O 2p Hubbard band is filled, so La_2CuO_4 should be an insulator. To explain the semiconducting behavior at low temperatures and the high positive value of the thermoelectric power, *Goodenough* introduced a slight oxidation of copper involving holes localized at the top of the filled Hubbard band [3.24]. A mechanism for this possible oxidation was proposed by *Nguyen* in terms of Schottky defects [3.29]. These defects allow the formation of cationic and anionic vacancies, leading to $La_{2(1-\epsilon)}Cu_{1-\epsilon}^{II}O_{4(1-\epsilon)}$ with $\epsilon \geq (4-5) \cdot 10^{-3}$ [3.29], which allows a very small proportion of copper to be oxidized.

b) Superconducting Properties

As is evident from the preceding subsection, up to February 1987, $La_2 \cdot CuO_4$ was claimed to be a semiconductor or an insulator at very low temperatures [3.19-21]. Superconductivity of this oxide with cationic stoichiometry was first demonstrated by *Beille* and co-workers [3.30]. They measured a critical temperature of 37 K after annealing the oxide for a long time in 1 oxygen at 1 bar pressure. The T_c is very sensitive to the applied magnetic field. It can be increased up to 42 K by applying 19 kbar hydrostatic pressure. Magnetic susceptibility measurements show that the superconducting properties are very sensitive to the thermal treatment: 50%-90% diamagnetism and an imaginary susceptibility characteristic of bulk superconductivity were observed in a sample annealed at 900°C under 500 bar oxygen pressure. One interesting feature is that superconductivity is observed in spite of a semiconducting behavior above T_c and a relatively large resistivity ($\rho \sim 6\Omega \cdot cm$) just above the resistivity fall. This particular behavior is attributed to a grain surface superconductivity [3.30-33]. Superconducting material with a more metallic behavior can be prepared by plasma oxidation experiments [3.35]. These experiments and the influence of the oxygen pressure on the properties of La_2CuO_4 are in agreement with the assumption of the existence of anionic vacancies in the "stoichiometric" as-synthesized compound. Superconducting materials can also be obtained after annealing at 450°C in oxygen at 1 bar by using an excess of copper or lanthanum oxide, leading to the "nonstoichiometric" $La_2Cu_xO_{4-y}$ oxides [3.33-35]. The influence of the oxygen pressure during the thermal treatment is enhanced in the case of these materials [3.36, 37]. Samples with x ranging from 0.98 to 1.04 have been studied. The resistivity of all the samples is approximately the same for all the compositions after annealing in oxygen at 1 bar. For x = 1.04, annealing at high oxygen pressure (900°C, 500 bar) induces a metallic behavior and reduces the normal state resistivity ρ at T_c by a factor of $4 \cdot 10^4$ [3.36] (Fig. 3.4b).

From the study of the superconducting compounds, it appears that superconductivity is closely related to the average valence of copper above 2^+. In this scheme, the superconducting properties of La_2CuO_4 can be explained in terms of Schottky defects. As was stated before, stoichiometric La_2CuO_4 prepared in air contains only divalent copper and exhibits Schottky defects according to the formula $La_{2(1-\epsilon)}Cu^{II}_{1-\epsilon}O_{4(1-\epsilon)}$. Annealing this compound at higher oxygen pressure allows oxygen to be fixed on the anionic vacancies, leading to the formula $La_{2(1-\epsilon)}Cu^{II,III}_{1-\epsilon}O_{4(1-\epsilon)+\epsilon'}$. Above a critical value of ϵ', i.e. a critical value for the ratio Cu(III)/Cu(II) (or $[Cu-O]^+/Cu(II)$), superconductivity is obtained. In the same way, an excess of copper oxide favors the formation not only of lanthanum vacancies but also of anionic vacancies according to the hypothetical nonsuperconducting oxide $La_{2(1-x)}Cu^{II}O_{4-\epsilon''}$, which can easily be oxidized, even in an oxygen flow, leading to the mixed valence superconducting oxide $La_{2(1-x)}Cu^{II,III}O_{4-\delta}$. An excess of lanthanum also favors the superconducting state, leading again to the mixed valence in agreement with the limit formula $La_2Cu^{II,III}_{1-x}O_4$. However, in this latter case the critical temperature decreases drastically (Fig. 3.4b), probably owing to the presence of vacancies on the copper sites.

A recent investigation by neutron diffraction has revealed the existence of two forms of "La_2CuO_4" [3.38]. So another way to explain the presence of trivalent copper in "La_2CuO_4" is to consider an excess of oxygen in the rock-salt-type layers, leading to the formula $La_2CuO_{4+\delta}$. This model is supported by a neutron diffraction study of the isostructural $La_2NiO_{4+\delta}$ (δ = 0.18) [3.39]. But, in the case of the cuprate, the observed oxygen stoichiometry deviation is too low (δ = 0.032) to make a good case for the proposed model, which leads to abnormally short oxygen-oxygen distances [3.40]. These two possibilities do not necessarily conflict and the issue of oxygen nonstoichiometry in La_2CuO_4 still remains open.

3.2.2 The Oxides $La_{2-x}A_xCuO_{4-x/2+\delta}$ (A = Ba, Ca, Sr)

The replacement of a proportion of La^{3+} by a divalent cation in La_2CuO_4 induces oxygen vacancies favorable to the formation of formally trivalent copper. As synthesized alkaline-earth-doped lanthanum oxides show an important increase of the conductivity at room temperature [3.41, 42]. Electron transport properties of these compounds have been extensively studied above liquid nitrogen temperature by *Er-Rakho* [3.43] and *Nguyen* et al. [3.18], and down to 4 K by many other researchers since the discovery of high-temperature superconductivity in the La-Ba-Cu-O system [3.44].

a) Electron Transport Properties Above 77 K

The behavior of the barium- and calcium-substituted oxides is very similar to that of the strontium-substituted compounds, so the electrical properties of only the latter materials will be described, owing to their large homogeneity range (10% substitution for Ba and Ca, 65% for Sr). In the formula of these oxides: $La_{2-x}A_xCuO_{4-x/2+\delta}$ (A = Ba,Ca,Sr), "δ" indicates the excess of oxygen introduced by the formation of formally trivalent copper, i.e. the $[Cu-O]^+$ complex, which has been determined by thermogravimetric reduction under hydrogen [3.18].

The electron transport properties of these phases are very complex since they depend on the substitution fraction x, on the $[Cu-O]^+$ concentration (and consequently on the δ value) and on the number of oxygen vacancies $(x/2-\delta)$ that are located in the basal plane of the CuO_6 octahedra (Chap.2). Figures 3.5a and b show the substitution fraction dependence of δ and of the conductivity at room temperature for various thermal treatments, and Fig.3.5c illustrates the dependence of the Seebeck coefficient at room temperature for air-quenched samples. It can be seen that nonnegligible amounts of "trivalent" copper can be formed even by quenching the materials in air from 1100°C down to room temperature, and that annealing in vacuum or in a pure oxygen flow at 450°C leads to a significant decrease or increase of the δ value for a given x value.

Air quenched samples. The value of δ increases with x up to x~0.33 (δ ~ 0.12), then it decreases slowly, reaching zero for x~1 (see also Table 2.2). For these quenched samples, the temperature dependence of the electrical

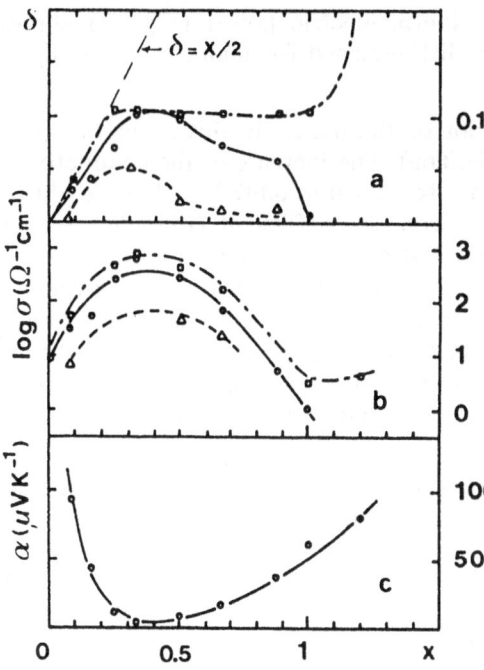

Fig.3.5. $La_{2-x}Sr_xCuO_{4-x/2+\delta}$ oxides. x dependence of δ (a), σ_{300K} (b) and α_{300K} (c) for various thermal treatments (after [3.18]). Air-quenched samples: (-); oxygen-annealed samples: (-·-·) vacuum-annealed samples: (-----). In (a) the numbers of oxygen vacancies can be obtained from the difference between the different curves and the line $\delta = x/2$

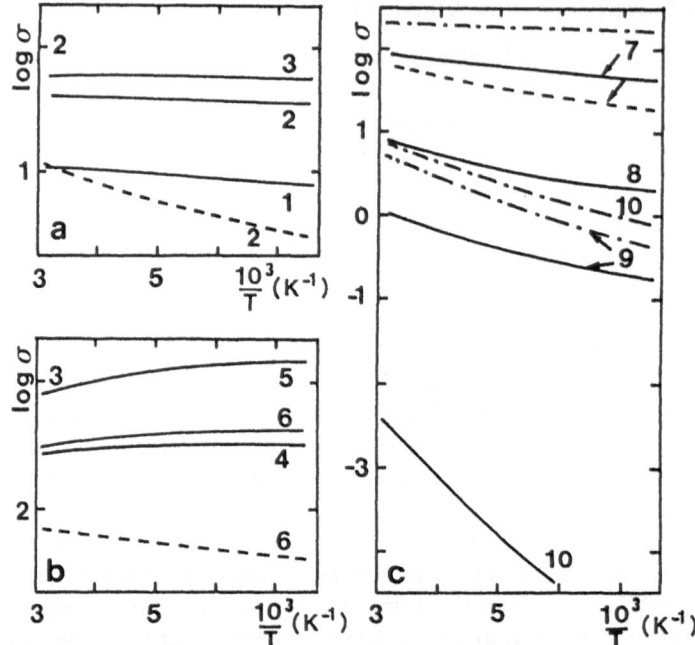

Fig.3.6. Thermal variation of the conductivity [$\Omega^{-1}cm^{-1}$] (logarithmic scale) of some compositions for the $La_{2-x}Sr_xCuO_{4-x/2+\delta}$ oxides obtained from various thermal treatments (after [3.18]). *1*: x = 0; *2*: x = 0.08; *3*: x = 0.16; *4*: x = 0.25; *5*: x = 0.33; *6*: x = 0.50; *7*: x = 0.66; *8*: x = 0.88; *9*: x = 1.0; *10*: x = 1.2; (——) air-quenched samples; (—·—) oxygen-annealed samples; (---) vacuum-annealed samples

conductivity (Fig. 3.6) and of the thermoelectric power (Fig.3.7) allows three domains of composition to be distinguished for temperatures ranging from 77 K to 300 K [3.18].

$0 \leq x \leq 0.16$: The electrical behavior of the oxides in this range of composition is like that of La_2CuO_4 (Fig.3.6a). The increase of the conductivity at room temperature from 10 $\Omega^{-1}cm^{-1}$ for x = 0 to 60 $\Omega^{-1}cm^{-1}$ for x = 0.16 (Fig.3.5b) and the decrease of the Seebeck coefficient (Fig.3.5c) are in agreement with the increase of δ, which is directly correlated to the number of charge carriers (the number of holes is equal to 2δ). It can be assumed that the number of anionic vacancies is too weak to modify the carrier mobility significantly, so the electron transport properties depend only on the hole density in the conduction band. However, some observations suggest that holes may be trapped in localized states at the top of the conduction band following the Mott model:

The thermal variations of σ are very weak and do not correspond to the metallic model:

$$\rho = \rho_0(1 + \gamma t) . \tag{3.2}$$

The $|\alpha|$ values are too high for metals.

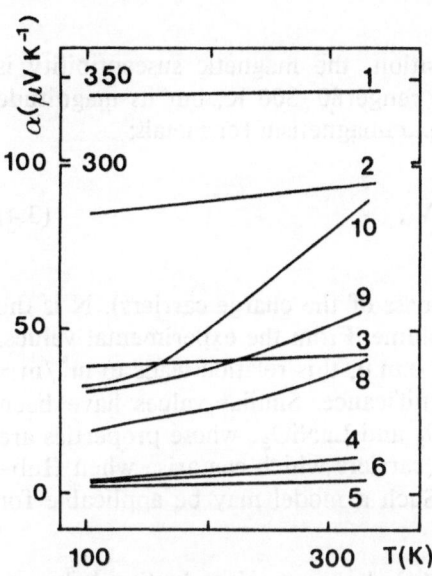

Fig.3.7. Thermal variation of the thermo-electric power for some compositions of the oxides $La_{2-x}Sr_xCuO_{4-x/2+\delta}$ [3.18]. The curves are numbered as in Fig.3.6

In this model [3.45, 46] a long range disorder involves the formation of localized states at the band edge on which charge carriers move by a variable range hopping. This supposes that the anionic vacancies are disordered in this range of composition, as can be expected from their low concentrations. In these conditions, α increases with temperature, following the relation

$$\alpha = \left(\frac{\pi^3}{3}\right)\left(\frac{k^2T}{e}\right)\frac{d\ln\sigma}{dE} \tag{3.3}$$

in agreement with experiments.

$0.16 < x \leq 0.5$: This range of composition corresponds to the highest values of δ and at the same time to the highest values of the conductivity (Fig.3.6b). The dependence, at 300 K, of σ and α on x agree well with that of δ. The thermal variations of the conductivity show a metallic behavior characterized by a slope $d\sigma/dT < 0$. Moreover, a linear variation of the resistivity with temperature is observed for $x = 0.33$. The value of the γ coefficient is $4.05 \cdot 10^{-3}$ °C^{-1}, this value is very close to the theoretical value for free electrons ($\gamma = 3.66 \cdot 10^{-3}$ °C^{-1}). For $x = 0.5$ the law for metals is verified only above room temperature. This metallic behavior is also in agrement with the weak values and near temperature independence of $|\alpha|$. The positive sign of the thermoelectric power shows that the conduction band is always more than half-filled. This result contradicts Uchida et al. [3.23] who predicted a negative sign for the Seebeck coefficient for $x > 0.23$ using two-dimensional tight-binding band calculations in the rigid-band approximation. This contradiction can be considered as only apparent since the calculations were carried out for materials without oxygen vacancies, i.e. with a larger number of charge carriers.

61

In these two domains of composition, the magnetic susceptibility is nearly temperature independent in the range 80–300 K, but its magnitude is too high to be interpreted as a Pauli paramagnetism for metals:

$$\chi_{Me} = 4\,\frac{\pi m}{\beta h^2}\,\mu_B^2 \left(\frac{3N}{\pi}\right)^{1/3} \left[1 - \frac{\beta^2}{3}\right]V \,, \tag{3.4}$$

where $\beta = m/m^*$ (m^* is the effective mass of the charge carriers), N is the density of holes, and V is the molar volume. From the experimental values, which range from $330 \cdot 10^{-6}$ to $835 \cdot 10^{-6}$ e.m.u, this relation leads to $m^*/m > 10$, which is without any physical significance. Similar values have been obtained for the metallic oxides $LaCuO_3$ and $LaNiO_3$, whose properties are explained by a gas of highly correlated carriers which can arise when Hubbard bands do not overlap (W < U). Such a model may be applicable for these oxides.

$x > 0.5$: The curves $\sigma = f(1/T)$ (Fig.3.6c) show a semiconducting behavior with a dramatic decrease of the conductivity values (associated with a decrease of the δ values) as x increases. For example, the oxide $La_{0.8}Sr_{1.2} \cdot CuO_{3.4}$ ($\delta = 0$) exhibits a conductivity of about $10^{-3}\ \Omega^{-1}cm^{-1}$ at room temperature. In the same way, the $|\alpha|$ value increases with x, in agreement with the decrease of δ. In this range of composition, the electron transport properties are characteristic of Mott semiconductors. The compositions corresponding to x = 0.88, 1 and 1.2 exhibit a variation of the conductivity according to Mott's relation [3.45, 46]

$$\sigma = A\exp[-(Q/kT)^{1/4}] \tag{3.5}$$

with $Q = (1.5/\beta^3)N(E_F)$ and where β denotes the rate of fall-off of the envelope of the wave function, $N(E_F)$ is the density of states near the Fermi level and A is a constant characterizing the electron-phonon interaction. This relation characterizes a variable range hopping of charge carriers localized in the conduction band near the Fermi level. The localization of the charge carriers may be related to the decrease of δ and to the increase of the number of oxygen vacancies and also to their possible ordering. The influence of the oxygen vacancies upon the conductivity is particularly evident for samples annealed in pure oxygen (see the following subsection).

Above 300 K, for small x values (corresponding to an orthorhombic cell) the same thermal variation of σ as for La_2CuO_4 is observed. It is attributed to variations in the oxygen content of the samples during the phase transition orthorhombic $\rightleftarrows$ tetragonal. In the case of strontium as well as for calcium and barium, the temperature of the end of the σ fall and the importance of this fall decrease as x increases, i.e. as the orthorhombic distortion of the tetragonal cell and the temperature of transition O $\rightleftarrows$ T decrease [3.18, 43]. For the tetragonal oxides, nonlinear variations of log $\sigma = f(1/T)$ are observed, which are again attributed to oxygen exchanges with the atmosphere.

Effects of oxygen intercalation - de-intercalation. Except for $x = 0$ and $x = 1.2$, for which $\delta = 0$, in quenched samples, annealing in vacuum leads to an important decrease of δ (Fig.3.5a) and consequently of σ. This decrease of δ can involve a change in the conduction mode: for example, the oxide corresponding to $x = 0.5$, which was metallic, becomes a semiconductor after oxygen de-intercalation (Fig.3.6b).

Conversely, oxygen intercalation and consequently an increase in the formally trivalent copper content are observed by annealing at 450°C in an oxygen flow, but the δ value remains dependent on the substitution fraction x (Fig.3.5a). For a given x value, the increase of δ with respect to the quenched samples leads to an increase of the conductivity, and the importance of the variation in the conductivity is a function of the importance of the variation of δ. So, for small x values the variation of δ is weak and σ is not very sensitive to annealing in oxygen. In contrast, for $x = 1.2$ the gain of oxygen is very important ($\delta \sim 0.33$), leading to a large increase of the conductivity at room temperature from 10^{-3} to $4 \; \Omega^{-1} cm^{-1}$. It is worth noting that the oxygen intercalation in this oxide is accompanied by the disappearance of the superstructures, i.e. the disappearance of the ordering of the oxygen vacancies, which may also be a parameter which influences the mobility of the charge carriers. From thermogravimetric analysis in hydrogen [3.18] and chemical analysis by redox titration [3.47, 48], it appears that for $x \leq 0.15-0.20$ there are no anionic vacancies in the annealed compounds. These results were confirmed by Hall coefficient measurements for $La_{2-x}Sr_xCuO_4$ with x ranging from 0 to 0.3 [3.47-50]. For all Sr concentrations R_H is positive, nearly temperature independent, and decreases linearly up to $x = 0.15$ [3.47-49] or $x = 0.20$ [3.50], giving approximately 1 hole per strontium cation. Hole mobilities ranging from 1.8-3.5 $cm^2 V^{-1} s^{-1}$ at 50 K to 0.5-0.7 $cm^2 V^{-1} s^{-1}$ at 300 K have been measured.

Influence of the anionic vacancies on the mobility of the charge carriers is illustrated by the behavior of compositions ranging from $x = 0.33$ to $x = 1.0$. In this domain, δ remains approximately constant and equal to 0.11, the number of oxygen vacancies increases from 0.04 ($x = 0.33$) to 0.39 ($x = 1.0$) and the room temperature conductivity decreases from 10^3 ($x = 0.33$) to 3 $\Omega^{-1} cm^{-1}$ ($x = 1.0$).

The study of the oxides $LaSrCuO_{3.5+\delta}$ provides another illustration of the complexity of the electron transport properties of these phases. The conductivity at room temperature increases from 0.8 ($\delta \sim 0.0$) to 4 $\Omega^{-1} cm^{-1}$ ($\delta \sim 0.11$) and its temperature dependence shows a semiconducting behavior with an activation energy of about 0.05 eV. $LaSrCuO_4$ ($\delta = 0.5$) crystallizes in the K_2NiF_4 structure (a=3.765Å, c=13.27Å) and also exhibit a semiconducting behavior ($\Delta E \simeq 0.003 eV$), but a lower σ value at 300 K is measured ($\sigma \sim 3 \cdot 10^{-2} \Omega^{-1} cm^{-1}$) [3.51], showing that σ passes through a maximum as δ increases for a given composition. This may be related to a change in the number of the charge carriers in the conduction band when δ varies. The fully oxidized oxide theoretically exhibits only formally trivalent copper, but magnetic studies provide evidence for small amounts of divalent copper. In a pure Cu(III) compound, the Cu $3d_{x^2-y^2}$-O 2p band in the model of Fig.3.3d must be empty and the $3d_{z^2}$ level must lie far below this band due to the large axial distortion of the copper octahedra, i.e. the c/a ratio, leading to an insulator. Thus it can be assumed that Cu^{2+} ions, as an impurity, act as donor levels lying just below the bottom of the conduction band, so that, in this naive model, $LaSrCuO_4$ would be an n-type semiconductor; unfortunately there are no data concerning the sign of the Seebeck voltage.

Anisotropy. Anisotropy in the electron-transport properties results from the conductive perovskite layers being interleaved with insulating rock-salt-type layers along the crystallographic c axis. Resistivity measurements on square plate shaped single crystals of composition $La_{1.95}Sr_{0.05}CuO_4$ with typical dimensions $8\times8\times2$ mm³ show that the resistivity along the c axis is 20 times large than that along the a axis at room temperature [3.52], a value comparable to that of the 90 K superconductor $YBa_2Cu_3O_7$ (Sect.3.4.2a).

b) Superconducting Properties

The oxides $La_{2-x}Ba_xCuO_{4-x/2+\delta}$ were identified by *Takagi* and co-workers [3.53] as the phase responsible for superconductivity in the system Ba-Cu-La-O [3.44,54]. The midpoint of the resistive transition is around 30 K for $x = 0.15$ [3.53,55] with a width of transition (defined by the temperature separation of the 10% and 90% drops in resistivity) of 3.7 K [3.55]. By magnetic susceptibility measurements a T_c (onset) of 29 K is observed [3.52]. Under pressure T_c increases from 32 to 40.2 K at 13 kbar [3.56] (T_c onset determined by resistivity measurements). This value of 30 K represents the maximum T_c observed in the barium-substituted phases. The variation of the critical temperature as a function of the fraction of substitution shows that T_c increases drastically with x from 6 K for $x = 0.05$ to 29 K for $x = 0.08$ then remains approximately constant up to $x \sim 0.15-0.20$ [3.55].

Calcium-substituted phases exhibit lower critical temperatures: the maximum of T_c is observed at $18-20$ K for $x = 0.15$ [3.56-58]. Annealing in air or in pure oxygen does not influence the midpoint T_c of the resistiv-

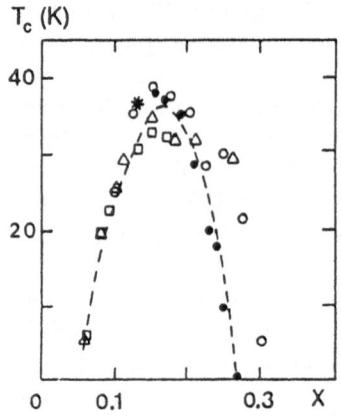

Fig.3.8. Dependence of T_c on substitution fraction in the Sr-doped La_2CuO_4 series according to (○) [3.61], (△) [3.47,48], (□) [3.31], (●) [3.71] and (✳) $La_{1.95}CuO_4$ [3.36]. The dashed line corresponds to the variation of T_c for stoichiometric oxides (% $[Cu-O]^+ = x$)

ity drop but it does affect the width of the transition (T_c onset: 31.5 and 26 K after annealing in air and in O_2, respectively) [3.59].

The strontium-doped materials have been investigated more, owing to the values of the critical temperature observed in this system, which are 7 K higher than in the system with barium. First *Cava* et al. [3.60] studied the variation of the superconducting properties as a function of Sr concentration (x ranging from 0 to 0.3 in steps of 0.1) and found that T_c is highest and the transition is narrowest near x = 0.2. For this composition, the resistive transition, for a sample slowly cooled from 900°C in an O_2 flow, is 1.4 K wide (10% to 90%) with a midpoint at 36.2 K. The associated dc diamagnetic susceptibility is a large fraction (60%–70%) of the ideal value indicating that superconductivity is essentially a bulk property. In fact, the maximum superconducting temperature (T_c at midpoint: 39.3K) is reached at x = 0.15 [3.61] and a careful investigation of the system shows that superconductivity is observed from x ~ 0.06 [3.31, 47, 48] to x ~ 0.3 [3.61] (Fig. 3.8).

The phases with the highest critical temperatures exhibit other interesting characteristics. At 4.2 K substantial fractions of the materials remain superconducting in a pulsed field of 45 T [3.62] and critical fields H_{c2} of 50 T to 125 T can be estimated for 0 K [3.63, 64]. Moreover, critical currents of $2 \cdot 10^3$ A/cm² and $4 \cdot 10^2$ A/cm² are measured at 4.2 and 18 K, respectively, in a magnetic field of 5 T [3.64].

Superconducting properties such as the critical temperature and width of the resistive transition are greatly influenced by the thermal history of the materials [3.60, 65, 66]. Typical curves of the temperature dependence of resistivity for several thermal treatments are given in Fig. 3.9 for $La_{1.84}$· $Sr_{0.16}CuO_{3.92+\delta}$ samples prepared in oxygen [3.65]. The specimen prepared at 1100°C and quickly cooled shows an onset at about 38 K and a broad transition with a tail that prevents zero resistance being reached above 24 K. Specimens annealed at lower temperatures exhibit, in addition to lower resistivity, an increase in T_c and appreciably sharper transitions. This behavior undoubtedly correlates with the oxygen content of the sample. In particular, the tail structure and the broad transitions can be associated with variations in oxygen concentration between the different grains of the ceramic or (and) with concentration gradients of oxygen in the grains. Thus, the δ value and consequently the number of $[Cu-O]^+$ complexes have an influence on the T_c value. For well-synthesized materials, and small x values

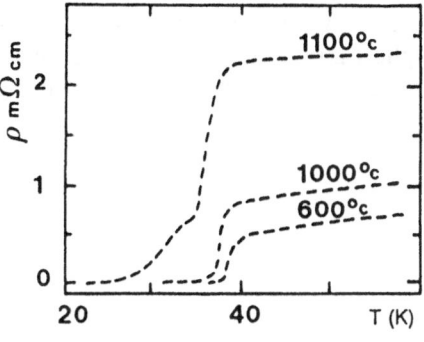

Fig. 3.9. Thermal variation of the resistivity of the oxide $La_{1.84}Sr_{0.16}CuO_{3.92+\delta}$ quenched from different temperatures. (After [3.65])

(up to x ~ 0.15), oxygen stoichiometry is reached [3.18, 46, 48], and the number of charge carriers is equal to x [3.46-50].

Plotting T_c as a function of x (or the percentage of formally trivalent copper) shows that T_c increases as $[Cu-O]^+$ content increases (Fig.3.8).

The decrease of T_c for x greater than 0.15 was first associated with a decrease of the $[Cu-O]^+$ concentration [3.47.48] which occurs as the oxygen vacancies appear. This disagrees with the data of *Nguyen* et al. [3.18], which shows an increase of δ, i.e. an increase of $[Cu-O]^+$, in this range of composition, associated with an increase of the electrical conductivity [3.18,61]. In the light of the results obtained by several groups concerning the evolution of T_c with doping or oxygen stoichiometry in bismuth and thallium superconductors [3.67-73], it appears that an optimum exists in the hole concentration which corresponds to a maximum for T_c. This agrees with the results obtained by *Torrance* et al. [3.74] on samples of the series $La_{2-x}Sr_xCuO_4$ prepared under 100 bar oxygen pressure, which do not exhibit oxygen vacancies. These authors show that for x > 0.15 the progressive disappearance of T_c is induced by high carrier concentration. Their data are plotted in Fig.3.8 (solid circles), they differ from those of *Tarascon* (open circles) and *Shafer* et al. (open triangles), whose samples are not oxygen stoichiometric, so that in the latter cases the substitution fraction x is greater than the hole concentration for x > 0.15. The dashed curve then represents the variation in T_c for stoichiometric $La_{2-x}Sr_xCuO_4$ oxides, i.e. for compounds where the carrier concentration is equal to x; it shows a maximum for x ~ 0.15. So, it appears that the only factor that seems to govern the critical temperature is the hole (or $[Cu-O]^+$) concentration. However, a critical value must be reached for the appearance of superconductivity, since no superconductivity is observed for x < 0.06. It is worth noting that the data obtained for $La_{1.95}CuO_4$ [3.36] agree with the curve $T_c = f([Cu-O]^+)$ (Fig.3.8).

The lack of data concerning the barium- and calcium-substituted La_2CuO_4 oxides and the difficulty of determining with accuracy the number of oxygen vacancies for small x values prevent us from understanding why these oxides exhibit lower T_c's than the strontium series. *Kishio* et al. [3.75] have studied the quasi-ternary composition $La_{1.8}(Ba,Ca,Sr)_{0.2}CuO_4$ with various mixtures of the alkaline earth cations. They found that the maximum T_c is obtained for the oxide containing only Sr and the minimum for that containing Ca only, and that the variation of the critical temperature can be related to the variation of the tetragonal cell by a linear function (the lowest T_c corresponds to the highest a parameter). A similar influence of the a parameter upon T_c was also pointed out by *Ganguly* et al. [3.76].

3.2.3 Substitutions for Lanthanum in the Superconducting Oxide $La_{1.8}Sr_{0.2}CuO_{4-y}$

a) Rare-Earth Substitutions

The substitution of Ln^{3+} magnetic ions for La^{3+} in La_2CuO_4, in contrast to that of strontium, does not change the oxidation state of copper. The effect of the Ln ions on the properties of the superconducting phases may help in understanding the interplay between magnetism and superconductivity. The rare earth oxides Ln_2CuO_4 show a structure slightly different from the La_2CuO_4 structure, so that the substitution of Ln ions for La is limited to low Ln concentrations. The $La_{1.6}Ln_{0.2}Sr_{0.2}CuO_{4-y}$ series is obtained as a tetragonal single phase for all the rare earths, with the exception of the heavy ones (Er, Dy,...). The samples are superconductors and T_c decreases when the atomic number Z of the lanthanide increases (Fig.3.10) [3.77]. With the decrease of T_c is associated an increase of the resistivity of the samples, and for ions with $Z > Z_{Nd}$ the resistivity above T_c does not vary linearly with temperature but goes through a broad minimum just above T_c, like in the nonmagnetic phase $La_{2-x}Sr_xCuO_4$ ($x < 0.15$). The magnetic measurements above T_c up to room temperature are consistent with trivalent rare earth cations. For example, an effective magnetic moment of $8\mu_B$ per Gd atom is determined in the Gd-doped sample, in good agreement with the value expected for a free Gd^{3+} ion ($7.94\mu_B$). If the charge of the lanthanide cations is really 3+, the decrease in T_c cannot be due to the introduction of magnetic impurities. In the latter case, the depression of T_c is generally described by the *Abrikosov* and *Gor'kov* formula [3.78].

$$T_c = n(N(E_F))I^2(g-1)^2 J(J+1) , \qquad (3.6)$$

where n is the concentration of the rare-earth ions, $N(E_F)$ is the density of states at the Fermi level, I is the exchange integral, and $(g-1)^2 J(J+1)$ is the de Gennes factor. So, the monotomic and smooth decrease of T_c through

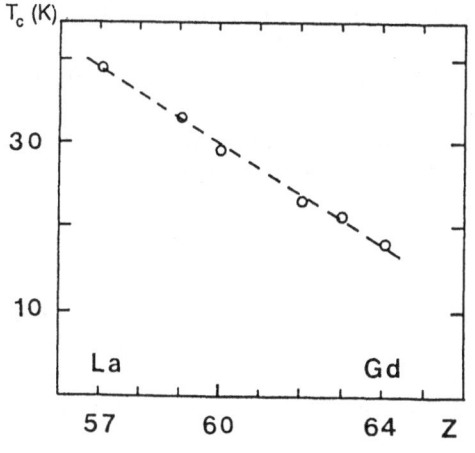

Fig.3.10. Critical temperature T_c as a function of lanthanide atomic number for $La_{1.6}Ln_{0.2}Sr_{0.2}CuO_{4-y}$. (After [3.77])

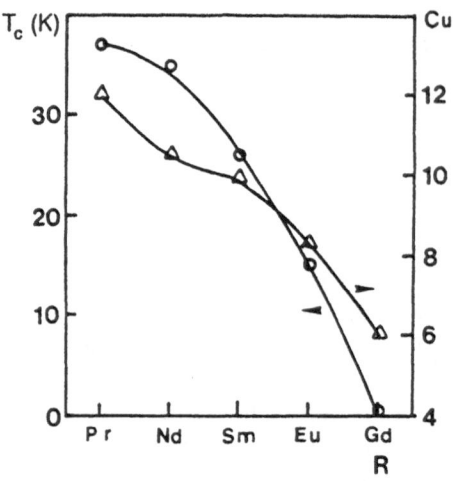

Fig.3.11. Variation of T_c and carrier concentration [Cu(III) %] as a function of Ln for the series $La_{1.7}Ln_{0.15}Sr_{0.15}CuO_4$ [3.79]

the Ln series cannot be a magnetic effect [3.35] since the de Gennes factor is not monotonic across the series and, in particular, is a minimum and a maximum for the neighboring elements Eu^{3+} (j=0) and Gd^{3+} (j=7/2).

The decrease in T_c could be due to a decrease in the $[Cu-O]^+$ concentration, which could be influenced by the nature of the rare earth cation.

Subramanian et al. [3.79] have studied the evolution of T_c and of the carrier concentration as a function of Ln for several compositions of the series $La_{2-x-y}Ln_xSr_yCuO_4$. They find that the reduction of T_c corresponds to a decrease of the hole concentration as illustrated in Fig.3.11 for x = y = 0.15.

The case of sustitution for praseodymium is more complex since it can exhibit both trivalent and tetravalent states. The $(La_{1-y}Pr_y)_{1.8}Sr_{0.2}CuO_4$ series has been studied by *Ganguly* et al. [3.76] in the range $0 \le y \le 0.25$. The T_c values at midpoint are lower than those measured by *Tarascon* et al. [3.77] for the compositions corresponding to y = 0 and 0.1, probably due to the method of preparation. T_c decreases from about 33K (y = 0-0.05) to 21K (y = 0.25). These results could be interpreted as resulting from the formation of Pr^{4+} ions, which reduced the hole concentrations, but they disagree with those of *Subramaniam* et al. [3.79] for the series $La_{1.85-x}$· $Pr_xSr_{0.15}CuO_4$. These latter results show that over the whole substitution range $(0 \le x \le 0.80)$, T_c and the percentage of formally trivalent copper remain approximately constant and that praseodymium would be in the trivalent state. Thus the nature of the valence of Pr ions in these oxides has not been elucidated.

b) Bismuth Substitution

Substitution of bismuth for lanthanum in the strontium-doped superconductors also modifies the superconducting properties of these oxides, as was shown by *Michel* et al. [3.80], who studied the oxides $La_{2-x}Bi_x'Sr_{x-x'}$· CuO_{4-y}. Pure phases with a tetragonal cell are obtained for $x \le 0.3$ and $x' \le$

0.2 without detectable variations of the cell parameters, and superconductivity was mainly observed for $x-x' = 0.1-0.2$. The composition $La_{1.8} \cdot Bi_{0.1}Sr_{0.1}CuO_{4-y}$ exhibits a resistive transition with a T_c at midpoint of about 23 K after annealing in air. This value is very close to that observed for $La_{1.9}Sr_{0.1}CuO_4$ synthesized in a flow of pure oxygen [3.58]. For a given strontium concentration ($x-x' = 0.2$), a bismuth concentration can cause T_c to reach an optimum value, since for an air-annealed sample corresponding to $x' = 0.05$, T_c at the midpoint of the resistive transition is 41 K, which should be compared with $T_c \simeq 36$ K measured for a sample with $x' = 0$ prepared in the same conditions, and for $x' = 0.1$ T_c is only around 29 K. Thus, the introduction of small amounts of bismuth is able to improve the superconducting properties of the strontium doped oxides, however, higher bismuth contents tend to destroy superconductivity. This phenomenon may be due to the ability of Bi(III) to be partially oxidized into Bi(V) due to the presence of the alkaline earth, as already observed for the superconducting mixed valence perovskite $Ba(Pb_{1-x}Bi_x)O_3$ [3.81]. As was assumed above for Pr substitutions, this possible partial oxidation of the bismuth ions may reduce the hole concentration. As a result, the oxidation conditions (annealing time and oxygen pressure) should have a great influence on the superconducting properties of these materials. This influence is shown in Fig.3.12 for the composition $La_{1.7}Bi_{0.1}Sr_{0.2}CuO_{4-y}$ prepared under various conditions (for details, see figure caption). From this figure it appears immediately that the use of pure oxygen destroys the superconducting properties (samples C and D). Nevertheless a fall of the resistivity is observed at about 42 and 35 K, respectively, indicating that superconducting regions are still present. In contrast, the samples A and B heated in air are superconductors with a T_c at midpoint of 29 K (sample A) and 42 K (sample B). This shows that the transition is displaced towards higher temperatures as the annealing time in air decreases. In the same way, an increase of the resistivity is observed as the annealing time or p_{O_2} increases: $\rho \sim (3-5) \cdot 10^{-3}$ $\Omega \cdot cm$ for A and B samples, and $\rho \sim 2 \cdot 10^{-2}$ and $2 \cdot 10^{-1}$ $\Omega \cdot cm$ for samples C and D, respectively. This behavior is in complete contrast with that of the oxides $La_{2-x}Sr_xCuO_{4-x/2+\delta}$, whose superconducting properties and electron

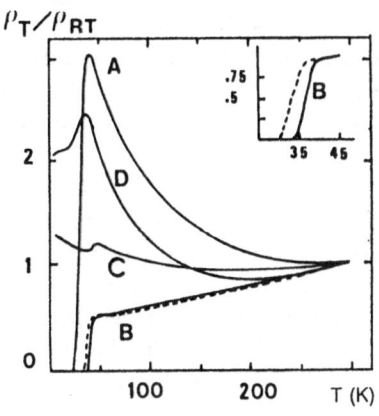

ρ_T/ρ_{RT}

Fig.3.12. Resistivity as a function of temperature, normalized to ρ_{300K}, for the oxide $La_{1.7}Bi_{0.1}Sr_{0.2}CuO_{4-y}$. (*A*) Annealed in air at 400°C for 18 h. (*B*) Quenched in air from 1000°C. (*C*) Annealed in pure oxygen at 400°C for 18 h. (*D*) Synthesized, sintered and annealed in pure oxygen. *Inset*: expanded view of the resistive transition for sample B, and $La_{1.8}Sr_{0.2}CuO_{4-y}$ (*dotted line*) for comparison. (After [3.80])

transport properties in the normal state are improved by oxygen thermal treatments. However, it is consistent with a preferential oxidation of bismuth with respect to copper in the presence of oxygen. The fact that the resistive transition of sample B ($T_c \sim 42K$) is 4 K higher than that of $La_{1.8}Sr_{0.2}CuO_{3.9+\delta}$ ($T_c \sim 38K$) has not yet been explained.

These results showed quite early on that bismuth was a potential element for superconductivity in copper oxides (Chap.5) but also drew attention to its particular property with respect to the oxidation atmosphere, which may be due to its tendency to be oxidized from Bi(III) to Bi(V), killing the superconductivity by a possible partial occupation of the copper sites by Bi(V) (see Chap.4 concerning the influence of substitutions on copper sites).

3.2.4 Other Cuprates with the K_2NiF_4 Structure

The phases Ln_2CuO_4 (Ln = Pr, Nd, Sm, Eu, Gd) exhibit a structure different from that of La_2CuO_4 (Chap.2) which is characterized by copper in square-coplanar coordination leading to the formation of $[CuO_2]_\infty$ layers perpendicular to the c axis of the tetragonal cell. The electrical properties are also quite different since they are characteristic of semiconductors. For example, the conductivity of Sm_2CuO_4 and Nd_2CuO_4 at room temperature are three to four orders of magnitude lower than that of La_2CuO_4 with an activation energy of about 0.15 eV below 600 K, and the Seebeck coefficients are highly negative, $\alpha_{300K} \sim -1000 \ \mu V \cdot K^{-1}$ [3.15], indicating that the charge carriers are electrons and that copper is probably strictly divalent.

Substitution of strontium for lanthanum in Ln_2CuO_4 allows the synthesis of oxides with the K_2NiF_4 structure for the compositions $Ln_{2-x}Sr_x \cdot CuO_{4-x/2+\delta}$ (Ln = Pr, Nd; $1 \le x \le 1.2$ and $\delta = 0$) [3.29]. The thermal dependence of the conductivity for as-synthesized samples is quite similar to that of the strontium-doped La_2CuO_4 oxide with the same strontium content prepared in the same conditions. It is characteristic of a semiconducting behavior that can be interpreted as a variable range hopping according to the linear variation of log σ vs $T^{-1/4}$ proposed by *Mott*, see (3.5). The Seebeck coefficients are positive and increase with temperature in agreement with this model. The high values of α ($\alpha_{300K} \sim 100$ and $180 \ \mu V \cdot k^{-1}$ for Pr and Nd, respectively) can be related to the low values of the conductivity [$\sigma_{300K} = 2 \cdot 10^{-2} \Omega^{-1} cm^{-1}$ (Ln = Pr) and $6 \cdot 10^{-4} \Omega^{-1} cm^{-1}$ (Ln = Nd)] and to the magnetic measurements, which can be interpreted in terms of the lone presence of Ln^{3+} and Cu^{2+} ions.

The orthorhombic solid solution $La_{2-x}Pr_xCuO_4$, $0 \le x \le 0.5$ exhibits a thermal variation of the conductivity similar to that of the nonsuperconducting La_2CuO_4 [3.19]. At room temperature, σ slightly increases with increasing x. This behavior has been interpreted by *Goodenough* as an increase of the width of the conduction band associated with a decrease of the unit cell parameters. But this interpretation fails with the $Ln_{0.8} \cdot Sr_{1.2}CuO_{3.4}$ oxides; the unit cell parameters of the Pr and Nd oxides are lower than those of the La phase, whereas σ_{300K} is higher and lower for Ln = Pr and Nd, respectively, than σ_{300K} for Ln = La ($\sigma_{300K} \sim 10^{-3}\ \Omega^{-1}cm^{-1}$).

3.3 Oxides with the Oxygen-Deficient $Sr_3Ti_2O_7$-Type Structure

The oxides that correspond to the general formula $Ln_{2-x}A_{1+x}Cu_2O_{6-x/2+\delta}$ are obtained for Ln = La with $0 \leq x \leq 0.14$ for A = Sr and x = 0.1 for A = Ca and for Ln $\neq$ La only in the case of A = Sr with x = 0.14 (Ln = Pr), $0 \leq x \leq 0.2$ (Ln = Nd), $0.7 \leq x \leq 0.9$ (Ln = Sm, Gd), and $0.6 \leq x \leq 0.9$ (Ln = Eu). The structure is an intergrowth of double oxygen-deficient perovskite layers with rock-salt-type monolayers. The oxygen vacancies are located at the junction of the two perovskite monolayers, leading to the formation of CuO_5 square pyramids (Fig.2.22).

3.3.1 The Oxides $La_{2-x}A_{1+x}Cu_2O_{6-x/2+\delta}$ (A = Sr, Ca)

The thermal dependence of the electrical conductivity is plotted in Fig.3.13 for several compositions. They are characterized by a sharp and continuous increase at low temperature (T < 300K) for the less-oxidized compounds - i.e., for strontium, quenched specimens - and by nearly temperature-independent behavior for the most-oxidized samples. The value of σ at room temperature increases with the concentration of formally trivalent copper from 7 $\Omega^{-1}cm^{-1}$ for $La_2SrCu_2O_6$ [Cu(III) ~ 0%] to $3 \cdot 10^2\ \Omega^{-1}cm^{-1}$ for $La_{1.86}Sr_{1.14}Cu_2O_{6.22}$ [Cu(III) ~ 29%]. These features correspond to a semi-conducting mode. The Seebeck coefficient, studied in the range 120-330 K, is positive, indicating a p-type conduction in agreement with the presence of holes introduced by the $[Cu-O]^+$ species. The quenched samples exhibit α values at 300 K ranging from 100 to 120 $\mu V \cdot K^{-1}$, which decrease continuously when T decreases, while the specimens annealed in oxygen are characterized by a small and nearly temperature-independent thermoelectric power ($\alpha_{300K} \sim 20 \mu V \cdot K^{-1}$). The room temperature conductivity of the cal-

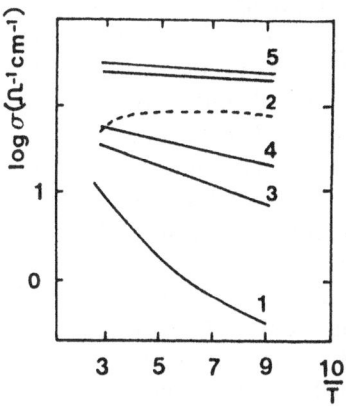

Fig.3.13. $La_{2-x}A_{1+x}Cu_2O_{6-x/2+\delta}$ oxides. Conductivity (logarithmic scale) vs T^{-1} for different compositions of the strontium series (solid lines): 1: x = 0, δ = 0; 2: x = 0, δ = 0.20; 3: x = 0.1, δ = 0.20; 4: x = 0.14, δ ~ 0.04; 5: x = 0.14, δ ~ 0.29; and for the calcium compound (dotted line). (After [3.82])

cium-quenched sample is twice as large as the strontium one for the same x value and its thermal variation is quite different since it is close to that of a metal in spite of the high α value at room temperature ($\alpha \sim 140 \mu\text{V·K}^{-1}$). The variation of σ vs T^{-1} for the oxygen-annealed calcium compound [Cu(III) $\sim$ 8%] is quite similar to that of the quenched sample and its value can be deduced from the previous one by using a multiplicative factor of about 1.4.

In these compounds the environment of the copper ions is mainly characterized by four short Cu-O distances of 1.95 Å in the "ab" plane and one long Cu-O distance close to 2.30 Å along the c axis, so that the splitting of the 3d orbitals (Fig.3.2) leads to a localization of the hole of the $3d^9$ configuration of the Cu^{2+} ion in the $3d_{x^2-y^2}$ orbital. It results that, in a first approximation, the band structure near the Fermi level can be considered to be the same as in Fig.3.3d.

For strontium compounds the fact that the conductivity depends sensitively on the hole concentration and that $La_2SrCu_2O_6$ exhibits a high value of the conductivity in spite of the absence (or only very small concentration) of $[Cu-O]^+$ species suggests that the separation between the upper and lower subbands is very weak. The increase of the Seebeck coefficient with T for the less-oxidized compounds and the fact that σ follows a $T^{-1/4}$ law for $La_2SrCu_2O_6$ are consistent with holes trapped at the top of the filled band. For the most-oxidized compounds [20%-30% of formal Cu(III)], whose conductivities are nearly temperature independent, the density of holes is large enough to produce a good mobility, although the case of metallic behavior has not yet been reached. The small value of the thermoelectric power is characteristic of a high density of degenerate-gas carriers.

The quenched calcium compound $La_{1.9}Ca_{1.1}Cu_2O_{5.97}$ exhibits surprisingly high and near-metallic conductivity for a small Cu(III) concentration ($\sim$2%). This shows the important effect of the band width W and of the intra-atomic Coulomb energy U on the electrical properties due to the nature of the A cation. In this compound, W becomes larger owing to the decrease of the unit cell parameter compared to the Sr oxides so that the upper and lower bands probably overlap slightly, leading to higher carrier mobility [3.82].

Above room temperature the thermal variations of the electron transport properties can easily be explained by oxygen concentration variations, which are observed by ThermoGravimetric (TG) measurements [3.82]. The influence of the oxygen stoichiometry on the electrical conductivity and the reversibility of the phenomenon has been studied for the oxide, $La_{1.9}Ca_{1.1}Cu_2O_{5.97}$. The thermal variation of the conductivity in argon reveals a dramatic fall of σ (Fig.3.14) associated with a loss of oxygen near 570 K (curve 1). At this temperature, heating is stopped and the sample is cooled progressively down to 77 K. In this latter temperature range a semiconducting behavior is observed (curve 2) owing to a lower oxygen content. Heating up to 500 K again in argon leads to the same curve. However, beyond 500 K, σ decreases again owing to the fact that thermodynamical equilibrium had not previously been reached. At 570 K argon is replaced by air

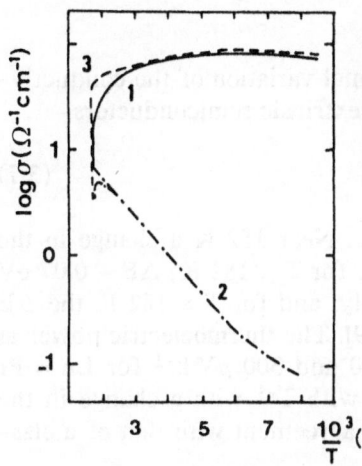

Fig.3.14. Conductivity vs T^{-1} for the oxides $La_{1.9}Ca_{1.1}Cu_2O_{5.97}$ under different conditions; *1*: first heating in an inert atmosphere; *2*: first cooling and second heating in an inert atmosphere; *3*: air introduction and second cooling (in air). (After [3.82])

and oxygen is intercalated, in agreement with the increase of σ. When σ becomes about constant, the sample is cooled and shows a behavior down to 77 K quite similar to that of the starting material.

Surprisingly, neither the strontium compounds nor the calcium one are superconductors [3.83] down to 4.2 K, in spite of the low dimensionality of the structure and of the electron transport properties. This behavior can be partly explained by structural considerations and the oxidation of copper. Copper in pyramidal coordination can be considered as forming $[CuO_2]_\infty$ planes perpendicular to **c**, owing to the long Cu-O apical distance. These $[CuO_2]_\infty$ planes are assumed to be responsible for superconductivity. Such a situation occurs for compounds containing only six or fewer oxygen atoms per formula. For higher oxygen contents extra oxygen atoms are located between two copper atoms, leading to the formation of octahedra with a Cu-O distance along **c** of about 1.85 Å, so that $[CuO_2]_\infty$ planes can no longer be considered to exist. Then superconductivity should only be observed in samples containing six (or fewer) oxygen atoms per formula. In such compounds the highest $[Cu-O]^+$ concentration would be for the oxide $La_{1.86}\cdot Sr_{0.14}CuO_6$, with a value of 7%. This value corresponds to the lower limit value (6%-8%) below which superconductivity does not exist in the Sr-doped La_2CuO_4 oxides.

Very recently, *Cava* et al. [3.84] have been successful in the search for superconductivity in this system. They synthesized under 20 atm axygen pressure in a sealed quartz tube, oxides with the composition $La_{2-x}Sr_xCa\cdot Cu_2O_6$ and showed that these oxides can be superconductors with a maximum T_c (60K) corresponding to about 0.2 holes per copper atom.

3.3.2 The Oxides $Ln_{2-x}Sr_{1+x}Cu_2O_{6-x/2}$

Only air-quenched samples of $Ln_{2-x}Sr_{1+x}Cu_2O_{6-x/2}$ have been studied and no trace of Cu(III) has been detected by thermogravimetric analysis in these conditions.

a) Ln = Pr, Nd; x = 0.14

These two oxides are characterized by a thermal variation of the conductivity which follows the classical expression for extrinsic semiconductors,

$$\sigma = \sigma_0 \exp(-\Delta E/kT) , \tag{3.7}$$

with a σ value of about 1 $\Omega^{-1}cm^{-1}$ at 300 K. Near 182 K a change in the slope of the curves $\log\sigma = f(T^{-1})$ is observed: for T < 182 K, $\Delta E \sim 0.07$ eV and 0.09 eV for Ln = Pr and Nd, respectively, and for T > 182 K the ΔE values are approximately 0.02 eV higher [3.29]. The thermoelectric power at room temperature is highly positive ($\alpha \sim 100$ and 300 $\mu V \cdot k^{-1}$ for Ln = Pr and Nd, respectively) and increases linearly with T^{-1} with a change in the slope around 190 K. This behavior of α is in agreement with that of a classical extrinsic semiconductor:

$$\alpha = - \frac{k}{e} \left[\frac{K_2}{K_1} - E_F \right] / kT , \tag{3.8}$$

where E_F is an activation energy and K_i are the values of the standard transport integrals. The positive sign of α and the E_F values calculated from the previous relation argues that the hole conduction is created by acceptor levels at around 0.03 eV (E_F) above the $3d_{x^2-y^2}$-O 2p filled subband, which does not overlap the empty $3d_{x^2-y^2}$ subband. These levels are assumed to be due to Ln impurities present in the starting materials.

b) Ln = Eu, Sm, Gd; x = 0.9

At room temperature the values of the conductivity are very weak, since they range from $6 \cdot 10^{-5}$ $\Omega^{-1}cm^{-1}$ for Ln = Eu to $1.3 \cdot 10^{-4}$ $\Omega^{-1}cm^{-1}$ for Ln = Gd and follow the relation for semiconductors [see (3.7) with $\Delta E \sim 0.34eV$]. The α values are positive (at 300 K, 170 $\mu V \cdot K^{-1} \leq \alpha \leq 250$ $\mu V K^{-1}$) and increase approximately linearly with T. This behavior is characteristic of Mott semiconductors. By annealing the europium compound in oxygen, σ_{300K} increases up to $8.4 \cdot 10^{-4}$ $\Omega^{-1}cm^{-1}$ owing to an increase of the oxygen content [Cu(III) $\sim 6\%$]. The poor electron transport properties of these oxides with regard to the previous ones are attributed to the greater number of oxygen vacancies.

3.4 Oxides with the Oxygen-Deficient Perovskite Structure

3.4.1 Nonsuperconducting Oxides

a) LaCuO₃

Among the cuprates which crystallize with the perovskite structure, La·CuO₃ [3.85] appears to be an exception since it exhibits only formally trivalent copper and no oxygen vacancies. This particularity is due to the method of synthesis, which needs a very high oxygen pressure (65 kbar). $LaCuO_3$ is metallic, but its conductivity is very low for a metal ($\sigma_{300K} \sim 0.2\,\Omega^{-1}cm^{-1}$). The magnetic susceptibility is weak and decreases slowly as temperature increases, however, its magnitude ($\chi_M = 3.2 \cdot 10^{-4}$ e.m.u. at room temperature) is too high to be interpreted as a Pauli paramagnetism characteristic of delocalized charge carriers, see (3.4). To raise the Pauli susceptibility to the experimental value requires $m^*/m \sim 10$. This value is characteristic of a gas of strongly correlated carriers (or degenerate spin polaron gas) [3.45, 46, 86]. $LaCuO_3$ crystallizes in a rhombohedral cell. The copper octahedra are regular [$d(Cu-O) = 1.94\,\text{Å}$] so the Cu $3d_{x^2-y^2}$ and Cu $3d_{z^2}$ orbitals lie at the same energy level (Fig.3.2). Then both orbitals will participate in the conduction band, leading to a three-dimensional band structure near the Fermi level, so that $LaCuO_3$ is not a superconductor.

b) $BaLa_4Cu_5O_{13+\delta}$ and $La_{8-x}Sr_xCu_8O_{20-\epsilon}$

The oxygen-deficient perovskites $BaLa_4Cu_5O_{13+\delta}$ ($\delta \sim 0.1$) and $La_{8-x}Sr_x \cdot Cu_8O_{20-\epsilon}$ ($\epsilon \sim 0.15$, $1.28 \leq x \leq 1.92$), whose structures are depicted in Figs. 2.14,17, exhibit very similar electron transport properties [3.87,88]. Both are metallic at least up to 700 K. The σ values at room temperature are very high since they are greater than $10^3\ \Omega^{-1}cm^{-1}$ for well-sintered materials. Typical curves $\rho = f(T)$ are given in Fig.3.15 for the barium compound and one of the strontium series ($x = 1.6$). Above room temperature, the curves agree well with the characteristic law for metals (3.2); the deviation from linearity observed at low temperatures is attributed to the influence of defects and especially to oxygen vacancies. The γ values deduced from these curves in the high temperature range are $4.1 \cdot 10^{-3}$ and $2.2 \cdot 10^{-3}\ ^\circ C^{-1}$ for the barium and strontium oxides, respectively. These values are close to the theoretical value for free electrons ($\gamma = 3.7 \cdot 10^{-3}\ ^\circ C^{-1}$). At room temperature the thermoelectric power is very weak, in agreement with a metallic behavior, and positive: $\alpha \sim 9\ \mu V \cdot K^{-1}$ for $BaLa_4Cu_5O_{13+\delta}$ and $2\ \mu V \cdot K^{-1} \leq \alpha \leq \mu V \cdot K^{-1}$ for $La_{8-x}Sr_xCu_8O_{20-\epsilon}$; it increases slightly with temperature. At temperatures lower than 200 K, the sign of α becomes negative for all com-

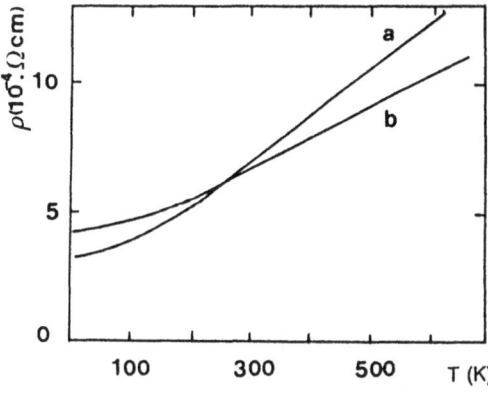

Fig.3.15. Thermal dependence of resistivity for (a) $BaLa_4Cu_5O_{13+\delta}$ ($\delta \sim 0.1$) and (b) $La_{6.4}Sr_{1.6}Cu_8O_{20-\epsilon}$ ($\epsilon \sim 0.15$)

positions, whereas no change in the X-ray diffraction patterns versus temperature is observed. This change in sign was first attributed to a lack of accuracy in the e.m.f. measurements due to the very weak α values [3.88], but it was also observed by other researchers [3.89]. The molar magnetic susceptibility is very weak. It decreases slightly with increasing temperature up to 250 K and then becomes nearly temperature independent. The χ_M values [$\chi_M = (110-120)\cdot 10^{-5}$ e.m.u. at room temperature] are however too high to be interpreted as a Pauli paramagnetism of an electron gas, so a strongly correlated gas as in $LaCuO_3$ is assumed. No superconductivity has been observed in these compounds down to 4.2 K [3.83,90]. This is in agreement with the three-dimensionality of the structures (Chap.2) and with the electronic band structure calculations [3.91] for $BaLa_4Cu_5O_{13+\delta}$, which show that the bands near the Fermi level are more nearly three dimensional than two or one dimensional. Considering the structure of these compounds, two σ^* conduction bands can be distinguished, owing to the anisotropy induced by the oxygen vacancies lying in the (001) plane of the tetragonal cell: a band $\sigma^*_\parallel$ parallel to the c axis involving the Cu $3d_{z^2}$ orbital and a band $\sigma^*_\perp$ perpendicular to the c axis involving the Cu $3d_{x^2-y^2}$ orbital. The mean Cu-O distances in the (001) plane are slightly larger than the mean out-of-plane Cu-O distances [3.92,93], so the copper polyhedra can be considered as slightly compressed along c, leading to the $3d_{z^2}$ level lying above the $3d_{x^2-y^2}$ (Fig.3.2). This results in the $\sigma^*_\parallel$ band being higher in energy and probably broader than the $\sigma^*_\perp$ band. Moreover, in the latter, oxygen vacancies may act as traps for the charge carriers. Assuming that the bands $\sigma^*_\parallel$ and $\sigma^*_\perp$ contribute to the conductivity and that a change in the sign of α occurs, a band model can be proposed involving a splitting of each band into two Hubbard bands, σ^{*1} and σ^{*2}, due to strong electron-electron interactions (Fig.3.16). In this model, the conduction in the $\sigma^{*2}_\parallel$ band is of n-type and in the $\sigma^{*2}_\perp$ band, p-type. The experimental thermoelectric power can be written as for intrinsic semiconductors [3.1]

$$\alpha = [(\alpha\sigma)_{\sigma^{*2}_\perp} + (\alpha\sigma)_{\sigma^{*2}_\parallel}]/(\sigma_{\sigma^{*2}_\perp} + \sigma_{\sigma^{*2}_\parallel}) . \qquad (3.9)$$

The sign of α then depends on the relative values of the products ασ. It will depend only on the relative values of σ if $|\alpha_{\sigma^{*2}_\perp}|$ is assumed to be of the

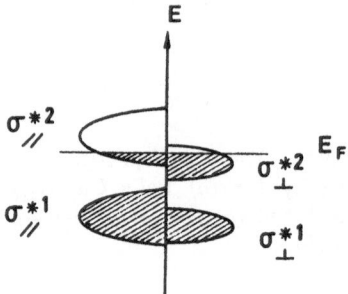

Fig.3.16. Proposed band model for $BaLa_4Cu_5O_{13+\delta}$ and $La_{8-x}Sr_xCu_8O_{20-\epsilon}$. (After [3.88])

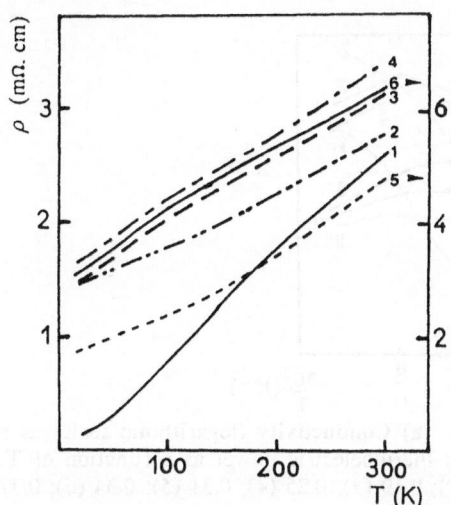

Fig.3.17. Variation of the resistivity as a function of temperature for the oxides $La_3LnBaCu_5O_{13+\delta}$: Ln = La(1), Nd(2), Sm(3), Gd(4), Dy(5) and Y(6). (After [3.95])

same order of magnitude but slightly greater than $|\alpha_{\sigma_{\parallel}}*2|$. Then at low temperatures, $\sigma_{\sigma_{\perp}}*2 \ll \sigma_{\sigma_{\parallel}}*2$, owing to the trapping of the charge carriers, and $\alpha\sigma_{\sigma_{\parallel}}*2$ dominates in the expression of α, which will be negative. On the other hand, at high temperatures, all the charge carriers are delocalized, leading to an increase of $\sigma_{\sigma_{\perp}}*2$, which allows $(|\alpha|\sigma)_{\sigma_{\perp}}*2 > (|\alpha|\sigma)_{\sigma_{\parallel}}*2$ and α to become positive. Oxygen can be removed under reducing conditions (N_2) without changing the structure but leading to less conductive samples: $BaLa_4CuO_{12.9}$ shows a semimetallic behavior [3.94].

Rare earths and yttrium can partly substitute lanthanum in $La_4Ba \cdot Cu_5O_{13}$ [3.95]. For Ln = Nd, Sm, Gd, Dy the maximum substitution is 25%, whereas it can reach 50% for yttrium. All the compositions exhibit a metallic character without any trace of superconductivity down to 20 K (Fig. 3.17).

c) $Ba_3La_3Cu_6O_{14+\delta}$

The first investigation of the influence of the oxygen stoichiometry upon the electron transport properties of the oxygen-deficient perovskite cuprates was carried out by *Provost* and co-workers [3.96] on the oxides $Ba_3La_3Cu_6O_{14+\delta}$. As a function of the oxygen pressure during annealing at 400°C, δ may vary in the range 0.05–0.43 (Fig.3.18) leading to 18%–31%

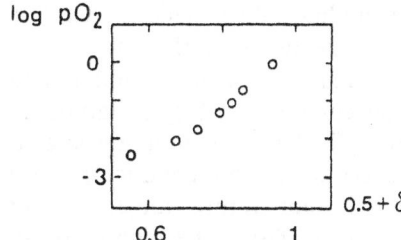

Fig.3.18. Dependence of the extra oxygen $(0.5+\delta)$ compared to a pure divalent oxide on the oxygen pressure during annealing at 400°C for 24 h for $Ba_3La_3Cu_6O_{14+\delta}$. (After [3.43])

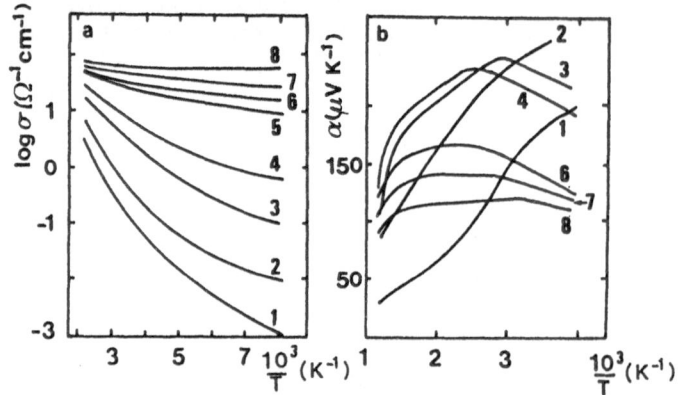

Fig.3.19a,b. The oxides $Ba_3La_3Cu_6O_{14+\delta}$. (a) Conductivity (logarithmic scale) as a function of reciprocal temperature and (b) thermoelectric power as a function of T, for different δ values: $\delta = 0.05$ (*1*); 0.10 (*2*); 0.19 (*3*); 0.25 (*4*); 0.31 (*5*); 0.34 (*6*); 0.37 (*7*) and 0.43 (*8*). (After [3.96])

formal Cu(III) from electroneutrality calculations. With increasing δ, the room-temperature conductivity increases continuously from 0.1 $\Omega^{-1}cm^{-1}$ (δ ~ 0.05) to 70 $\Omega^{-1}cm^{-1}$ (δ ~ 0.43) and a smooth transition from a semiconducting behavior to a semimetallic behavior is observed for the curves $\sigma = f(1/T)$ (Fig.3.19a). The molar magnetic susceptibilities are weak, are nearly independent of the oxygen concentration and their values decrease from 80 to 300 K by a factor of 1.5. In contrast, the values and the thermal variation of the thermoelectric power are strongly dependent on δ (Fig.3.19b). Three different types of curves are observed, depending on the oxygen composition.

$\delta = 0.05$ and 0.10. At low temperatures (T < 200K), α exhibits small values ($\alpha < 70\mu V\cdot K$ at 120K) and increases linearly with T. These values and those of the conductivity at the same temperature (σ ~ 10^{-3} and $10^{-2}\Omega^{-1}cm^{-1}$ at 120K for $\delta = 0.05$ and 0.10, respectively) are characteristic of a polaron hopping and probably of variable range hopping since the activation energy deduced from the curves log σ vs T^{-1} shows a continuous variation from 100 to 300 K. At higher temperatures (T > 300K), α assumes greater values ($\alpha > 150\mu V\cdot K^{-1}$ at 300K), which, combined with the linear variation of logσ = $f(T^{-1})$, are characteristic of the tendency towards a classical extrinsic semiconductive behavior.

$\delta = 0.18$ and 0.25. The Seebeck coefficient also increases with T at low temperatures (T < 250K) but exhibits much higher values. This increase of α with the oxygen content corresponds to a decrease in the number of charge carriers near the Fermi level, so the increase of σ may be understood as a progressive delocalization of the charge carriers. At higher temperatures, 250-300 K, α decreases linearly with 1/T. This behavior, associated with the linear variation of logσ = $f(1/T)$, is typical of semiconductors with electron excitation in a conduction band. In this temperature range, a direct comparison between activation energies derived from the conductivity

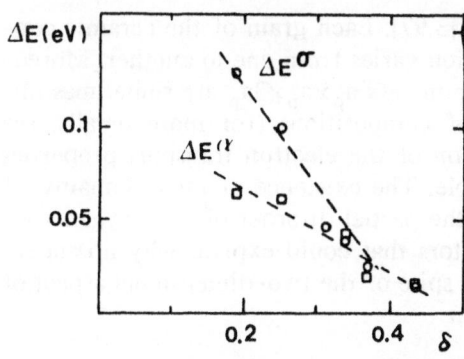

Fig.3.20. Activation energies (conductivity and Seebeck coefficient) as a function of δ for $Ba_3La_3Cu_6O_{14+\delta}$. (After [3.96])

(ΔE^σ) and from the Seebeck coefficient (ΔE^α) (Fig.3.20) reveals that they are both small. However, ΔE^α still remains lower than or equal to ΔE^σ, which suggests that the hole mobility is still activated with an energy $\Delta E^\mu = \Delta E^\sigma - \Delta E^\alpha$, when the density of states near the Fermi level is low.

$\delta = 0.34 - 0.43$. The conductivity varies very slightly with temperature from 80 to 400 K and becomes more and more independent of temperature as δ increases. At the same time, α decreases but remains much higher than for classical metals. Above 300 K, $\log \sigma$ increases linearly as T^{-1} decreases, with a very weak slope (Figs.3.19, 20), while α decreases linearly with T^{-1} with the same slight slope as σ, indicating a very degenerate semiconductor with a large number of charge carriers in the vicinity of the conduction band.

For $\delta \geq 0.18$, ΔE^α decreases approximately linearly with δ. In the same range of composition, the linear decrease of α with (1/T) observed at high temperatures follows the known relation for extrinsic semiconductors when a gap exists:

$$|\alpha| = \frac{k}{e}\left(\frac{\Delta E^\alpha}{kT} + \text{const.}\right). \tag{3.10}$$

If electron transport properties governed by the Mott model [3.45, 46] are assumed, ΔE^α should be equal to the difference between the Fermi level E_F and the mobility edge E_C, below which holes would be delocalized, and the conduction would be due to excitation into the mobility edge. The decrease of ΔE^α, as the number of holes increases indicates that a critical δ value may exist which corresponds to $\Delta E^\alpha = 0$ (E_F crosses the E_C energy level), giving rise to metallic conduction, i.e. when the density of holes at the top of the conduction band is large enough. This critical δ value is close to 0.50.

The great similarity between the structure of these oxides and that of the "123" superconductors, as revealed by neutron diffraction studies (Chap.2), allows a Cu $3d_{x^2-y^2}$-O $2p$ band to be proposed as the conduction band. But the reality is much more complex, as was shown by a high resolution electron microscopy study [3.97]. Each grain of the ceramic consists of microdomains whose orientation varies from one to another. Moreover, deviations from the unit cell parameters $a_p \times a_p \times 3a_p$ are sometimes observed, leading to local variations of compositions (for more details, see Chap.6), so an extensive interpretation of the electron transport properties of these oxides would be unreasonable. The existence of microdomains, of local variations of composition, and the partial disorder of the oxygen vacancies and La and Ba atoms are factors that could explain why no superconductivity is observed [3.90,98], in spite of the two-dimensional aspect of the structure, as will be discussed later.

3.4.2 Superconducting Oxides: Properties of $YBa_2Cu_3O_{7-\delta}$

The superconducting oxides with an oxygen-deficient perovskite structure are known as the 90 K superconductors. As was shown in Chap.2, they correspond to the general formula $LnBa_2Cu_3O_{7-\delta}$ (Ln stands for rare earths and yttrium) and their structure corresponds to that of the perovskite with ordering of oxygen vacancies and Ln and Ba atoms. In these oxides, copper is fivefold (square pyramidal) and fourfold (square planar) coordinated. Of this series, $YBa_2Cu_3O_{7-\delta}$ is the most studied compound, owing to the fact that it was the first to be synthesized [3.99-102].

"123" superconductors containing exactly seven oxygen atoms per formula unit are not easily obtained in ceramics and generally a slight oxygen deficiency is observed, even after annealing in a pure oxygen flow at a low temperature, as shown in the recent study of the oxygen content as a function of the annealing temperature and of the oxygen pressure by *Karpinski* et al. [3.103]. Nevertheless, in the following part of this chapter, we shall use the term "stoichiometric oxide" to mean a carefully synthesized compound with a δ value (where δ is the oxygen deficiency with respect to seven) as low as possible, even if this value is not strictly equal to zero.

a) The "Stoichiometric" Oxide

Well-synthesized samples are known to be superconductors below 92K with a very sharp resistive transition of about 1-2K width. In contrast to the 40K superconductors, the critical temperature at the midpoint of the resistivity fall is almost pressure independent [3.104], even up to 170 kbar [3.105], while the transition width increases.

The critical field is close to 50T at 77K and it is estimated to be higher than 100T at 0K [3.106-108]. A critical current of 140 kA/cm^2 is calculated at 1.5K in a magnetic field of 2T from magnetic hysteresis cycles. This value decreases with increasing H or T [3.109]. Values of Jc lower than 10^3 A/cm^2 are generally measured on ceramic samples, without an applied magnetic field [3.109-111]. But these values can be increased up to $(2-4)\cdot10^4$ A/cm^2 at 77K (H = 0) for textured ceramics [3.112-114] and to around $5\cdot10^6$ A/cm^2, in the same conditions, for single crystalline thin films [3.114,115].

In the normal state and up to 500-550 K, a roughly metallic behavior is observed on sintered samples ($\gamma \sim 2.04\cdot10^{-3}$ °C^{-1}), and for higher tem-

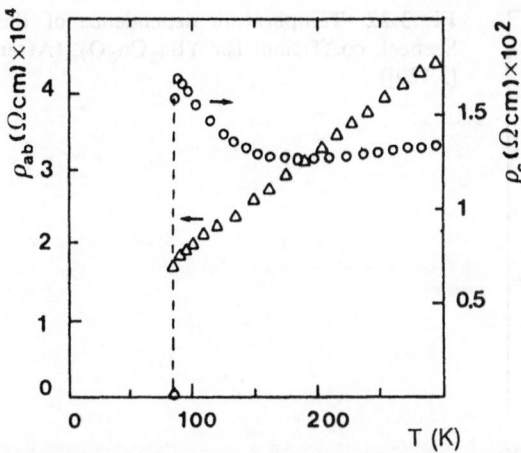

Fig.3.21. Resistivity tensor components parallel to the Cu-O planes (ρ_{ab}) and perpendicular to them, i.e., along the c axis (ρ_c), measured for a single crystal of $YBa_2Cu_3O_7$. (After [3.117])

peratures the resistivity increases dramatically with T [3.116], owing to the removal of oxygen ($\rho_{800K}/\rho_{300K} \sim 10$).

The electron transport properties of this oxide are highly anisotropic, as depicted in Fig.3.21 for the thermal dependence of the resistivity measured on single crystals in two perpendicular directions, i.e. along the c axis of the unit cell (ρ_c) and perpendicular to this axis (ρ_{ab}). At room temperature the ratio ρ_c/ρ_{ab} is in the range 30-35 and increases with decreasing temperature to reach a value of about 100 just above T_c [3.117-120], following two very different behaviors: metallic in the "ab" plane, i.e. parallel to the triple oxygen-deficient perovskite layers, and semiconducting along the c axis. Thin film measurements reveal a different behavior: ρ_c does not exhibit semiconducting behavior and the ratio ρ_c/ρ_{ab} varies from 4-7 at room temperature to about 5-10 just above T_c [3.121, 122].

The sign of the thermoelectric power is positive above the critical temperature; its value is very small ($\alpha_{300K} \sim 5\mu V \cdot K^{-1}$) and decreases linearly when temperature decreases in the range 200-300 K ($\alpha_{200K} \sim 3 \mu V \cdot K^{-1}$), but below 150 K it departs from the linear behavior and shows an anomalous increase as T_c is approached [3.123]. This behavior, which was deduced from ceramic measurements, is confirmed by a single-crystal study [3.120], which also reveals strong anisotropy of the thermopower (Fig.3.22). In the "ab" plane, the Seebeck coefficient exhibits the same behavior as for ceramic samples, while along the c axis it increases continuously with temperature, following an expression in the form $\alpha = aT+bT^2$, and exhibits higher values (at room temperature $\alpha_c \sim 10\mu V \cdot K^{-1}$, $\alpha_{ab} \sim 3\mu V \cdot K^{-1}$). In contrast with that for La_2CuO_4-type superconductors, the Hall coefficient is highly temperature dependent and decreases with increasing T; for ceramic samples, at room temperature, R_H ranges from $+1 \cdot 10^{-3}$ to $+2 \cdot 10^{-3}$ cm^{3}C^{-1} [3.49, 124, 125]. One interesting feature is the results of single-crystal measurements: for H parallel to the c axis, R_H is found to be positive, corre-

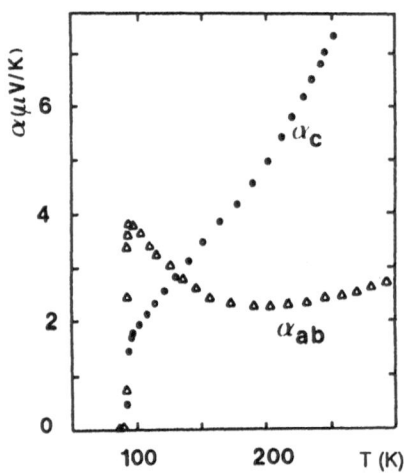

Fig.3.22. Temperature dependence of the Seebeck coefficient for $YBa_2Cu_3O_7$. (After [3.120])

sponding to hole conduction, but it is found to be smaller and negative when H is parallel to the "ab" plane [3.117, 118, 126]. These results are consistent with the predictions of band-theory analysis for La_2CuO_4-type cuprates [3.127, 128] and extended to $YBa_2Cu_3O_7$, but the positive sign of α_{ab} does not agree with the theory which predicts a negative value. This contradiction between theory and experiments stems from the difficulty of establishing an accurate model of the complex electronic structure of this oxide. The Coulomb repulsion energy is $U_{dd} \sim 5\text{-}6$ eV [3.25-28, 129], close to the 3d electron band width (W $\sim$ 5eV [3.28]).

The band structure calculations show that the upper valence band crossing the Fermi energy level is formed by the antibonding Cu-O orbitals: Cu $3d_{x^2-y^2}$-O $2p_x, 2p_y$. Only this two-dimensional band is generally considered in the theories, and the holes induced by the mixed valence should be in this band. However, spectroscopic study [3.27] of a single crystal reveals that hole states are found in the "ab" plane as well as along the c axis, but at lower energy ($\sim$ 1.5eV). Moreover, *Bianconi* et al. [3.131] have studied the symmetry of the charge carriers $3d^9\underline{L}$ by measuring the polarized L_3 X-ray absorption spectra of a single crystal. They observed that the ligand holes are formed not only in the two-dimensional band derived from the Cu $3d_{x^2-y^2}$ but also in a band derived from the Cu $3d_{z^2}$ orbital orthogonal to the "ab" plane. These results indicate that the ligand holes are in a band involving also the Ba-O layers, which must be taken into account in the theories of high T_c superconductivity.

b) Influence of the Oxygen Nonstoichiometry

$YBa_2Cu_3O_7$ begins to lose oxygen at about 350-450 °C even if it is heated in a pure oxygen flow (Fig.3.23), leading to the nonstoichiometric oxides $YBa_2Cu_3O_{7-\delta}$, whose electron transport properties have been studied by several teams [3.35, 129, 131-138]. Although not all the results coincide, it is well established that T_c decreases drastically from 92 K for $\delta \simeq 0$, to around 28 K for $\delta \simeq 0.7$ [3.131], a value beyond which the samples are semicon-

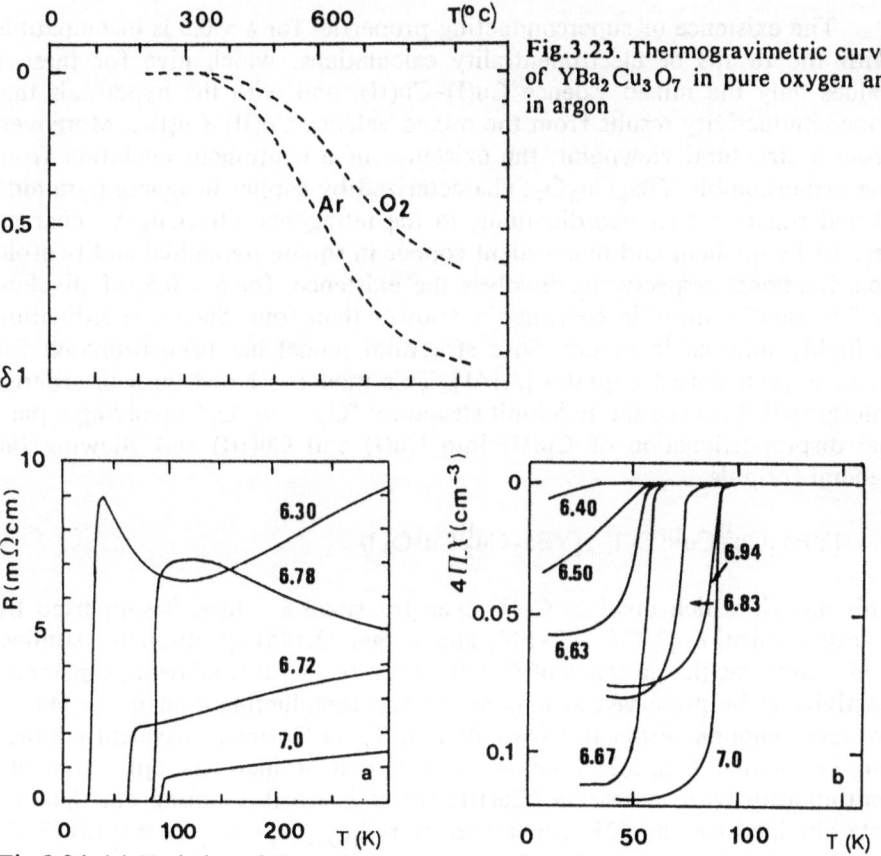

Fig.3.23. Thermogravimetric curves of $YBa_2Cu_3O_7$ in pure oxygen and in argon

Fig.3.24. (a) Variation of the resistance as a function of temperature for some δ values of the $YBa_2Cu_3O_{7-\delta}$ series (after [3.131]), and (b) Meissner effect measurements. (After [3.132])

ductors. As an example, some curves showing the temperature dependence of the resistivity and of the magnetic moment, for several δ values, are plotted in Figs.3.24a,b. At the same time as superconductivity disappears, the unit cell symmetry becomes tetragonal. As T_c decreases, a decrease of the percentage of diamagnetism is observed, indicating a decrease of the superconducting volume. The conductivity of the semiconducting oxides shows a thermally activated conduction. However, the activation energy remains temperature dependent [3.139], indicating a variable range hopping conduction mode, as confirmed by a well-defined $T^{-1/4}$ dependence for the conductivity. As the oxygen content deviates from seven per formula unit, the Seebeck coefficient increases to reach values greater than 100 $\mu V \cdot K^{-1}$ for the semiconducting compositions, corresponding to a decrease of the density of charge carriers near the Fermi level. The evolution of the Hall coefficient as a function of δ shows an increase from $2 \cdot 10^{-3}$ ($\delta = 0$) to $4 \cdot 10^{-3}$ cm^{3}C^{-1} ($\delta \sim 0.1$), then it remains constant up to $\delta \sim 0.4$ and drastically increases for higher δ values ($R_H \sim 22 \cdot 10^{-3}$ cm^{3}C^{-1} for $\delta \sim 0.7$) [3.140].

The existence of superconducting properties for $\delta > 0.5$ is incompatible with the results of electroneutrality calculations, which give for these δ values only the mixed valence Cu(I)-Cu(II), and with the hypothesis that superconductivity results from the mixed valence Cu(II)-Cu(III). Moreover, from a structural viewpoint, the existence of a continuous evolution from the orthorhombic $YBa_2Cu_3O_7$, characterized by copper in square pyramidal and square planar coordinations, to the tetragonal $YBa_2Cu_3O_6$, characterized by divalent and monovalent copper in square pyramidal and twofold coordinations, respectively, involves the existence, for $\delta < 0.5$, of divalent or "trivalent" copper in coordinations lower than four. Such a coordination is highly unusual in oxides. So a structural model has been proposed for these oxygen-defect cuprates [3.141]. This model is based on a disordered intergrowth between the two limit structures "O_7" and "O_6" involving a partial disproportionation of Cu(II) into Cu(I) and Cu(III) and allowing the general formula

$$[YBa_2Cu_2^{II}Cu^{III}O_7]_{1-\delta}[YBa_2Cu_2^{II}Cu^IO_6]_\delta \ .$$

This model, which involves Cu(I) even for small δ values, is supported by X-ray absorption [3.136, 142-144] and Auger [3.145] spectrometry studies. It also involves the presence of Cu(III), even for $\delta > 0.5$, allowing superconductivity to be preserved as long as the superconducting domains or chains are large enough. From this formulation, T_c vs "formally trivalent copper" can be plotted (Fig.3.25), demonstrating that it increases approximately continuously with increasing "Cu(III) %". It is worth pointing out that the data obtained for the 40K superconductors $La_{2-x}Sr_xCuO_4$ ($x < 0.15$) [3.47,

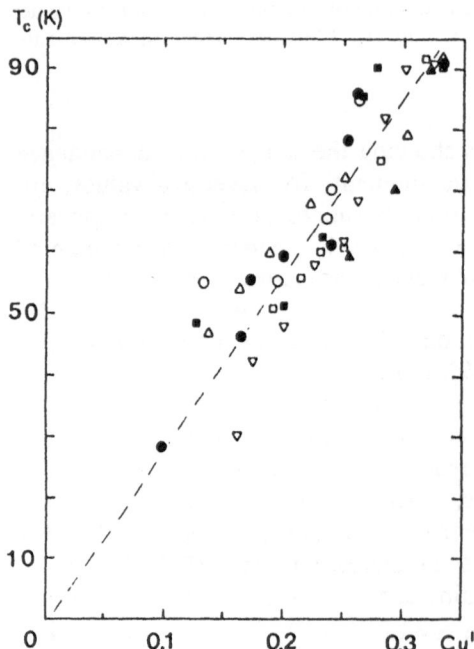

Fig.3.25. T_c vs Cu(III)% (or [Cu-O]$^+$ %) according to different researchers [black symbols represent results of resistivity measurements (T_c offset)]: (•) [3.131]; (△) [3.134]; (■) [3.135]. Open symbols represent results of magnetic measurements (T_c onset): (Δ) [3.35]; (∇) [3.134]; (○) [3.133]; ( ) [3.132]

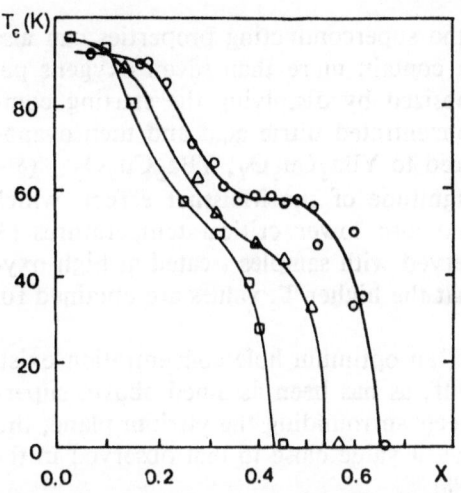

Fig.3.26. Variation of T_c as a function of δ for long-annealed $LnBa_2Cu_3O_{7-\delta}$ samples: Ln = Y (o), Gd ($\triangle$), and Nd ($\square$). (After [3.166])

48, 61] and $La_{2-x}CuO_4$ [3.36], which are plotted in Fig.3.8, show that there is no evidence of discontinuity in the evolution of T_c.

The experimental values of T_c as a function of δ differ according to the method of preparation. For example, when samples are obtained from short annealings at high temperatures at a low oxygen pressure [3.35, 132], a monotomic decrease is observed, whereas plateaus with $T_c \simeq 90$ K ($\delta \leq 0.2$) and 50-60 K ($0.3 \leq \delta \leq 0.5$) are observed for samples resulting from long annealing times (for an example, see Fig.3.26). These plateaus result from an ordering of the oxygen vacancies. Such an ordering, which involves superstructures, was observed by electron diffraction microscopy for samples with $\delta \sim 0.5$, and structural models were proposed for the orthorhombic phase $YBa_2Cu_3O_{6.5}$, called the orthorhombic phase II (for more details, see Chap.6). The most probable model for this latter phase results in the doubling of the a parameter [3.146, 147]. Susceptibility measurements of vacuum annealing of $YBa_2Cu_3O_{7-\delta}$ samples at 400 °C for various periods of time provide evidence for the presence of two superconducting phases whose fraction changes with oxygen stoichiometry [3.148]. The "orthorhombic phase II" is assumed to be responsible for superconductivity at 50-60 K. Superconductivity in this oxide is assumed to be due to the CuO_2 planes orthogonal to the c axis and belonging to the pyramidal layers, whereas T_c is enhanced to 90 K in the orthohombic phase I ($\delta \leq 0.1$) due to an increase of electronic coupling of the CuO_2 planes by the interleaving CuO_4 square planar groups [3.35, 135, 149]. This idea is firmly supported by the recent work of *Tolentino* et al. [3.150], who studied the symmetry of the holes as a function of δ by polarized X-ray absorption spectroscopy on a single crystal. They found that, when δ decreases, the first holes which appear have xy symmetry, corresponding to the conductive planes and chains, and that, for $\delta \leq 0.3$, there appear in addition holes with z symmetry, which might correspond to a coupling between the different superconducting planes and chains inside the triple oxygen-deficient perovskite layers, increasing T_c to 92 K. The importance of the nature of the Cu-O polyhedra in the middle of the triple perovskite layers for the superconducting properties can also explain the behavior of oxides that contain more than seven oxygens per unit cell. Such oxides can be synthesized by dissolving the starting compounds (Y_2O_3, CuO, $BaCO_3$) in concentrated nitric acid and then evaporating and calcining [3.151]. Compared to $YBa_2Cu_3O_7$, $YBa_2Cu_3O_{7+\delta}$ ($\delta \sim$ 0.1-0.2) oxides exhibit a lower magnitude of the Meissner effect, which decreases as δ increases [3.151], and also lower critical temperatures [3.152, 153]. The same behavior is observed with samples treated at high oxygen pressure [3.154], so it appears that the higher T_c values are obtained for the "stoichiometric" oxide ($\delta = 0$).

As in the 40K superconductors, an optimum hole concentration exists which is, in this case, around 33%. If, as has been assumed above, superconductivity is due top the CuO_2 sheets surrounding the yttrium plane, this leads to 16.5% holes per copper sheet, a value close to that observed in the K_2NiF_4-type superconductors ($\sim$15%).

c) Substitution for Yttrium and Barium

Yttrium can be replaced by many rare earth ions without changing the orthorhombic structure and the superconducting properties, T_c remains, for most of them, in the range 88-92 K [3.35, 120, 155-160]. This occurs for the oxides $LnBa_2Cu_3O_7$ with Ln denoting Nd to Lu (except Pm, Tb). The fact that T_c is not significantly affected, even by the presence of magnetic rare earth cations, is in contrast with the classical superconductors such as the A15s for which traces of magnetic impurities kill the superconductivity. Thus it appears that the interactions between the two-dimensional copper-oxygen framework and the planes of Ln^{3+} ions are very weak, bringing out the importance for superconductivity of the role of the Cu-O square planar groups, which are about 6.5 Å [Cu(1)-Ln distance] from the rare earth ions. The magnetic moments, extracted from the magnetic susceptibility measurements, generally agree with the theoretical values expected for free trivalent rare earth ions, suggesting that all the rare earths are in the trivalent state. In the normal state, all these oxides exhibit a metallic behavior with ρ_{300K} increasing with Ln^{3+} size.

Yttrium can be partly replaced by a divalent cation such as calcium, leading to the solid solution $Y_{1-x}Ca_xBa_2Cu_3O_{7-x/2}$ with $0 \le x \le 0.3$ [3.161, 162]. As x increases, the oxygen content, and consequently T_c decreases: $T_c \sim$ 20K (offset in resistivity measurements) for x = 0.3; moreover, the resistive transition is broadened [3.161]. To preserve oxygen stoichiometry, cross substitutions, like the series $Y_{1-x}Ca_xBa_{2-x}La_xCu_3O_7$, have been investigated, but the results do not agree. Some researchers [3.162-164] find that there is a complete homogeneity range ($0 \le x \le 1$) and describe the upper limit $CaBaLaCu_3O_7$ as a tetragonal superconductor with $T_c \sim$80K, whereas others [3.165] find the solid solution limited to x $\sim$ 0.5 with 75K $\le T_c \le$ 90K.

When yttrium is replaced by a rare earth, T_c is also dramatically affected by oxygen stoichiometry, as shown in Fig.3.26 for $NdBa_2Cu_3O_{7-\delta}$ and $GdBa_2Cu_3O_{7-\delta}$ samples obtained after long annealings [3.166]. The region of δ where samples are superconducting and the width of plateaus become narrower as the ionic radius of the rare earth ion increases.

The lanthanum-substituted oxide requires particular attention. $LaBa_2 \cdot Cu_3O_7$ was prepared for the first time as a superconductor ($T_c \sim 75K$) by *Michel* et al. [3.167], then it was reported to exhibit a critical temperature ranging from about 50 K [3.168, 169] to 80-90 K [3.155, 170-172] and also not to be a superconductor [3.35, 158, 173], depending on the thermal treatment. This strange behavior results from the fact that this oxide can be obtained either in a tetragonal form or in an orthorhombic form. The synthesise of "good" $LaBa_2Cu_3O_{7-\delta}$ (orthorhombic form) requires two important procedures. The first is sintering above 950 °C in N_2 gas atmosphere, the second is annealing at 300 °C in dried O_2 gas atmosphere [3.174], below the temperature of the transformation from tetragonal to orthorhombic, which occurs at about 350° C [3.172]. The superconducting form results from an ordering of the La^{3+} and Ba^{2+} ions, whereas a partial disorder occurs in the tetragonal form. The smaller difference in size between lanthanum and barium compared to the other rare earths allows such a partial disorder, which may also induce disordering of the oxygen vacancies and deviation from the oxygen stoichiometry. Moreover, if the nature of the Ln^{3+} planes seems to be insensitive to the superconducting properties, this disorder, which involves the presence of lanthanum in the BaO planes, leads to a change in the cationic charge of these planes, which are found to be active in the formation of the holes [3.130] and could influence the superconducting and normal state properties. A good example is the oxide $La_3Ba_3 \cdot Cu_6O_{14+\delta}$, which has been described earlier: the cell symmetry is tetragonal, the barium sites are partly occupied by lanthanum (25%), the oxygen vacancies are weaker and are delocalized at the level of Cu(1) atoms. In spite of its high trivalent copper content (~33%), close to that of $YBa_2 \cdot Cu_3O_7$, the oxygen-annealed compound is not a superconductor and exhibits a semimetallic behavior. On the other hand, the study of the series $La_{3-x}Ba_{3+x}Cu_6O_{14+\delta}$ shows that a solid solution exists for $0 \leq x \leq 1$ and that a resistive transition occurs as soon as $x \sim 0.2$ ($T_c \sim 40K$) [3.175] with a minimum of the curve $\rho = f(T)$ at $T \sim 100$ K. But the results of high-resolution microscopy for $x = 0$ (Chap.6) and the cell parameter values for the two limit compositions $x = 0$ and $x = 1$ allow one to assume that, in fact, domains of a superconducting phase with composition close to $La \cdot Ba_2Cu_3 \cdot O_{7-\delta}$ and whose dimensions and number are large enough to allow them to percolate are formed in the bulk.

Other examples can be found to illustrate the influence of the nature of the "Ba-O" planes on the superconducting properties. The first is that of the series $YBa_{2-x}Sr_xCu_3O_{7+\delta}$ [3.176] which, a priori, does not involve any change in the oxygen content. Single phases are obtained for $x \leq 1$. In this range of composition, all the samples exhibit a metallic behavior and a resistive transition, with the critical temperature at midpoint decreasing from 92 K ($x = 0$) to 82 K ($x = 1$), with which is associated a decrease of the orthorhombic distortion of the unit cell. Refinement of the structure for $x = 1$ shows that the oxygen content does not vary in a significant way, so that the decrease of T_c may be attributed to a local perturbation of the structure in the neighborhood of the Sr sites due to the smaller size of Sr^{2+} compared to Ba^{2+} ($\Delta R \sim 0.2 \text{Å}$). These results should be compared with those from the study of the series $YBa_{2-x}La_xCu_3O_{7+\delta}$: the difference in size between La^{3+} and Ba^{2+} ions is like that above ($\Delta R \sim 0.24 \text{Å}$) but the difference of charge may involve oxygen contents greater than seven, as shown for the compound $YBa_{1.6}La_{0.4}Cu_3O_{7+\delta}$ [3.168, 177], which is near the limit of the orthorhombic domain in this series. For this composition, T_c is 40 K [3.168], 45 K [3.178] and 52 K ($\delta = 0.05$) [3.177], i.e. at much lower values than observed with strontium. Since La^{3+} and Sr^{2+} ions have comparable size and close electronegativities, the sharp decrease in T_c can be attributed, in a first approximation, to the presence of extra oxygen atoms (i.e., more than seven per formula unit) induced by the difference of charge. *Manthiram* et al. [3.177] have attributed the decrease of T_c to the formation of peroxide-like species $(O_2)^{2-}$ within the perovskite layers to preserve the local charge, leading to a reduction of the number of charge carriers. Such species have been observed in $YBa_2Cu_3O_{7-\delta}$ samples by X-ray photoelectron spectroscopy [3.145]. This hypothesis is supported by studies of the series $LnBa_{2-x}Ln_xCu_3O_{7+\delta}$ [3.175, 179] (Ln = Nd, Sm, Eu, Gd and Dy), which also show a sharp decrease of T_c as x increases and correlatively an increase of the Hall constant, which is evidence for a decrease of the effective charge carrier concentration [3.179]. Some results are plotted in Fig. 3.27.

From the rare earths that do not lead to superconducting oxides (cerium, praseodymium and terbium) only the oxide $PrBa_2Cu_3O_{7\pm\delta}$ can be prepared as a single phase [3.35, 180, 181] but with a tetragonal cell symmetry, which is known to be unfavorable for superconductivity. In addition to the absence of superconducting properties, this oxide exhibits a semiconducting behavior; however, the resistivity at room temperature depends on the thermal treatment and decreases with increasing oxygen content: $\rho_{300K} \sim 3.6$, 0.12 and 0.02 $\Omega\cdot$cm for 6.8, 7.0 and 7.1 oxygen atoms per unit cell, respectively [3.180].

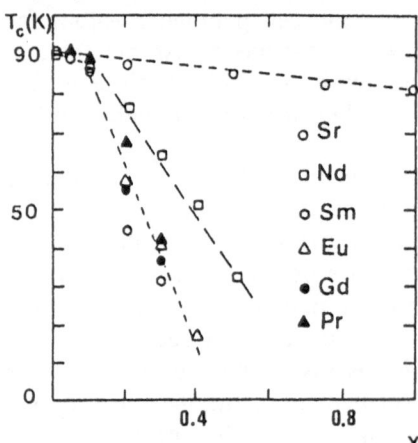

Fig.3.27. Midpoint T_c vs x for various oxides of the series $YBa_{2-x}A_xCu_3O_{7\pm\delta}$ (A = Sr, Nd, Sm, Eu, Gd) and $Y_{1-x}Pr_xBa_2Cu_3O_{7-\delta}$

The study of the series $Y_{1-x}Pr_xBa_2Cu_3O_{7+\delta}$ [3.181, 182] reveals that as x increases, T_c decreases up to x = 0.5 ($T_c \sim 30K$), then no transition is observed. At the same time, ρ_{300K} increases and its thermal dependence becomes more and more like that of a semiconductor. The Seebeck coefficient remains positive over all the range of concentrations but its value increases with x to reach a value close to 125 $\mu V \cdot K^{-1}$ at room temperature for x = 1. The increase in magnitude of α indicates a decrease of the density of states at the Fermi level; moreover, the shape of the curves $\alpha = f(T)$ is close to that observed for the oxygen-deficient oxides $YBa_2Cu_3O_{7-\delta}$ with large δ values. These results show that the influence of praseodymium on T_c may be attributed to a decrease of the Cu(III) content, in which case Pr should be mainly in the tetravalent state.

Magnetic measurements, carried out for the composition $PrBa_2Cu_3 \cdot O_{7.1}$, show effectively the presence of Pr^{4+} ions with a ratio $Pr^{4+}/Pr^{3+} \sim 8.1$ [3.180]. Using this value, the average valence of copper can be calculated as 2.1, leading to 10% "trivalent" copper, which corresponds to a value where no superconductivity is observed in the "123" phases.

This is in complete disagreement with the results of spectroscopic studies by X-ray absorption [3.183, 184], which indicate that the Pr valence is close to III in $PrBa_2Cu_3O_{7+\delta}$ as well as in the solid solution $Y_{1-x}Pr_x \cdot Ba_2Cu_3O_{7+\delta}$ and that hybridization between the valence-bond state and Pr ions may possibly occur. The latter could be at the origin of the apparent reduced valence state for praseodynium, as has been observed for bismuth in bismuth superconductors [3.185]. The study of the series with constant oxygen stoichiometry $Y_{1-x-y}Ca_yPr_xBa_2Cu_3O_{6.95\pm0.02}$ ($0\leq x\leq 0.3$, $0\leq y\leq 0.2$) sheds some light on this problem [3.186]. The critical temperature (50% of the resistivity drop) vs Ca concentration y for four Pr concentrations x is plotted in Fig.3.28. Maxima are clearly visible in both the x = 0.15 (curve 3)

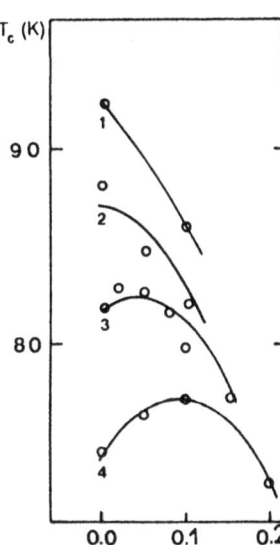

Fig.3.28. Variation of T_c as a function of calcium concentration y for different Pr contents in the oxides $Y_{1-x-y}Ca_yPr_xBa_2Cu_3O_{6.95\pm0.02}$: x = 0(1), 0.1(2), 0.15 (3) and 0.20(4). (After [3.186])

and $x = 0.20$ (curve 4) series, and can be expected for lower x values. The horizontal shift of the maxima vs y shows that increasing Pr concentration requires a larger calcium content to obtain the maximum. Since Ca contributes holes, this implies that Pr contributes electrons and hence that its formal valence is greater than III. In addition, the decrease in the magnitude of the maximum value of T_c when the Pr concentration increases, evinces the destructive effect on superconductivity of Pr ions. This is explained as a hybridization of the Pr 4f states with the valence-bond states associated with the Cu-O planes, which would generate an appreciable exchange interaction between the Pr magnetic moments and the spin of the mobile holes [3.186].

The presence of a maximum in the evolution of T_c vs calcium concentration for constant x is more direct evidence for an optimal hole concentration in the superconductive Cu-O sheets.

Substitution of alkaline-earths for barium will be discussed in Chap.4.

4. Substitutions in La$_2$CuO$_4$-Type and YBa$_2$Cu$_3$O$_7$-Type Superconductors

A considerable number of papers have been published on the effects of substitution at either the rare earth or copper sites on the superconducting properties of the La$_2$CuO$_4$- and YBa$_2$Cu$_3$O$_7$-type structural families. It is, of course, not possible to discuss here all these results, in spite of their great importance from the physical viewpoint. It should, however, be emphasized that it is very difficult to interpret many of these papers, in spite of their great interest, because the materials are very often not sufficiently characterized from the chemical and crystallographic viewpoints. Indeed, it is easy to understand that if the oxygen content has not been determined accurately, a rigorous interpretation is prevented. In the same way, knowledge of the distribution of cations in the structure, or of their influence on the structural evolution, is important for an understanding of the superconducting properties. Thus, this chapter will focus above all on the chemical and crystallographic aspects of these substitutions and on their relationships with the superconducting properties, especially the evolution of the critical temperature. For practical reasons, most of the references are taken from two conferences that were held in Berkeley, California, in June 1987 and in Interlaken in February 1988 (published in Physica C, Vols.153-155).

4.1 Substitution on the Rare-Earth Sites

The substitution of a Rare-Earth (RE) magnetic ion for lanthanum or yttrium may help our understanding of the interplay between magnetism and superconductivity. For this reason, the first investigations which were performed concern the RE substitution.

4.1.1 La$_2$CuO$_4$-Type Oxides

Unlike La$_2$CuO$_4$, none of the rare-earth oxides exhibit a structure of the K$_2$NiF$_4$ type (Chap.2). Thus the substitution of RE ions for lanthanum in the La$_2$CuO$_4$-type superconductors is limited to low RE concentrations. Few investigations have been performed in these systems up to now, owing to the discovery of the 92 K superconductors very soon after the La$_2$CuO$_4$-type oxides.

Tarascon et al. [4.1] have focused their attention on the La$_{1.6}$Sr$_{0.2}$RE$_{0.2}$CuO$_{4-y}$ series. For these compositions they observed a single phase for all the rare earths, except for the heavier ones (Er, Dy). They charac-

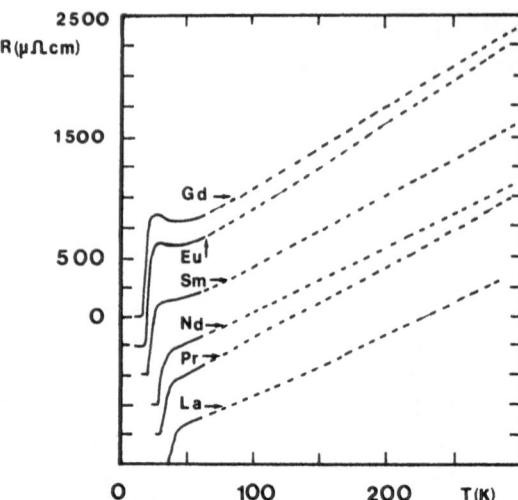

Fig.4.1. Resistivity as a function of temperature from below T_c to 300 K for the $La_{1.6}Sr_{0.2}RE_{0.2}CuO_{4-y}$ series. (After [4.1])

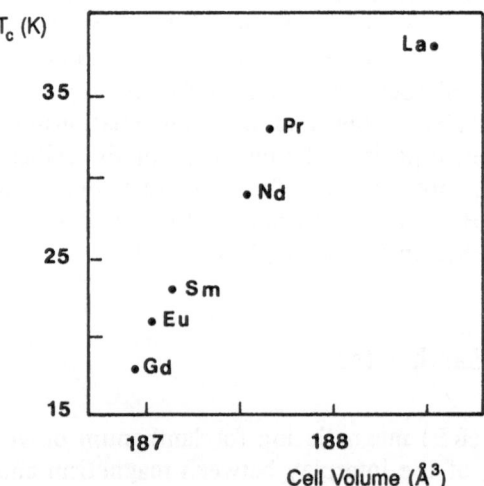

Fig.4.2. Superconducting critical temperatures and tetragonal unit cell volume for various rare earths for the $La_{1.6}Sr_{0.2}RE_{0.2}CuO_{4-y}$ series. (After [4.1])

terized a tetragonal cell for these oxides ($a \sim a_p$; $c \sim c_{La_2CuO_4}$), with a and c decreasing continuously as Z of the lanthanide increases. Thus these results show clearly that here the rare-earth ions are located on the lanthanum sites. The main effect observed by these researchers concerns the critical temperature T_c, which decreases with the substitution of rare earths of increasing Z from 39 K for lanthanum to 18 K for gadolinium (Fig.4.1). This depression of T_c appears to be correlated with the decrease of the tetragonal cell volume (Fig.4.2): T_c decreases smoothly through the rare-earth series. This effect of RE substitution upon superconductivity was presented in detail in Chap.3, and, as previously discussed, it should be pointed out that an effect other than the "volume effect" may be at the origin of this evolution, namely, the trivalent copper content, which may decrease with the size of the RE ions. A careful study of the oxygen content of these materials should allow the question of which is the responsible effect to be answered. The substitution of sodium for lanthanum in the La_2CuO_4-type structure by *Subramanian* et al. [4.2] is very interesting. The researchers showed that the oxides $La_{2-x}Na_xCuO_4$ were superconducting up to about 40 K for x values from 0.2 to 0.5.

Substitution of bismuth for lanthanum in the La_2CuO_4-type oxide has shown, for the first time, the particular behavior of bismuth, and especially its influence on the critical temperature in relation to the method of synthesis (Fig.3.12). Pure tetragonal phases ($a{\sim}a_p$; $c{\sim}c_{La_2CuO_4}$) with K_2NiF_4-type structure were synthesized with the compositions $La_{2-x}Bi_{x'}Sr_{x-x'}{\cdot}CuO_{4-y'}$ for $x \leq 0.3$ and $x' \leq 0.2$ with critical temperatures ranging from 23 to 41 K [4.3]. The spectacular influence of the experimental conditions of synthesis on the superconducting properties of these oxides has already been discussed in Chap.3.

4.1.2 $YBa_2Cu_3O_7$-Type Oxides

A great deal of work has been carried out concerning the substitution of rare-earth (RE) ions for yttrium. The references given here [4.1, 4-20] are obviously incomplete, and will not be discussed. A more detailed study, especially of the superconducting properties of these oxides, has been presented in Chap.3.

All the isostructural oxides $REBa_2Cu_3O_{7-\delta}$ (Table 4.1) have been prepared, with the exception of RE = Tb. The different researchers obtain very similar results for most of the oxides with regard to the crystal data.

Table 4.1. Meissner effect results, effective magnetic moments μ_{eff}, and paramagnetic Curie temperatures Θ, for the $REBa_2Cu_3O_{7-\gamma}$ series, along with their electrical properties (at 300K resistivity ρ_{300}, the transition temperature T_{c0} and midpoint (T_{cm}) of the superconducting transition]. (From [4.1])

Compound	Meissner effect [%]	μ_{eff}(calc) $[\mu_B]$ per RE atom	μ_{eff}(exp) $[\mu_B]$ per RE atom	Θ [K]	T_{cm} [K]	T_{c0} [K]	ρ_{300K} $[\mu\Omega{\cdot}cm]$
$NdBa_2Cu_3O_{7-y}$	48	3.62			95.3	92.0	4200
$SmBa_2Cu_3O_{7-y}$	32	0.85			93.5	88.3	2800
$EuBa_2Cu_3O_{7-y}$	35	0.00			94.9	93.7	2500
$GdBa_2Cu_3O_{7-y}$	36	7.94	7.75, 7.85[*]	- 2, - 3[*]	93.8	92.2	1650
$DyBa_2Cu_3O_{7-y}$	31	10.65	10.45,10.69[*]	- 7, - 8[*]	92.7	91.2	1550
$HoBa_2Cu_3O_{7-y}$	33	10.61	10.48,10.60[*]	-16, -19[*]	92.9	92.2	1200
$ErBa_2Cu_3O_{7-y}$	34	9.58	9.93, 9.63[*]	-15, -16[*]	92.4	91.5	1880
$TmBa_2Cu_3O_{7-y}$	33	7.56	7.53, 7.69[*]	-35, -38[*]	92.5	91.2	1110
$YbBa_2Cu_3O_{7-y}$	49	4.54			87.0	85.6	1000
$LuBa_2Cu_3O_{7-y}$	10	0.00			89.5	88.2	17000

[*] Samples annealed under vacuum

With the exception of RE = La and Pr, all the oxides are found to exhibit an orthorhombic cell similar to that of $YBa_2Cu_3O_7$ ($a \sim a_p$, $b \simeq a_p$, $c \simeq 3a_p$). Moreover, a systematic decrease of the cell volume is observed from Nd to Yb, but the variations of the lattice parameters are not absolutely continuous (Table 4.2). Lanthanum and praseodymium oxides exhibit distinctive behavior: they were found in a first step by many researchers to be tetragonal and nonsuperconductors. Subsequently $LaBa_2Cu_3O_{7-\delta}$ was prepared with orthorhombic symmetry (Chap.2) and superconducting properties. In the same way, orthorhombic $PrBa_2Cu_3O_{7+\delta}$ was able to be prepared by several researchers [4.21-24], but unlike the other oxides this phase is not superconducting. Like the pure yttrium compound, the tetragonal oxides can be obtained by annealing the superconducting samples under vacuum at 450°C. Thus there is no doubt that the limiting $REBa_2Cu_3O_7$ oxides can be synthesized. It has been observed from Seebeck-coefficient measurements that the tetragonal oxides are n-type semimetals or semiconductors. As for yttrium, it is noted that the orthorhombic phase with both p-type metal-like conductivity and superconductivity can easily be restored by reannealing the samples in an oxygen flow.

Several attempts to substitute cerium and also bismuth for yttrium in $YBa_2Cu_3O_7$ were unsuccessful. Several researchers report the possible substitution of scandium for yttrium without loss of superconducting properties. However, no structural study allowed the introduction of scandium in the $YBa_2Cu_3O_7$-type matrix to be proved: indeed, the X-ray data indicate multiphase oxides for $x \geq 0.25$ for $Y_{1-x}Sc_xCu_3O_{7-\delta}$ [4.25] and no variations of T_c [4.25]. Thus further work needs to be done to understand whether scandium has entered the lattice or not.

Substitution of calcium for yttrium in $YBa_2Cu_3O_6$ allows the superconducting phase $Y_{1-x}Ca_xBa_2Cu_3O_6$ to be isolated, with a T_c of 50 K [4.26]. Such an oxide is characterized by a disproportionation of copper, with the mixed valency $Cu(II)-Cu(III)$ in the pyramidal copper layer, whereas univalent copper forms layers of CuO_2 rods. Superconductivity up to 44 K is also induced by calcium in the oxide $Y_{1-x}Ca_xBa_2Cu_3O_{6+\delta}$ ($\delta \leq 0.20$) [4.27].

4.2 Substitution on the Copper Sites

Taking into consideration the fact that superconductivity occurs in the copper-oxygen layers, one expects that substitution for copper would dramatically influence the superconducting properties of these oxides. The elements that are suitable for substitution for copper should have a similar size and should present octahedral, pyramidal or square planar coordination. The main elements that were investigated were nickel, iron, cobalt, zinc, aluminum and palladium.

Table 4.2. Crystal data for $REBa_2Cu_3O_{7-y}$, RE = Y or a rare-earth element, for compounds as grown under 1 atm O_2 (orthorhombic) or annealed under vacuum (tetragonal). The oxygen content in the samples as-prepared and the change in oxygen y after annealing in vacuum are determined by thermogravimetric analysis as described in the text. a, b and c are the dimensions of the unit cell, and V is the volume of the unit cell. (From [4.1])

| RE | y | \multicolumn{4}{c}{As prepared (under O_2)} | | \multicolumn{3}{c}{Annealed in vacuum} |
		a [Å]	b [Å]	c [Å]	V [Å³]	Δy	a,b [Å]	c [Å]	V [Å³]
Y	6.72	3.8237 (8)	3.8874 (8)	11.657 (2)	173.28 (6)	0.53	3.8589 (8)	11.800 (4)	175.72 (10)
La		3.8562 (8)	3.9057 (10)	11.783 (3)	177.47 (7)		3.9007 (8)	11.858 (3)	180.42 (9)
Pr		3.922 (2)			180.7 (11)/3	0.69			
Nd	7.16	3.8546 (8)	3.9142 (12)	11.736 (2)	177.07 (7)	0.47	3.8930 (6)	11.882 (2)	180.13 (9)
Sm	7.11	3.855 (2)	3.899 (2)	11.721 (4)	176.11 (13)	0.51	3.884 (2)	11.822 (8)	178.40 (2)
Eu	7.10	3.8448 (8)	3.9007 (10)	11.704 (3)	175.53 (7)	0.56	3.8829 (8)	11.814 (4)	178.11 (10)
Gd	7.08	3.8397 (12)	3.8987 (18)	11.703 (3)	175.19 (11)	0.54	3.8770 (6)	11.810 (2)	177.51 (6)
Dy	6.82	3.8284 (8)	3.8888 (8)	11.668 (2)	173.71 (5)	0.58	3.8656 (10)	11.783 (5)	176.07 (11)
Ho	6.71	3.8221 (8)	3.8879 (8)	11.670 (2)	173.42 (6)	0.66	3.8601 (8)	11.791 (3)	175.69 (9)
Er	6.82	3.8153 (8)	3.8847 (12)	11.659 (2)	172.80 (7)	0.58	3.8540 (8)	11.796 (4)	175.22 (11)
Tm	6.65	3.8101 (8)	3.8821 (12)	11.656 (2)	172.41 (7)	0.70	3.8491 (10)	11.788 (4)	174.65 (10)
Yb	6.71	3.7989 (8)	3.8727 (10)	11.650 (2)	171.40 (6)		3.8411 (4)	11.828 (2)	174.51 (5)

4.2.1 La$_2$CuO$_4$-Type Oxides

Few investigations were carried out on the La$_2$CuO$_4$ oxides, as previously explained. One attractive point was the replacement of copper by nickel, owing to the existence of the La$_2$NiO$_4$ oxide, which exhibits the K$_2$NiF$_4$-type structure and a metallic behavior but does not superconduct. This investigation, performed [4.1] on the 39 K superconductor La$_{0.85}$Sr$_{0.15}$CuO$_4$, showed that the oxides La$_{1.85}$Sr$_{0.15}$Cu$_{1-x}$Ni$_x$O$_4$ were single phase over a large range of x values ($0 \leq x \leq 1$) and could be indexed in a tetragonal cell ($a \sim a_p$, $c \simeq c_{La_2CuO_4}$). It was also observed (Table 4.3) that the a parameter increases and c decreases with increasing x. The substitution of nickel for copper lowers the critical temperature dramatically (Fig.4.3). Indeed, the latter decreases from 39 to 21 K if just 0.025 Cu is replaced by nickel. As soon as x reaches 0.075, a pronounced semiconducting behavior is observed. The same dramatic change is obtained for the magnetic properties (Fig.4.4): a Pauli paramagnetism is observed up to x = 0.05, whereas at higher x values the susceptibility increases with decreasing temperature according to the Curie-Weiss law. Thus, as nickel replaces copper the carrier density decreases, leading to a semimetal-semiconductor transition. The substitution of zinc, cadmium or mercury for copper in this structural type was studied by *Bauer* and co-workers [4.28-30]. They found that copper could be replaced statistically by zinc to a large extent in La$_{1.85}$Sr$_{0.15}$(Cu$_{1-x}$Zn$_x$)O$_4$ with $0 \leq x \leq 0.45$. They obtained a monotonic increase of the unit cell, with a structural transition from tetragonal symmetry for x < 0.15 to orthorhombic symmetry of the La$_2$CuO$_4$ type for 0.15 < x < 0.45. They also pointed out that the tetragonal pyramidal coordination of copper, due to its Jahn-Teller effect, is progressively changed into a more regular octahedral coordination as the Zn/Cu ratio increases. Like for nickel, the replacement of copper by zinc dramatically affects T$_c$, which decreases from 39 K to less than 20 K as soon as 0.02 zinc per formula unit is substituted for copper (Fig.4.5). The substitution of the couple "Ni$_{0.5}$Zn$_{0.5}$" for copper leads to very similar results (Fig.4.5). Attempts made by these researchers [4.30] to

Table 4.3. Room temperature resistivity, transition temperature and lattice parameters for La$_{0.85}$Sr$_{0.15}$Cu$_{1-x}$Ni$_x$O$_{4-y}$. (From [4.1])

x	a [Å]	c [Å]	V [Å^3]	T$_{cm}$ [K]	ρ_{300K} [$\mu\Omega\cdot$cm]
0	3.777 (3)	13.226 (1)	188.60(1)	39.3	2300
0.025	3.7749(2)	13.222 (1)	188.42(2)	22.6	2380
0.05	3.7761(2)	13.2078(9)	188.33(2)	<4.2	3500
0.075	3.7789(4)	13.187 (1)	188.31(4)	no	4000
0.1	3.7915(3)	13.1671(3)	188.29(6)	no	4200
0.125	3.7833(3)	13.1582(3)	188.36(3)	no	9300
0.20	3.7873(4)	13.1138(2)	188.09(6)	no	
0.30	3.7971(3)	13.0460(1)	188.10(3)	no	

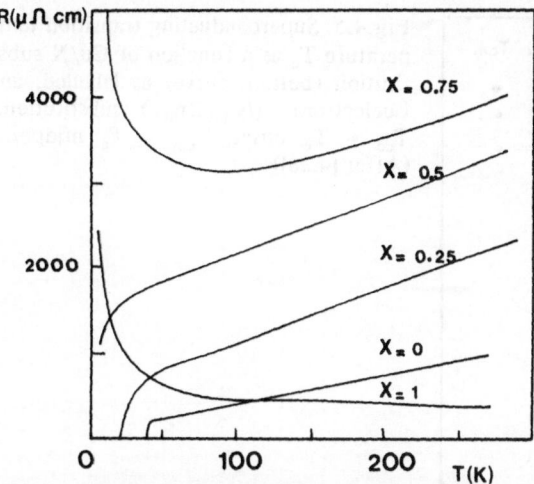

Fig.4.3. The temperature dependence of the resistivity for several members of the $La_{1.85}Sr_{0.15}Cu_{1-x}Ni_xO_{4-y}$ series. (After [4.1])

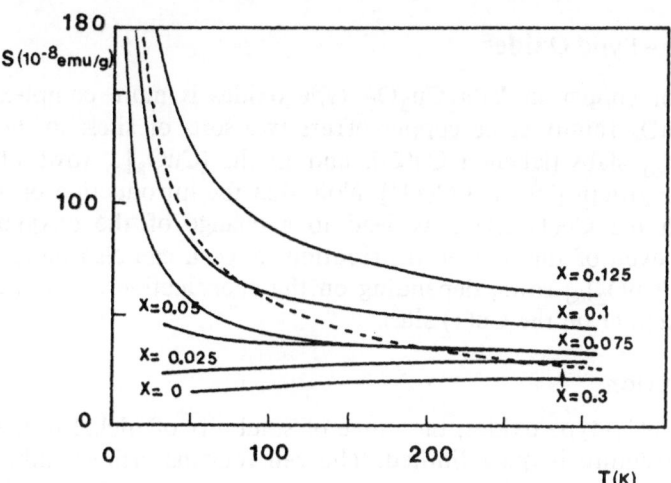

Fig.4.4. The observed magnetic susceptibility between 4.2 K and room temperature for the $La_{1.85}Sr_{0.15}Cu_{1-x}Ni_xO_{4-y}$ series. (After [4.1])

replace copper by mercury and cadmium led to multiphase $La_{0.85}Sr_{0.15}\cdot$ $(Cu_{1-x}M_x)O_4$ samples as soon as x > 0.02. Moreover, the parameters of the tetragonal cell (Fig.4.6) did not vary, and neither did the T_c values for x < 0.02 (Fig.4.5). This suggests that Hg and Cd do not substitute for copper, in agreement with their much greater size.

Similar results were observed for the substitution of nickel and zinc in $La_{1.85}Ba_{0.15}(Cu_{1-x}M_x)O_{4-\delta}$ (M=Ni, Zn) by *Maeno* and *Fujita* [4.31], who observed a dramatic decrease of T_c as x increases (Fig.4.7). In addition, they pointed out that doping with nonmagnetic Zn has a stronger effect on the lowering of T_c than doping with nickel.

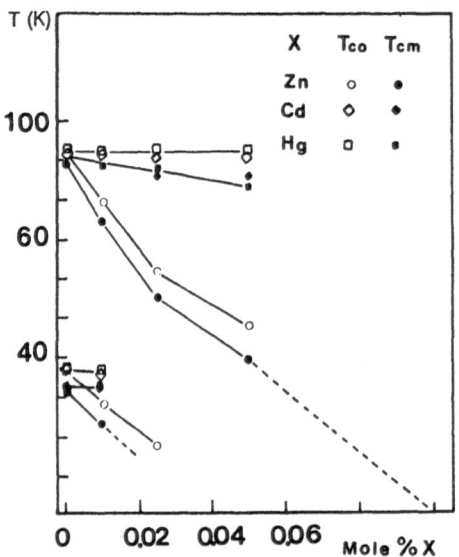

Fig.4.5. Superconducting transition temperature T_c as a function of Cu/X substitution (bottom curve) as labelled, and isoelectronic ($Ni_{0.5}Zn_{0.5}$) substitution. $T_{co} = T_c$ onset, $T_{cm} = T_c$ midpoint. (After [4.28])

X	T_{co}	T_{cm}
Zn	o	•
Cd	◇	◆
Hg	▢	▪

4.2.2 $YBa_2Cu_3O_7$-Type Oxides

The substitution for copper in $YBa_2Cu_3O_7$-type oxides is more complex than for the La_2CuO_4 family since copper offers two sorts of sites: in the pyramidal $[CuO_{2.5}]_\infty$ slabs [labelled Cu(2)], and in the $[CuO_2]_\infty$ rows of CuO_4 square planar groups [labelled Cu(1)]. Note that the introduction of a transition metal on the Cu(1) site may lead to a change of the oxygen stoichiometry, and even of the oxygen distribution: oxygen can also be located between the $[CuO_2]_\infty$ rows, depending on the coordination of the M element replacing copper on the Cu(1) sites.

a) Nickel Substitution

Unlike in the La_2CuO_4-type oxides, the range of solubility of nickel in the $YBa_2Cu_3O_7$ type structure is quite limited. The cell remains orthorhombic with a slight decrease of the cell parameters a,b,c, upon nickel substitution (Table 4.4). A single-phase $YBa_2Cu_{3-x}Ni_xO_{7-\delta}$ can be synthesized only for $0 \leq x \leq 0.50$. Moreover, electron probe microscopy analysis (EPMA) tends to indicate [4.31] that, starting from the nominal composition $YBa_2 \cdot (Cu_{1-x}Ni_x)O_{7-\delta}$, not all the nickel enters the $YBa_2Cu_3O_7$-type matrix (Fig.4.8). Nevertheless, the different researchers [4.1,31-33] agree that T_c decreases rather continuously as x increases. However, the T_c values differ from one publication to another: for instance, *Tarascon* et al. [4.1] observed that the superconducting transition decreases from 93 K at x = 0 to 50 K at x = 0.50, whereas the measurements performed by *Adrian* et al. [4.32] (Fig.4.9) using resistive and inductive methods suggest that resistive T_c's may be significantly wrong in the case of substitution experiments. *Adrian* et al. explained these results as due to a random distribution of the impurity atoms in combination with a small coherence volume. They rejected the

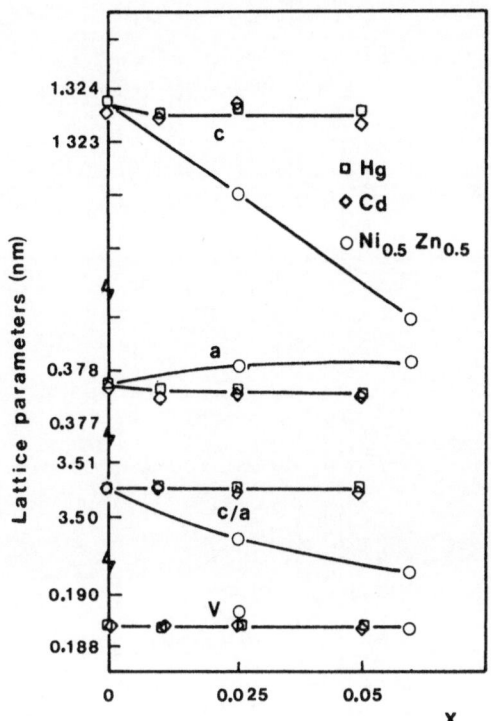

Fig.4.6. Lattice parameters as a function of x. Samples containing Cd and Hg appear to be multiphase for x > 0.02. (After [4.30])

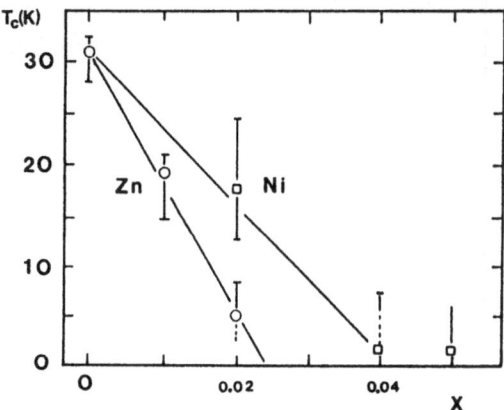

Fig.4.7. Dependence of T_c on x for $La_{1.85}Ba_{0.15}(Cu_{1-x}M_x)O_{4-\delta}$ with M = Ni and Zn. (After [4.29])

hypothesis of inhomogeneity due to the method of synthesis. This important feature of a random distribution of nickel between the Cu(1) and Cu(2) sites has not been confirmed so far by structural investigations. Whether nickel acts on the oxygen framework causing a local distortion or a non-stoichiometry is not known. Nevertheless, the fact is that T_c increases with increasing pressure for the nickel-substituted oxide, contrary to the pure $YBa_2Cu_3O_7$ oxide (Fig.4.10). This phenomenon is interpreted by *Adrian* et al. [4.32] as the nickel impurities causing disturbances of the electronic correlations that are reduced under pressure.

Table 4.4. Room temperature resistivity, transition temperature and lattice parameters for $YBa_2Cu_{3-x}Ni_xO_{7-\delta}$. (From [4.1]). Furthermore, we list the Curie–Weiss constant θ, the temperature-independent susceptibility χ_0, the effective magnetic moment μ per Ni atom derived from the Curie constant C, the temperature interval over which the fit was determined, and the rms deviation of the data from the Curie–Weiss law (in percent)

x	a [Å]	b [Å]	c [Å]	V [Å³]	T_{cm} [K]	ρ_{300K} [μΩ·cm]	μ_{eff} per mole of 3d metal	Θ [K]	$10^{-4}\chi_0$ emu/mole of 3d metal	r [%]	T_{fit} [K]
0	3.8237(8)	3.8874(3)	11.657(3)	173.28(6)	91.6	1800	0.29	-21	0.88	0.5	100-300
0.25	3.8191(8)	3.8857(8)	11.6571(3)	172.98(7)	63.9	1360	0.62	14	1.17	0.13	80-300
0.50	3.8197(7)	3.8832(4)	11.6470(5)	172.75(8)	52.3	2500	0.87	-4	1.45	0.23	60-300

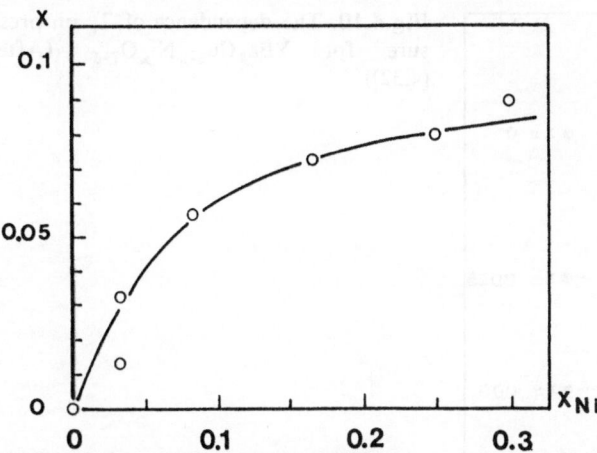

Fig.4.8. Concentration of Ni in $YBa_2(Cu_{1-x}Ni_x)_3CuO_{7-\delta}$ determined by EPMA, plotted vs the nominal concentration x_{Ni}. (After [4.31])

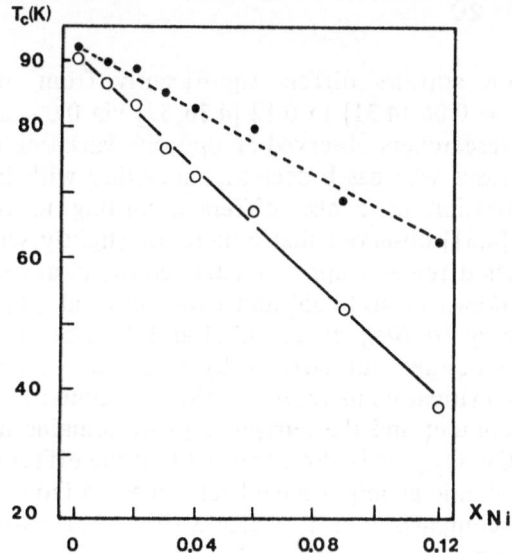

Fig.4.9. T_c versus nominal x for $YBa_2Cu_{3-x}Ni_xO_{7-\delta}$ measured resistively (•) and inductively (○). (After [4.32])

b) Iron Substitution

The substitution of iron for copper in the "123" structure could bring important information about the mechanism of superconductivity in this oxide. This is why several investigations were carried out almost simultaneously. The fact that doping with iron induces an orthorhombic-to-tetragonal transition in $YBa_2Cu_{3-x}Fe_xO_{7+\delta}$ was observed very early by *Maeno* and *Fujita* [4.31]. This phenomenon was confirmed by several researchers [4.34-37]. The reported critical concentration of iron at which the ortho-

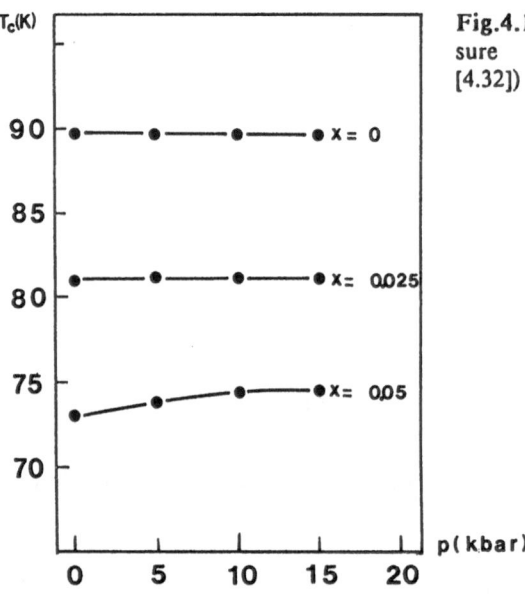

Fig.4.10. The dependence of T_c on pressure for $YBa_2Cu_{3-x}Ni_xO_{7-\delta}$. (After [4.32])

rhombic-to-tetragonal transition appears differs significantly from one publication to another: from x = 0.06 [4.31] to 0.12 [4.36, 37] via 0.09 and 0.07 [4.35] (Fig.4.11a). All the researchers observed an opposite variation of the a and b parameters: a increases whereas b decreases smoothly with increasing x up to x_c. The variation of c also differs according to researcher(s): *Maeno* and *Fujita* [4.31] observed that c increases slightly with increasing x (Fig.4.11b), whereas it remains approximately constant, at least for x ≤ 0.10, according to *Ishikawa* et al. [4.35] and *Ullmann* et al. [4.34] and decreases smoothly according to *Bieg* et al. [4.36] and *Raveau* et al. [4.37]. Another feature that was pointed out early on by *Maeno* and *Fujita* concerns the dependence of the critical temperature on the iron content: T_c decreases with increasing iron content and the tetragonal phase remains superconducting, unlike in $YBa_2Cu_3O_{7-\delta}$. It is also observed that the effect of Fe doping upon T_c is not as dramatic as might have been expected from its magnetic properties. But again, comparison of T_c versus x (Fig.4.12) reveals significant discrepancies between reported values. For instance, for x = 0.10, *Maeno* and *Fujita* observed that T_c has decreased to 60 K (Fig. 4.12a), whereas it has only decreased to 85 K according to *Ishikawa* et al. (Fig. 4.12b) and to 80 K according to *Ullmann* et al. (Fig.4.12c) and *Raveau* et al. (Fig.4.12e), and does not decrease significantly according to *Bieg* et al. (Fig.4.12d). These differences, which are due to the method of synthesis, result from the oxygen nonstoichiometry. It has indeed been shown by *Deslandes* [4.38] that the oxygen content was closely related to the temperature and duration of synthesis for the same nominal composition, leading to dramatically different T_c's. Unfortunately, most of the results described in the various publications [4.31, 34-36] cannot be used for a rigorous interpretation, since the oxygen content of the samples has not been determined. This is why the synthesis of "stoichiometric" phases YBa_2Cu_{3-x}.

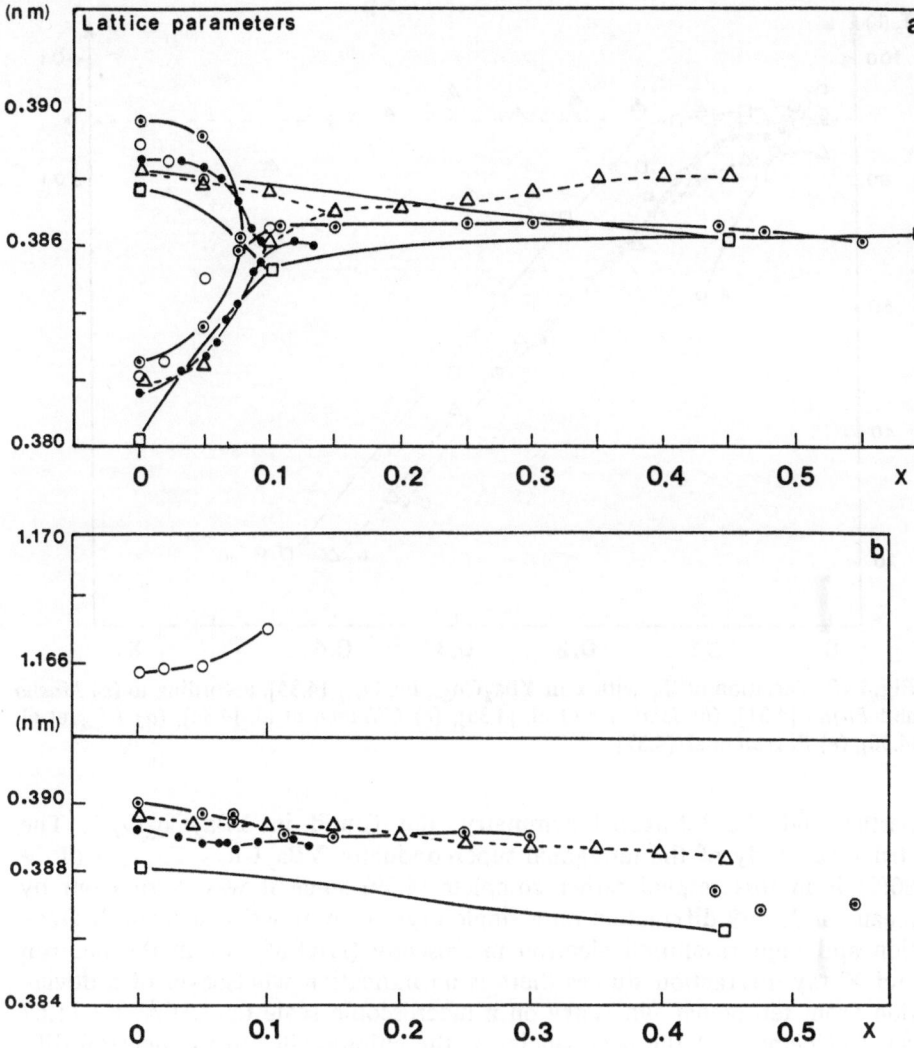

Fig.4.11a,b. Lattice parameters of $YBa_2Cu_{3-x}Fe_xCuO_{7-\delta}$ versus x: (a) a and b parameters, (b) c parameter,

- o according to *Maeno* and *Fujita* [4.31],
- • according to *Ishikawa* et al. [4.35],
- ⊙ according to *Ullmann* et al. [4.34],
- □ according to *Bieg* et al. [4.36],
- △ according to *Raveau* et al. [4.37]

Fe_xO_7 was undertaken [4.37], the oxygen content being determined chemically. Thus, the data given in Fig.4.12 correspond to an oxygen deviation from stoichiometry δ smaller than 0.05, and will be used later for the interpretation of the variation of T_c in terms of the mixed valence of copper.

The structural study of these oxides is of central importance in attempting to understand the compatibility between the superconducting pro-

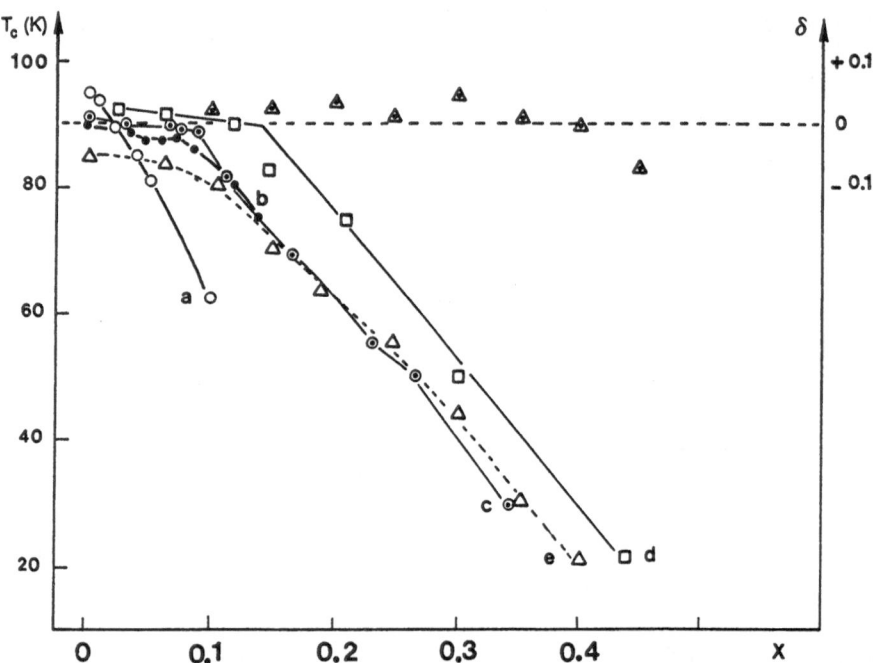

Fig.4.12. Variation of T_c with x in $YBa_2Cu_{3-x}Fe_xO_{7-\delta}$ [4.35], according to (a) *Maeno* and *Fujita* [4.31], (b) *Ishikawa* et al. [4.35], (c) *Ullmann* et al. [4.34], (d) *Bieg* et al. [4.36], (e) *Raveau* et al. [4.37]

perties and the tetragonal symmetry, not found in $YBa_2Cu_3O_{7-\delta}$. The structural study of the tetragonal superconductor $YBa_2Cu_{2.85}Fe_{0.15}O_7(T_c \neq$ 80K) is in this respect rather complete [4.39] since it was carried out by means of X-ray diffraction on a single crystal, by powder neutron diffraction and high resolution electron microscopy (HREM). From the neutron and X-ray diffraction studies there is no indication whatsoever of a deviation from tetragonal symmetry on a macroscopic scale (a=3.86Å, c= 11.67 Å). Refinement of the occupancies of the anionic sites in the neutron diffraction study leads to the "O_7" stoichiometric composition, in agreement with the X-ray single crystal results. The distribution of copper and iron in the copper sites was carefully determined by combining information from the two different kinds of experiments (X-ray and neutrons). It was found that iron is distributed over the two copper sites Cu(1) and Cu(2), in agreement with the study by Mössbauer spectroscopy of a sample of composition $YBa_2Cu_{2.82}Fe_{0.18}O_7$ by *Qiu* et al. [4.40]. However, it is observed that there is a preferential occupation of the Cu(1) sites by iron: of the 0.15 nominal Fe^{3+} ions, 0.10 Fe^{3+} ions are located on the Cu(1) sites (corresponding to the copper ions in square planar coordination in $YBa_2Cu_3O_7$), i.e., they occupy 10% of these sites, whereas only 0.06 Fe^{3+} ions are located in the Cu(2)-sites (in the pyramidal layers), i.e. they occupy less than 3% of these sites. These results are somewhat different from those obtained by *Chaudouët* et al. [4.41] by Mössbauer spectroscopy for the oxide $YBa_2Cu_{2.82}$·

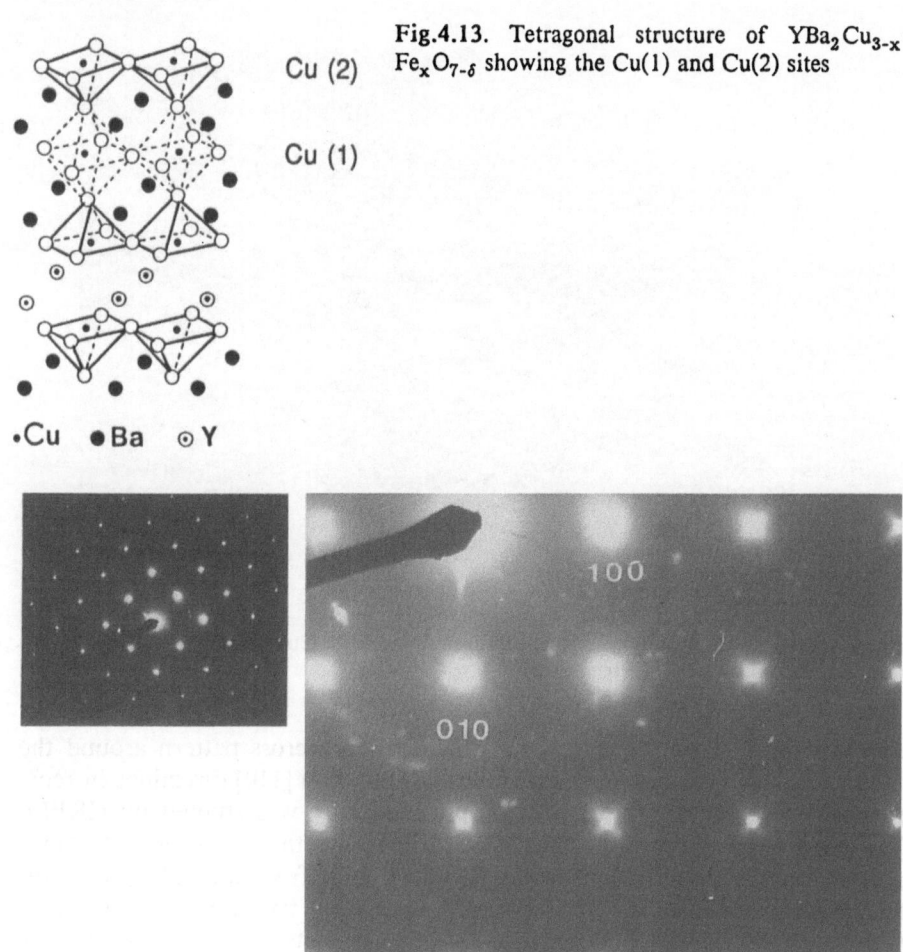

Cu (2)

Cu (1)

Fig.4.13. Tetragonal structure of $YBa_2Cu_{3-x} \cdot Fe_xO_{7-\delta}$ showing the Cu(1) and Cu(2) sites

•Cu ●Ba ⊙Y

100

010

Fig.4.14. Electron diffraction pattern of $YBa_2Cu_{2.85}Fe_{0.15}O_7$, a^*a^* plane

$Fe_{0.18}O_{7-\delta}$. They concluded that iron substitutes only on the Cu(1) site. Another important feature is that Cu(1) is underoccupied. Both single crystal X-ray diffraction and powder neutron diffraction show that the site appears to be deficient by about 4%, leading to the approximate formula $YBa_2[Cu_{1.94}Fe_{0.06}]_{py}$ $[Cu_{0.86}Fe_{0.10} \blacksquare_{0.04}]_{sp}O_7$. In this tetragonal structure (Fig.4.13) previously described in Chap.2, the oxygen vacancies are not ordered in rows along [010] but are randomly distributed at the Cu(1) level over the $0 \frac{1}{2} 0$ and $\frac{1}{2} 0 0$ positions as in the tetragonal $YBa_2Cu_3O_{7-\delta}$ oxide. In fact, this statistical distribution of the oxygen atoms over two sites gives only an average structure and does not reflect the local structure. The electron diffraction (ED) and high-resolution electron microscopy studies show that the structural arrangement is more complex than deduced from the X-ray and neutron studies. The ED patterns confirm without any ambiguity the apparent tetragonal symmetry of this phase, as shown for the a^*a^* plane in Fig.4.14. The reflections are unsplit, indicating that there is a macro-

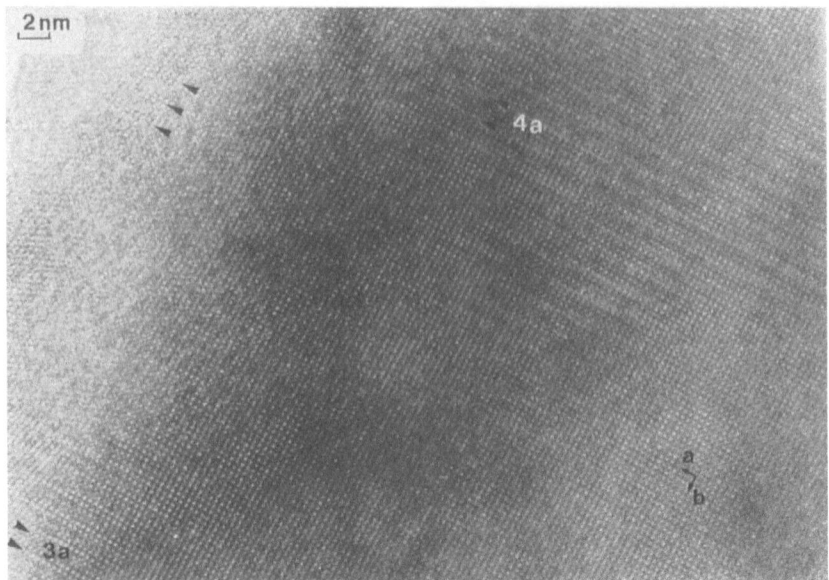

Fig.4.15. High-resolution electron micrograph of $YBa_2Cu_{2.85}Fe_{0.15}O_7$. Beam direction $||c_,$ "3a" and "4a" superstructures

scopic twinning, but instead they show a diffuse cross pattern around the reciprocal lattice points oriented along the [110] and [1̄10] directions in reciprocal space. The origin of this diffuse scattering was studied by HREM. The most striking feature of the HREM images is the occurrence of short-range structural modulations in the **aa** plane, most frequently along the axes and with periods of 3a or 4a (Fig.4.15). Also, twofold superstructures (Fig.4.16) and Moiré-like contrasts appear. The latter may indicate the existence of two-dimensional superstructures or a possible overlap of two differently oriented domains. From this it is evident that the local symmetry must be lower than tetragonal. We assume that this local structural distortion towards lower (probably orthorhombic) symmetry is caused by short-range ordering of oxygen atoms in the **aa** plane; also, Cu/Fe/■ ordering might play a role. It is the coexistence of several different superperiods together with the lack of long-range order that prevents the formation of superstructure reflections. In this sense the superconducting $YBa_2 \cdot (Cu_{0.95}Fe_{0.05})_3O_7$ crystals are tetragonal only in a statistical way, and the structure which was determined by X-ray and neutron diffraction constitutes an average structure. Due to the existence of a well-defined average lattice that is common to several domains (evidenced by the electron microscopy results as well as by the observation of a perfectly tetragonal X-ray diffraction pattern with quite narrow Bragg peaks) the averaging process occurring during diffraction involves structure factors (rather than intensities, as in the case of the incoherently twinned undoped crystals) and is therefore relevant also to the powder neutron diffraction studies on

106

Fig.4.16. High-resolution electron micrograph of $YBa_2Cu_{2.85}Fe_{0.15}O_7$. Beam direction $\|c$; "2b" superstructure

$YBa_2(Cu_{0.95}Fe_{0.05})_3O_7$. Different models have been proposed where several such "local superstructures" can coexist in the same crystal.

It is worth pointing out that the situation in the tetragonal Fe-doped $YBa_2Cu_3O_7$ oxides is quite different from that observed in the tetragonal undoped $YBa_2Cu_3O_{6.25}$. In the pure copper tetragonal oxide, the $[CuO_2]_\infty$ chains are interrupted or nonexistent, owing to the presence of a large amount of copper (I) in twofold coordination. In the Fe-doped tetragonal oxides, as δ remains close to zero, the amount of copper (I) can be considered as negligible. Moreover, in this latter oxide, iron can take a pyramidal coordination on the Cu(2) site and a pyramidal, octahedral or a tetrahedral coordination on the Cu(1) site. Thus, in these compounds the $[CuO_2]_\infty$ chains are probably formed of short segments linked together by iron polyhedra, inducing a change in the orientation of the segments. This change of orientation can be compared to that induced by twinning; but, contrary to microtwinning, it would appear in a random way and over short distances, implying a metrically tetragonal symmetry.

The disorder between iron and copper atoms in the **aa** plane explains why Mössbauer spectroscopy studies on Fe-doped "123" compounds show (at least) three different coordinations for the Fe atoms. The coordination of Fe on the Cu(1) site alone may range from 2 to 6 nearest oxygen neighbors. The relative proportions of these different Fe environments will most likely depend sensitively on details of the local ordering of Fe/Cu/■ and thus on the thermal history of the individual sample. It is quite likely that the ratio of Fe occupation of the Cu(1) and Cu(2) sites is also a function of preparation parameters. Therefore, the observed variations of T_c between samples of identical overall composition and crystallographic (i.e., long-range-order sensitive) appearance is not astonishing.

The neutron and electron diffraction study of the tetragonal 40 K superconductor $YBa_2Cu_{2.77}Fe_{0.23}O_{7.13}$ by *Marezio* and co-workers [4.42] is also very interesting. Unlike in $YBa_2Cu_{2.85}Fe_{0.15}O_7$, they found that the Fe cations substitute only for Cu(1) cations. They confirm the Mössbauer results concerning the existence of several coordinations for iron - pyramidal, octahedral and tetrahedral - depending upon the oxygen content and distribution of the additional oxygen with respect to "O_7" between the $[CuO_2]_\infty$ rows on the same level as Cu(1). As for $YBa_2Cu_{2.85}Fe_{0.15}O_7$, the authors observe electron diffraction patterns with diffuse scattering planes parallel to $(1\bar{1}0)^*$ and $(110)^*$. They interpret this diffuse scattering in terms of linear clusters of iron cations extending along the $[1\bar{1}0]$ and $[110]$ directions. The models they propose (Fig.4.17) are directly deduced from the microtwinning models given for $YBa_2Cu_3O_{7-\delta}$ [4.43, 44] by just placing the Fe^{3+} cations at the domain boundaries, forming FeO_4 or FeO_5 or FeO_6 polyhedra. The oxidation state of iron in these compounds is not completely known. There is no doubt that the majority of iron exhibits the trivalent state, but the presence of tetravalent iron cannot be ruled out.

This smooth evolution of T_c with increasing iron content shows that the behavior of these high T_c superconductors is very different from that of classical superconductors, where one observes that superconductivity is

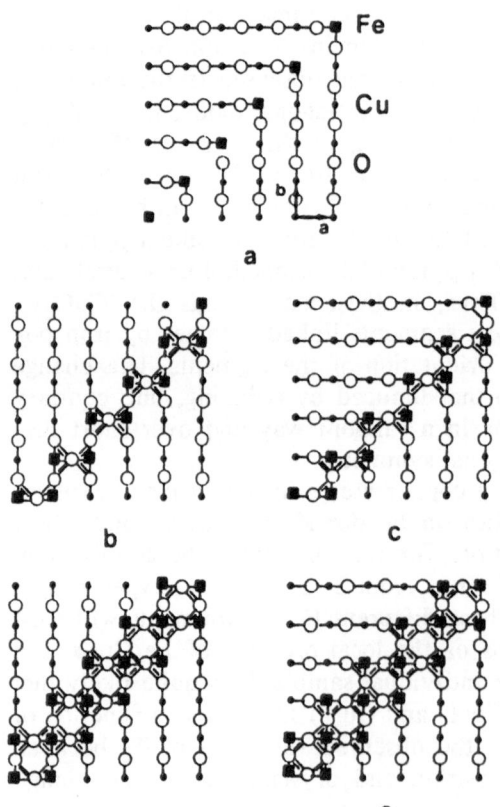

Fig.4.17. Models of iron clusters in $YBa_2Cu_{3-x}Fe_xO_7$. (After [4.42])

very rapidly killed by the introduction of small amounts of magnetic impurities. This suggests that, besides the magnetic properties, another factor plays an important role in superconductivity. This factor, which corresponds to the mixed valence $Cu(III)/Cu(II)$, or more exactly to the $Cu(III)$ content, will be discussed in the final chapter of this book, together with the evolution of oxygen nonstoichiometry in $YBa_2Cu_3O_{7-x}$. Nevertheless, it can already be mentioned that the ratio $Cu(III)/Cu$ (total) governs the superconductivity of these oxides.

c) Other Transition Element Substitutions

A systematic study of the substitution of different transition elements for copper has not yet been performed. Moreover, the oxygen stoichiometry in those oxides has not really been controlled, so that any interpretation of their superconducting properties is subject to controversy. For instance, *Veit* et al. [4.45] have investigated the T_c suppression in $YBa_2 \cdot (Cu_{3-x}M_x)O_{7-\delta}$ for M = Ti,V,Cr,Mn,Fe,Co,Ni, in the range $0.005 \leq x \leq 0.10$. From the T_c depression rate found from resistance measurements (Fig. 4.18) they find evidence for three mechanisms: a classical mechanism due to the localized magnetic moment of the M impurities, which is prominent for M in the 3d row; a mechanism due to oxygen loss induced by impurities such as iron and cobalt; and a third unexplained mechanism observed, for instance, in the case of nickel. However, *Tarascon* et al. [4.1] obtain contradictory results for cobalt, which they find to be similar to nickel. In the same way it is now known from the controlled stoichiometric $YBa_2Cu_{3-x} \cdot Fe_xO_7$ [4.37,38] superconductors that the deviation from oxygen stoichiometry is not the only factor which governs their properties.

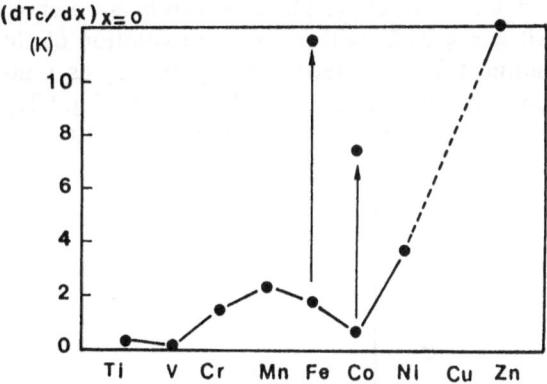

Fig.4.18. T_c depression rate $(dT_c/dx)_{x=0}$ for different transition metal (M) impurities $(x < 3\%)$ in $YBa_2Cu_3O_{7-y}$. (After [4.45])

The substitution of palladium for copper is also of interest owing to the $4d^8$ configuration of $Pd(II)$, which leads to a Jahn–Teller effect and imposes a square planar coordination for $Pd(II)$. The substitution of Pd for Cu in the "123" plane is indeed possible and leads to the orthorhombic phase

$YBa_2Cu_{2.5}Pd_{0.5}O_{7-\delta}$ [4.46]. In this phase one observes an ordering of palladium ions, which exclusively occupy the Cu(1) sites, i.e. they are located in the $[CuO_2]_\infty$ rows in agreement with their square planar coordination. The resistive measurements show a T_c of 49 K, but the magnetic measurements show only a 5% Meissner effect. Further investigation will be necessary to understand the behavior of palladium in this structure.

d) Other Substitutions

The substitution of zinc for copper in the $YBa_2Cu_3O_7$ oxides has a spectacular effect on the T_c depression. Indeed, it was shown [4.28-33,47] that T_c decreases more rapidly with Zn substitution than with Ni or Fe substitution (Fig.4.19). Unlike for transition elements such as nickel or iron, the solid solution of $YBa_2(Cu_{3-x}Zn_x)O_{7+\delta}$ was found to be limited ($0 \leq x \leq 0.3$). Moreover, in this homogeneity range no significant variation of the cell parameters has been observed. Nevertheless, the dramatic decrease of T_c caused by zinc doping confirms clearly that Zn^{2+} enters the copper lattice. This greater effect of Zn^{2+} on superconductivity may be due to the d^{10} electronic configuration of this cation, which does not allow any mobility of the d electrons, contrary to the transition metal ions, and breaks the conductivity. In this respect, isoelectronic cations Cd^{2+} and Hg^{2+} show a very different behavior: *Remsching* et al. [4.30] do not observe any variation of T_c for these cations for the "solid solution" $YBa_2Cu_{3-x}M_xO_{7-\delta}$ for $0 \leq x \leq 0.15$. In fact, it is likely that cadmium and mercury do not substitute for copper in the "123" matrix, because they are much larger than copper.

Substitution of aluminum for copper was studied by several research groups [4.31, 48-50] almost simultaneously. The results obtained by *Siegrist* et al. [4.50] give a good idea of the role of aluminum in the superconducting properties of the series $YBa_2Cu_{3-x}Al_xO_{7-\delta}$. These researchers observed a limited homogeneity range ($0 \leq x \leq 0.22$) with a smooth evolution of the cell parameters (Table 4.5) tending towards a tetragonal symmetry as x increases. One striking feature concerns the variation of T_c with x (Fig.4.20): one observes that relatively high aluminum concentration ($x \leq 0.10$) does

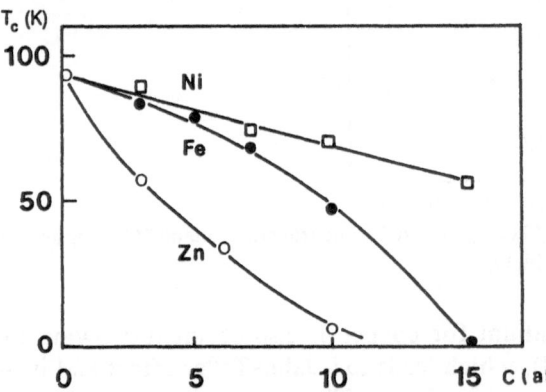

Fig.4.19. Superconducting transition temperatures versus concentration of Fe, Ni and Zn in $YBa_3Cu_2O_{7-\delta}$. (After [4.33])

Table 4.5. Aluminum concentration [4.50], lattice parameters, and superconducting transition temperature for $Ba_2YCu_{3-x}Al_xO_7$ samples. The estimated standard deviations (in parentheses) refer to the last digit printed. (The superscript a refers to ceramic samples)

x(Al)	a [Å]	b [Å]	c [Å]	T_c [K]
0.00	3.851 (3)	3.851 (3)	11.672 (9)	92
0.05[a]	3.835 (5)	3.884 (5)	11.688 (9)	88
0.10[a]	3.844 (2)	3.8947 (3)	11.657 (5)	75
0.11	3.847 (4)	3.851 (3)	11.659 (8)	82
0.12	3.855 (3)	3.859 (3)	11.684 (8)	51
0.13	3.843 (3)	3.859 (2)	11.667 (4)	45
0.135	3.856 (2)	3.860 (2)	11.674 (4)	45
0.18	3.854 (2)	3.856 (2)	11.668 (3)	20
0.22	3.854 (2)	3.861 (2)	11.672 (4)	<4.2

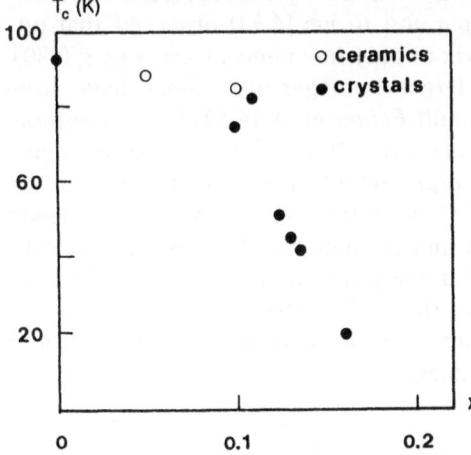

Fig.4.20. Superconducting onset temperatures as a function of aluminum content in $Ba_2YCu_{3-x}Al_xO_7$ samples. (After [4.50])

not significantly lower T_c, which remains close to 80 K for x = 0.10. But beyond x = 0.10, T_c decreases sharply with increasing aluminum concentration. An important problem in these phases is a knowledge of aluminum content. As was pointed out by the researchers, single phases are difficult to prepare and only some samples could be obtained as single crystals. In the same way, no systematic study of the oxygen stoichiometry has been performed in these oxides. Nevertheless, the results show that Al is incorporated into the "123" matrix on the copper sites and suggests that aluminum, like the transition metals, favors cross-linking of Cu-O chains, leading to a progressive suppression of superconductivity.

The substitution of gallium for copper is also possible, according to *Maeno* and *Fujita* [4.31]. The effect of gallium is very similar to that of iron and cobalt. A progressive transition from an orthorhombic symmetry (for x = 0) to a tetragonal symmetry (x = 0.10) is obtained for $YBa_2\cdot Cu_{3-x}Ga_xO_{7+\delta}$. In the same way the decrease in $T_c(x)$ is approximately regular, with a slope rather close to that of cobalt and iron compounds. Thus, this confirms that reduction in T_c is insensitive to whether the dopant is magnetic or nonmagnetic.

4.3 Substitutions on Other Sites

Two other sorts of sites are available for substitution in the "123" oxide: the barium sites and the anionic sites. Several studies have been carried out in this direction. Although not really fruitful, they are reviewed here.

4.3.1 Substitution for Barium

The substitution of potassium for barium in $YBa_2Cu_3O_{7-\delta}$ is able to increase the Cu(III) content and could be of interest for enhancing superconductivity. Several attempts have been made to synthesize the oxide $YBa_{2-x}K_xCu_3O_7$ [4.51-57]. The results obtained by the different researchers are sometimes contradictory. *Gupta* and *Braun* [4.53] observed that potassium is dissolved in the "123" matrix at only low concentration ($x \leq 0.30$), without changing the T_c, whereas larger homogeneity ranges have been found by *Saito* et al. [4.54] ($x \leq 1.2$) and *Felner* et al. [4.52] ($x \leq 1$) without a major change or with only a small change in T_c, which may not be significant. The results obtained for the replacement of barium by sodium are similar [4.53], except that the solubility of sodium seems to be much lower ($x < 0.10$). In both series, potassium and sodium, the fact that neither the lattice parameters nor the T_c value for the pure phase change significantly suggests that K^+ and Na^+ do not enter the "123" matrix.

The substitution of isovalent ions such as strontium [4.55] leads to a decrease of both T_c and lattice parameters.

4.3.2 Fluorination of $YBa_2Cu_3O_{7-\delta}$

Fluorine, because its size is close to that of oxygen, should be a potential candidate for the synthesis of new high-T_c superconductors: it should allow new oxyfluorides to be realised by juggling with the charge balance between the anions. Curiously, up to now, very few results have been obtained in this field. Several attempts have been made to introduce fluorine into the "123" matrix [4.56-59]. However, in several cases the materials were synthesized in air, [4.56, 57, 59], so that it is not really known how much fluorine was incorporated in the formula $YBa_2Cu_3O_{7-x}F_x$. Nevertheless, the results obtained show that T_c is not affected or else decreases in the presence of fluorine. The results obtained by *Perrin* et al. [4.58] are most significant, since they used different experimental methods for the synthesis of $YBa_2Cu_3O_{7-x}F_x$ in the absence of air, especially starting from $YBa_2Cu_3O_6$ in an NF_3 flow. They have definitely shown that fluorine was inserted in the "123" matrix and that it leads to a degradation of T_c. A recent study carried out by *Chevalier* et al. [4.60] has revealed that fluorination can be used for passivation of superconducting materials.

5. Bismuth, Thallium and Lead Superconducting Cuprates

The possibility of replacing the yttrium ions in $YBa_2Cu_3O_7$ by almost any trivalent lanthanide without producing any change of the critical temperature, even when the lanthanide is a magnetic cation, has encouraged the investigation of the related quaternary oxides involving copper, an alkaline earth element and a trivalent element other than a rare earth. Two families were discovered successively: the bismuth copper oxides and the thallium copper oxides. Some time later, the investigation of lead cuprates allowed new superconductors to be synthesized. The crystal chemistry of these systems has not yet been investigated in an exhaustive way, so that many phases, superconducting or not, remain to be isolated. Thus, this chapter will focus on the structure of the superconducting phases that have been isolated, and in particular on their electron transport properties.

5.1 Bismuth Alkaline-Earth Superconducting Cuprates

Trivalent bismuth was recognized quite early (in May 1987) as a potential cation for generating superconductors in the presence of copper [5.1]. The idea of searching for such materials was founded on the fact that bismuth, like lanthanides, exhibits a trivalent state and has a very similar size, but that, unlike these elements, its lone pair $6s^2$ can be stereochemically active. This latter property, which confers on bismuth (III) a nonspherical configuration, may lead to unique structural behavior. In this respect, the existence of the Aurivillius phases [5.2], such as $Bi_2BaNb_2O_9$ (Fig. 5.1), was attractive: such oxides exhibit a lamellar structure built up from $(Bi_2O_2)^{2+}$ layers of edge-sharing BiO_4 pyramids intergrown with $[Nb_2O_8]^{2-}$ double perovskite layers. This ability of bismuth to induce a two-dimensional character was considered an important feature for the generation of new high-T_c superconductors, in agreement with the theory developed by *Labbé* and *Bok* [5.3], and *Friedel* [5.4]. On the other hand, the fact that Bi(III), can be partially oxidized in air into Bi(V) in the presence of an alkaline earth cation such as barium or strontium had to be taken into consideration, since it could have either a positive or negative effect on the superconducting properties. These efforts were awarded with the discovery of the first signs of superconductivity at temperatures ranging from 7 to 22 K in the system Bi-Sr-Cu-O [5.5]. This phenomenon was observed for the composition $Bi_2Sr_2Cu_2O_7$, which was prepared by heating a mixture of

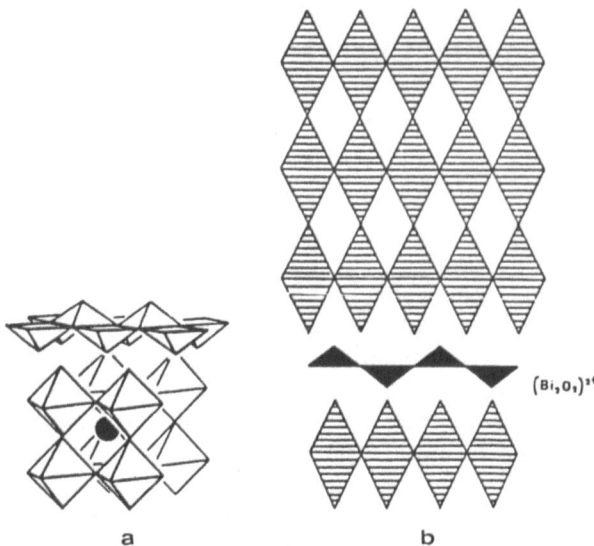

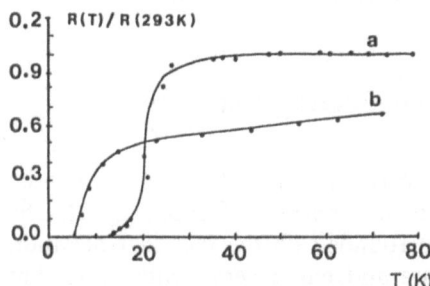

$(Bi_2O_2)^{2+}$

Fig.5.1. Structure of $Bi_2BaNb_2O_9$ (**a**) drawn in perspective and (**b**) schematized in terms of $[Bi_2O_2]^{2+}$ layers and $[Nb_2O_8]^{2-}$ double perovskite layers

Fig.5.2. Resistivity ratio versus temperature for the phase $Bi_2Sr_2CuO_6$ (nominal composition $Bi_2Sr_2Cu_2O_7$) showing the resistive transition at 22 K for the ultrapure oxide (*a*) and at 7 K for the "commercial" compound (*b*)

Bi_2O_3, Sr_2CuO_3 and CuO at 900 °C in air and quenching down to room temperature. It was also observed that the temperature variation of resistivity (Fig.5.2) depended upon the nature of the nominal compounds: for ultrapure starting materials the midpoint of the resistive transition corresponds to $T_c = 22$ K, the zero resistance being reached below 14 K (Fig.5.2, curve *a*), whereas for the oxides prepared from normal purity compounds, T_c is much lower, close to 7 K (Fig.5.2, curve *b*), owing to the presence of magnetic impurities. The close relationships between this structure and the perovskite structure were also shown at the same time from the electron diffraction and high resolution electron microscopy preliminary investigations (Fig.5.3), which showed that the orthorhombic cell of this phase was related to the perovskite in the following way: $a=5.32\simeq\sqrt{2}a_p$, $b=26.6$Å$\sim$ $5\sqrt{2}a_p$ and $c=2\times24.4$Å. From subsequent structural studies of other members of this series, it is now known that this oxide corresponds to the composition $Bi_2Sr_2CuO_6$.

114

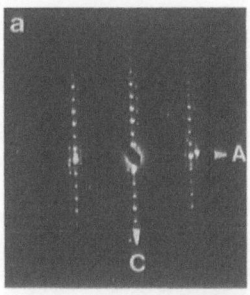

Fig.5.3. Electron diffraction pattern (a) corresponding to the (001) and (010) planes of $Bi_2Sr_2CuO_6$, and HREM image (b) of the corresponding (001) plane

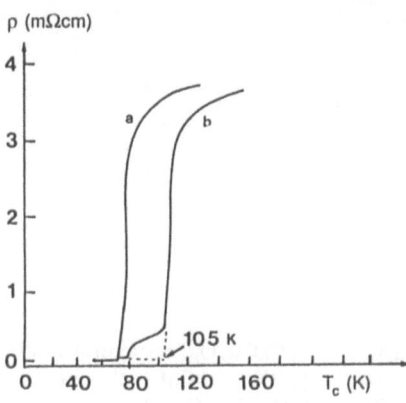

Fig.5.4. Temperature dependence of resistivities in $Bi_1Sr_1Ca_1Cu_2O_x$ oxides (a) sintered in air at 800°C for 8 h, then cooled in a furnace, and (b) sintered at 882°C for 20 min followed by annealing at 872°C for 9 h. (After [5.6])

The introduction of an additional element, calcium, into Bi-Sr-Cu-O allowed the first signatures of superconductivity to be observed by *Maeda* et al. [5.6] for the nominal composition $BiSrCaCu_2O_x$. For this composition the researchers observed two sorts of behaviors from the resistivity measurement (Fig.5.4): a sharp decrease of resistivity at 105 K, with zero resistance being reached only at 80 K, for samples sintered in air at 800 °C and cooled in a furnace (Fig.5.4a), and a sharp decrease of resistivity with zero resistance at 75 K for samples sintered at 882 °C followed by annealing at 872 °C in air. In the same way it was shown that the magnetization vs temperature curve (Fig.5.5) exhibits a Meissner effect characteristic of superconductivity below 100 K.

The structural studies [5.7-14] of the superconducting phases of the system Bi-Sr-Ca-Cu-O allowed the 80 K superconductor to be identified

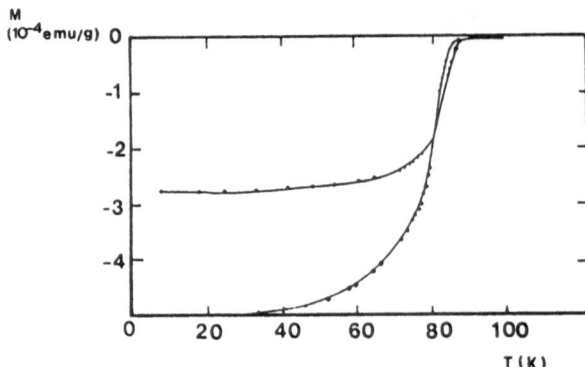

Fig.5.5. Magnetization vs temperature for $Bi_2Sr_2CaCu_2O_8$ according to *Tarascon* et al. [5.7] for warming in a field of 10 G (*lower curve*) and for cooling under a field of 10 G (*upper curve*)

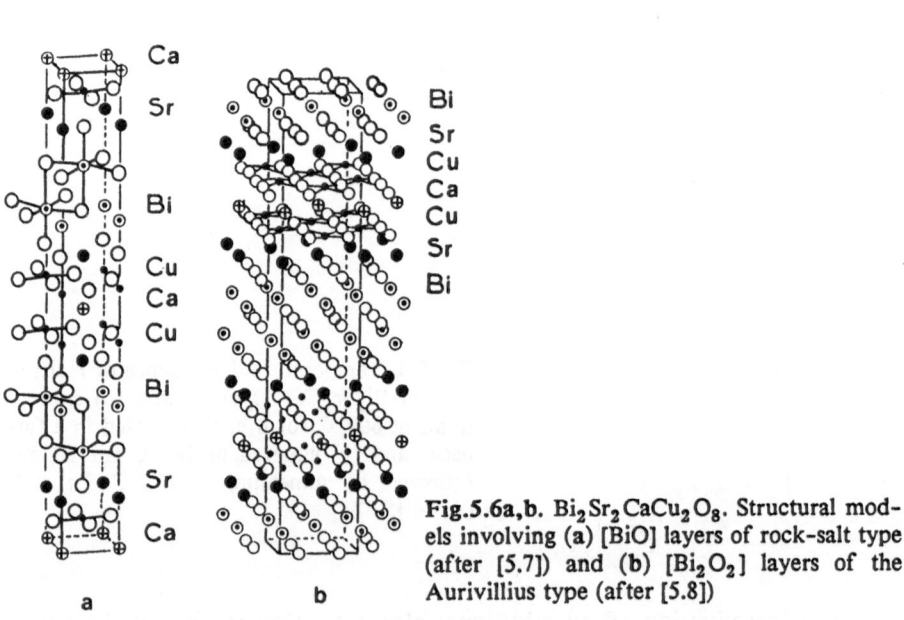

Fig.5.6a,b. $Bi_2Sr_2CaCu_2O_8$. Structural models involving (a) [BiO] layers of rock-salt type (after [5.7]) and (b) $[Bi_2O_2]$ layers of the Aurivillius type (after [5.8])

as $Bi_2Sr_2CaCu_2O_8$, and structural models to be established. From the single-crystal X-ray studies carried out by *Tarascon* et al. [5.7] and by *Subramanian* et al. [5.8] two possible structures can be considered for $Bi_2Sr_2CaCu_2O_8$. Using a tetragonal subcell (a=3.817Å, c=30.6Å). *Tarascon* et al. [5.7] proposed a structure formed of "neutral $Bi_2Sr_2CaCu_2O_8$ slabs 13 Å thick parallel to (001), of oxygen deficient perovskite-type, weakly bonded together by long Bi-O distances at z = 1/4 and 3/4" with a crystallographic shearing of the perovskite type at these latter levels. In fact, examination of their drawing of the structure (Fig.5.6a) shows that the given description is not absolutely correct, since the oxygen atoms located at the same level as the bismuth atoms do not exhibit the perovskite configuration but form BiO rock-salt-type layers, i.e. are turned through

45° with respect to the oxygen atoms located at the same level as the copper atoms. Thus, this first structural model can be described better as double oxygen-deficient perovskite layers $[SrCaCu_2O_5]_\infty$ intergrown with triple distorted rock-salt-type layers $[Bi_2SrO_3]_\infty$. Each $[SrCaCu_2O_5]_\infty$ layer is itself formed of two single $[CuO_{2.5}]_\infty$ layers of corner-sharing CuO_5 pyramids interleaved with a plane of calcium ions. Such layers, in which calcium exhibits pseudocubic coordination, are similar to those observed in $La_2CaCu_2O_6$ and in $YBa_2Cu_3O_7$. The triple rock-salt-type layers are formed of a bismuth bilayer sandwiched between SrO monolayers. The poor quality of the crystals did not allow the R factor to be lowered, so that the low accuracy of the atomic positions does not allow a significant discussion of the interatomic distances. Nevertheless, it will be seen later that this structural model is not far from the reality. *Subramanian* et al. [5.8] solved the structure in an orthorhombic subcell ($a=5.399\text{Å}\simeq\sqrt{2}a_p$, $b=5.414\text{Å}\simeq\sqrt{2}a_p$, $c=30.9\text{Å}$). The metallic atoms were found to be rather close to those observed by *Tarascon*, except that a higher precision is obtained, especially on the bismuth positions, which are distributed over two closely spaced sites. Thus, this structure (Fig.5.6b) also shows the double oxygen-deficient perovskite $[Sr_2CaCu_2O_5]_\infty$ layers. The researchers do not give any information about the positions of the oxygen atoms of the $[Bi_2O_2]^{2+}$ layers. However, from their drawing it can be seen that this additional oxygen is placed between the bismuth layers, like in Aurivillius phases, suggesting BiO_4 pyramids, so that this model corresponds to the intergrowth of $[Bi_2O_2]^{2+}$ layers and $[Sr_2CaCu_2O_5]_\infty$ perovskite layers.

Neutron powder diffraction studies [5.11, 12] lead to contradictory results: *Bordet* et al. [5.11] confirm the existence of rock-salt-type layers, as in *Tarascon's* model, whereas *von Schnering* et al. [5.12] observe that these oxygens are located at the same level as the bismuth ions along c, but turned through 45° around this direction with respect to *Tarascon's* model, forming true $[Bi_2O_4]_\infty$ oxygen-deficient perovskite layers, leading to the limiting formula $Bi_2Sr_2CaCu_2O_{10}$. According to *Bordet* et al. an intermediate composition is obtained, corresponding to the composition $Bi_2Sr_2\cdot CaCu_2O_{10-\delta}$, in which bismuth presents two oxidation states (V and III) simultaneously. Such a controversy is due to the complexity of the structure. Electron diffraction and electron microscopy studies [5.9, 10] show that all the crystals exhibit an incommensurate superstructure along the a axis. In the electron diffraction patterns, extra spots whose distance and intensity vary from one crystal to the other, appear systematically along $[100]^*$; the satellites generally settle in an incommensurate position (the commensurate position corresponds to a superstructure $10\sqrt{2}/2\cdot a_p \sim 27\text{Å}$). Figure 5.7a,b show [010] and [001] ED patterns, giving evidence of the conditions of reflection h0l ($h,l = 2n$) and hk0 ($h,k = 2n$), respectively. Figure 5.7c shows a [001] pattern obtained from a crystal where the superstructure is assumed to appear in two perpendicular directions ([110] of the perovskite cell) in two superimposed areas of the crystal. The high-resolution electron microscopy studies confirm the stacking sequence of the metallic atoms along c without any ambiguity, as shown by Fig.5.8, where

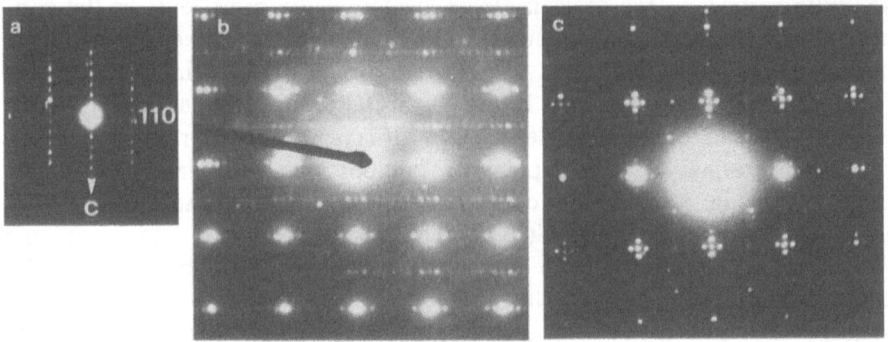

Fig.5.7a–c. ED patterns observed in $Bi_2Sr_2CaCu_2O_{8+\delta}$. (a) [010]; (b) [001] with incommensurate satellites along $[100]^*$, $q \simeq 4.6$; (c) superposition of two areas [001] with extra spots in two perpendicular directions

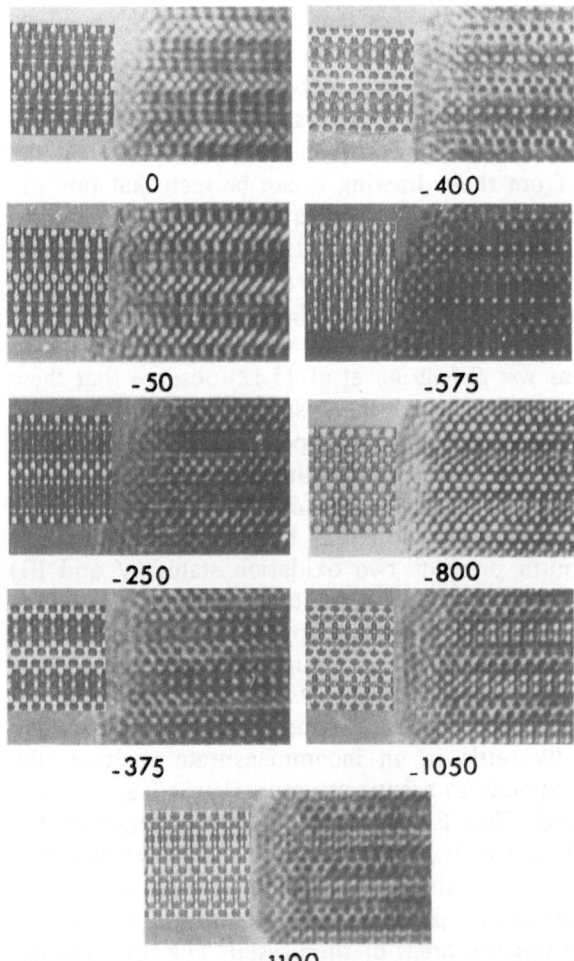

0

−400

−50

−575

−250

−800

−375

−1050

−1100

Fig.5.8. Experimental (*right part of each panel*) and calculated through-focus series for $Bi_2Sr_2(Ca_{0.5}Sr_{0.5})Cu_2O_8$ (zone axis [110], defocus values are in Ångstroms)

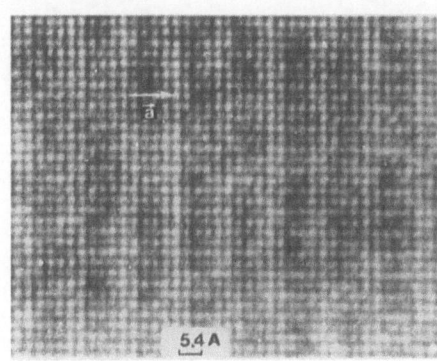

Fig.5.9. [001] HREM image of Bi_2Sr_2 $CaCu_2O_8$. Note the modulations of the contrast which appear systematically along **a** and in local areas along **b**

the observed and calculated through-focus series along [110] allow the sequence "Sr-Bi-Bi-Sr-Cu-Ca-Cu-Sr" to be identified (see, for instance, the -575Å defocus). But, on the other hand, incommensurability is observed in the [001] high-resolution electron microscopy images (Fig.5.9). In these images the typical contrast consists of square arrays of white dots spaced about 2.7 Å apart, but modulations of the contrast are clearly visible. A first type of modulation is observed along the a axis (Fig.5.9) with two or three (depending on the crystal area, thickness, and focus values) rows of bright dots alternating with three (or two) rows of dark dots. The periodicity is then five rows with about 2.7 Å spacing. This modulation might be due to bismuth displacements. The second type of modulation is less evident and appears only in some crystal areas (Fig.5.9 bottom right-hand corner) along the b direction: every second row of dots is brighter. This feature is also visible in some [100] images and cannot be ascribed to crystal or beam tilting, because of the random variation of the modulation from one dot to the next. Besides this phenomenon of incommensurability, another problem prevents an accurate structural determination. It concerns the lamellar morphology of these samples, which allows easy cleavage. Most [100] and [110] images show split lamellae, which are often thinner than 100 Å (Fig.5.10) and in which the breaking interface consists of amorphous layers of various thicknesses, sometimes with voids. Another consequence of this feature is the occurrence of bending of crystals, which can be really extensive. It sometimes causes one lamella to be at a small angle to the adjacent ones, making it difficult to ensure a perfect incidence over all the crystal. Compared to $YBa_2Cu_3O_{7-\delta}$, the stacking faults are not numerous (Chap.6).

The single-crystal studies performed by *Sleight* and co-workers [5.8] suggested a possible homogeneity range for this phase, due to the partial substitution of calcium for strontium, leading to the formula $Bi_2Sr_{3-x}Ca_x$· Cu_2O_8 with $0.4 \leq x \leq 0.9$. The value of the c parameter (30.9Å) obtained by these researchers, greater than those obtained by *Tarascon* et al. [5.7] and by *Hervieu* et al. [5.10] (30.6-30.7Å) is in agreement with this point of view. It is not realistic to discuss the interatomic distances in detail here, owing to all the difficulties which prevent a rigorous structural determina-

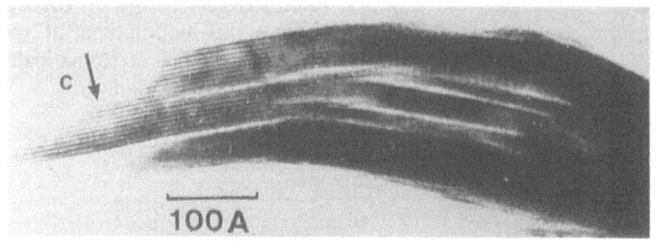

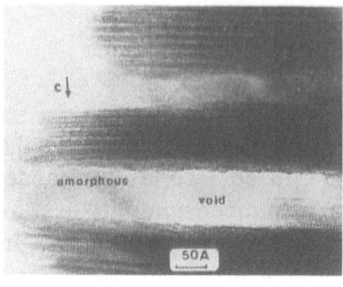

Fig.5.10. [100] images illustrating the lamellar character of the crystals of $Bi_2Sr_2CaCu_2O_8$

tion being performed. Nevertheless, there are some features which can be outlined with certainty.

The Cu-Cu distance between the pyramidal sheets in the double oxygen-deficient perovskite layer, close to 3.2 Å, is very similar to that observed in the $YBa_2Cu_3O_7$-type oxides, where the copper-oxygen sheets are separated by 3.3 Å. Though not accurate, the Cu-O distances of the CuO_5 pyramids remain close to those observed for other oxides, with four basal short Cu-O distances (close to 1.9Å) and one long apical Cu-O bond (2.4Å), indicating that copper is located almost at the center of a square, forming a CuO_4 group. A third remarkable characteristic deals with the distance between two successive bismuth sheets, close to 3 Å, which in the absence of oxygen between them would confer a unique lamellar character to this structure.

The 110 K resistive transition in the system Bi-Sr-Ca-Cu-O, first attributed to $Bi_2Sr_2CaCu_2O_8$ [5.6] was proposed by *Tarascon* et al. [5.15] to correspond to the oxide $Bi_2Sr_2Ca_2Cu_3O_{10}$. These authors were able to isolate crystals with this composition and with a 110 K transition (Fig.5.11) by heating a mixture of nominal composition "4334" ($4Bi_2O_3/3SrO/3CaO/4CuO$) to temperatures close to 885 °C, but below the melting point, and maintaining this temperature for about one week, then furnace cooling the product (6h). Magnetic measurements performed on such crystals (Fig.5.12) show Meissner values greater than 50% without taking into account the demagnetization factor. From the X-ray pattern of one of these crystals, *Tarascon* et al. determined the parameters of a tetragonal cell (a=5.39Å, c=37.1Å). However, they found that the diffraction lines were broad and not exactly at the 001 positions. Such phenomena can be attributed to the presence of stacking faults within the crystals (Chap.6) inherent to the method of synthesis. Recent studies carried out by *Kijima* et al. [5.16] confirmed that the 107 K superconductor corresponds to a pseu-

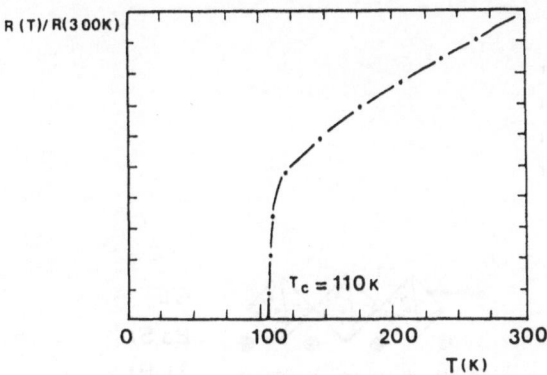

Fig.5.11. Resistivity normalized to the resistivity at 300 K for $Bi_2Sr_2Ca_2Cu_3O_{10}$. (After [5.15])

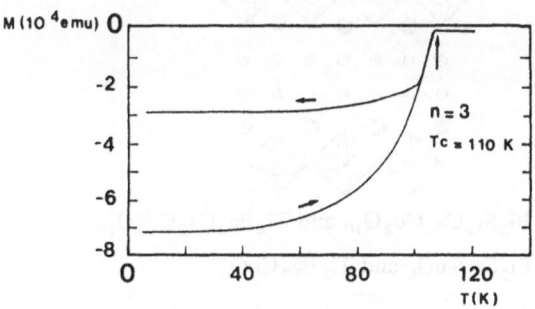

Fig.5.12. Magnetization versus temperature for a $Bi_2Sr_2Ca_2Cu_3O_{10}$, from [5.15]. The upper curve is for cooling in a field of 10 G (Meissner) and the lower curve is for warming in a field of 10 G (shielding)

dotetragonal phase (a=5.40Å, b=27.0Å and c=36.8Å). But these research-ers obtained this phase for the nominal composition "BiSrCa$_3$Cu$_4$O$_x$" pre-heated at 800 °C and then fired at 870 °C for 24-120 h and slowly cooled down to room temperature. The possible ideal composition of this 107 K superconductor, $Bi_2Sr_2Ca_2Cu_3O_{10}$, has been confirmed by *Sleight* and co-workers [5.17], and *Amelinckx* and co-workers [5.18]. Thus there is no doubt now that the structure of this phase (Fig.5.13) can be described as formed of triple distorted $[Bi_2SrO_3]_\infty$ rock-salt-type layers $[(Bi_2O_2)^{2+}$ Aurivillius-like layers cannot however be ruled out] intergrown with triple oxygen-deficient perovskite layers $[SrCa_2Cu_3O_7]_\infty$. In such a structure the latter layers are built up from two pyramidal $[CuO_{2.5}]_\infty$ sheets, as in $Bi_2Sr_2CaCu_2O_8$, and one $[CuO_2]_\infty$ sheet of corner-sharing CuO_4 square planar groups, interleaved with planes of calcium ions. The synthesis of this phase in the form of bulk pure samples is rather difficult, but has been really well done by *Endo* et al. [5.19]. Its stabilization by lead [5.20-22] has allowed bulk 108 K superconductors to be synthesized, even in the form of pure phases. The synthesis of the "2212"-type superconductors $Bi_{2-x}Pb_xSr_2Ca_{1-x}Y_xCu_2O_8$ [5.21] for $0 \le x \le 1$ confirms that the introduc-

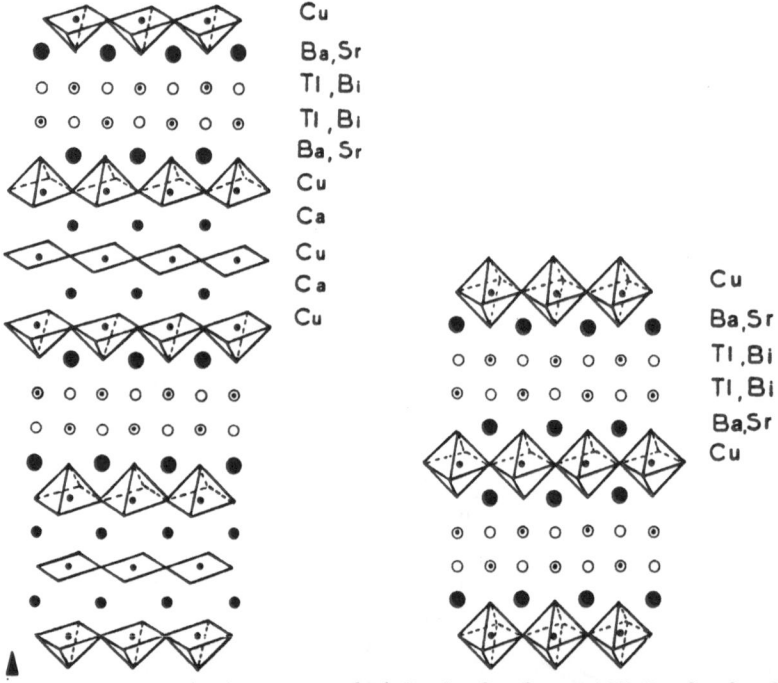

Fig.5.13. Schematized structure of $Bi_2Sr_2Ca_2Cu_3O_{10}$ and $Tl_2Ba_2Ca_2Cu_3O_{10}$

Fig.5.14. Schematized structure of $Bi_2Sr_2CuO_6$ and $Tl_2Ba_2CuO_6$

tion of lead does not greatly affect the superconducting properties of these oxides for low x values: T_c remains close to 85 K up to x = 0.5, and falls down abruptly to zero below this value.

The 22 K superconductor $Bi_2Sr_2CuO_6$ [5.5,24] also belongs to this structural family. A structural model (Fig.5.14) based on the intergrowth of triple distorted rock-salt-type layers $[SrBiO_3]_\infty$ and single octahedral perovskite layers is certainly very close to the reality. However, the exact structure, and especially the Bi-O distances, obtained from the single-crystal X-ray study cannot be considered as significant. This is due to the micaceous nature of the crystals, which were plagued with stacking faults and imperfections. Indeed satellites are observed along both a and b axes, due to a twinning or misorientation phenomena. Such modulations, which correspond to a period of approximately $5\sqrt{2}a_p$ are similar to those observed for $Bi_2Sr_2CaCu_2O_8$. In addition, modulations are observed along c, with a periodicity of approximately 3 times the 24.6 Å spacing. Thus the Bi-O distances, which range from 1.97 to 3.28 Å, can be considered as the result of strong distortion of the BiO_6 octahedra but cannot be discussed further.

Examination of the formula of the three superconductors deduced from the structural studies - $Bi_2Sr_2CuO_6$, $Bi_2Sr_2CaCu_2O_8$ and $Bi_2Sr_2Ca_2 \cdot Cu_3O_{10}$ - suggest that no trivalent copper would be present in those oxides. This is in fact inconsistent with the chemical analysis performed by several researchers, which shows the presence of either Cu(III) or Bi(V) in significant amounts. This confirms that all these structures are not well resolved, owing to the problems of modulations and the quality of the crystals. Several hypotheses can be proposed to explain the possible mixed valence Cu(II)/Cu(III) in these oxides. The first deals with an excess of oxygen, leading for instance to $Bi_2Sr_2CaCu_2O_{8+\delta}$ or $Bi_2Sr_2CuO_{6+\delta}$ formulations. Such an excess could result from the intercalation of additional oxygen in the bismuth bilayers. A second hypothesis concerns the possibility of bismuth deficiency, leading to bismuth vacancies or to a partial replacement of bismuth by a divalent cation Sr^{2+} or Ca^{2+}. A third attractive model has been proposed by *Cheetham* et al. [5.25], in which vacancies are present on the strontium sites. Moreover, the possibilities of partial replacement of strontium by calcium, leading to homogenity ranges such as $Bi_2Sr_{2-x}Ca_{1+x}Cu_2O_8$ or $Bi_2Sr_{2-x}Ca_{2+x}Cu_3O_{10}$ are not yet understood.

The X-ray absorption study of these phases at CuL_{III}-edge [5.26] has definitely shown the existence for copper of the $3d^9L$ configuration (L: ligand hole) characteristic of the mixed valence $Cu(II)-Cu(III)$ and for bismuth an electronic configuration $Bi(III-\delta)$ intermediate between $Bi(III)$ and metallic bismuth. This suggests that a deviation from the stoichiometric formula $Bi_2Sr_2Ca_{m-1}Cu_mO_{2m+4}$ is not necessary for the appearance of the mixed valence of copper in those oxides, but that this latter results from a dynamical transfer of holes from the bismuth-oxygen layers to the copper-oxygen layers. Thus, a rigorous structural resolution of the actual structure of these compounds will be absolutely necessary in order to understand their superconducting properties.

5.2 Thallium Alkaline-Earth Superconducting Cuprates

Studying the substitution of a trivalent cation, thallium (III) for yttrium, in the 92 K superconductor $YBa_2Cu_3O_{7-\delta}$, *Sheng* et al. [5.27,28] discovered superconductivity at 85 K in the Tl-Ba-Cu-O system. Two compositions were mainly investigated by these researchers: $TlBaCu_3O_{5.5+x}$ and $Tl_2Ba_2 \cdot Cu_3O_{8+x}$. In order to minimize decompositions and evaporation of Tl_2O_3, they used short periods of heating and quenching. The samples were prepared by first heating mixtures of $BaCO_3$ and CuO in air at 925° and then mixing the resulting Ba-Cu oxide with an appropriate amount of Tl_2O_3. Pellets of this latter mixture were then heated at 880°-910°C for 2-5 min in an oxygen flow. In these conditions the variation of resistance with temperature (Fig.5.15) shows a rather sharp transition with onset temperatures above 90 K and zero resistance at 81 K. The magnetization measurements made by this group confirmed that the diamagnetism-onset temperature was about 85 K, but the Meissner expulsion at 10 K was found to be weak, 15% of diamagnetic shielding, representing about 5% of perfect superconductivity. At this stage of the investigations, the phase responsible for superconductivity could not be identified by X-ray diffraction owing

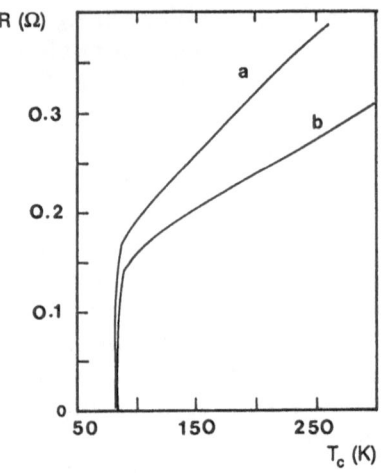

R (Ω)

0.3

0.2

0.1

0

50   150   250

T_c (K)

Fig.5.15. Variations of resistance with temperature for samples of $TlBaCu_3O_{5.5+x}$ (a) and $Tl_{1.5}Ba_2Cu_3O_{7.3+x}$ (b). (After [5.27-29])

to its presence in very small amounts. Using similar methods of preparation, *Sheng* et al. [5.29, 30] then detected bulk superconductivity at 120 K in the Tl-Ca-Ba-Cu-O system. But this method, which involves a slight melting of the mixture, always leads to multiphase samples. Subsequent investigations of the Tl-Ca-Ba-Cu-O system allowed two phases to be identified as responsible for superconductivity by several groups almost simulataneously [5.31-33]: $Tl_2Ba_2CaCu_2O_8$, called the "2212" phase, was assumed to exhibit a critical temperature close to 108 K, whereas the highest-yet T_c of 125 K was attributed to a second phase, $Tl_2Ba_2Ca_2Cu_3O_{10}$, called "2223".

It is worth pointing out that the methods of preparation used by *Sheng* et al. lead to the formation of mixtures of superconductors, owing to the very short times of reaction used in order to avoid thallium oxide volatilization. A pure phase [5.34] $Tl_2Ba_2CaCu_2O_8$ can be synthesized by reaction in a sealed quartz tube: an appropriate mixture of Tl_2O_3, BaO_2, CaO and CuO is pressed into the form of a pellet, placed in a small alumina crucible, and then allowed to react in a quartz tube at temperatures ranging from 750 to 900 °C for durations of 3-12 h, which is followed by furnace cooling. Simultaneously, single crystals were synthesized by *Sleight* and coworkers [5.35].

The crystal structure of this latter superconductor is closely related to that of $Bi_2Sr_2CaCu_2O_8$, but it is less complex in that the symmetry is tetragonal ($a=3.86Å\simeq a_p$, $c=29.38Å$) and that no superstructures or incommensurable extra-spots along a are observed. Thus the structure (Fig.5.6a) can essentially be described as the intergrowth of double sheets of corner-sharing CuO_5 pyramids interleaved with calcium ions, forming double oxygen-deficient perovskite layers $[BaCaCu_2O_5]_\infty$ and triple distorted rock-salt-type layers $[BaTl_2O_3]_\infty$. Unlike $Bi_2Sr_2CaCu_2O_8$, the crystals are not mica-like, and thus the interatomic distances are much more significant. The CuO_5 pyramids always exhibit the same geometry as encountered in the preceding oxides, with four short Cu-O distances (1.93Å) in the basal plane, and one long Cu-O distance along c (2.69Å). Copper tends to take a square planar coordination, and the copper distances between the $[CuO_{2.5}]_\infty$ sheets remain close to 3.2 Å, as in $YBa_2Cu_3O_7$. It is also interesting to notice that the planarity of the copper-oxygen sheets forming the basal planes of the CuO_5 pyramids increases from $YBa_2Cu_3O_7$ to $Bi_2Sr_2CaCu_2O_8$ and to $Tl_2Ba_2CaCu_2O_8$. In contrast to the bismuth oxides, the rock-salt-type nature of the thallium bilayers is unambiguous. Also, the very short distance between two adjacent thallium layers (2.2 Å) does not leave any space for additional oxygen. However, the $[BaTl_2O_3]_\infty$ slabs seem to be strongly distorted, especially the TlO_6 octahedra, which exhibit Tl-O bonds ranging from 1.98 to 2.46 Å. An important point concerns the possible nonstoichiometry or "solid solution" of the thallium and calcium distributions. One observes a high B thermal factor for thallium, and a low one for calcium. Refining the occupation factors of these sites allowed the proposal of a partial occupation of the calcium sites by thallium and either a thallium nonstoichiometry or a partial substitution of calcium for thallium on thallium sites. Thus stoichiometric compositions like $Tl_{2-x}Ca_xBa_2 \cdot Ca_{1-y}Tl_yCu_2O_8$ with x and y ranging from 0.1 to 0.2 can be proposed. An alternative to this hypothesis is $Tl_{2-x}Ba_2Ca_{1-y}Tl_yCu_2O_8$ with x < 0.1 and y $\simeq$ 0.1. The electron diffraction and high-resolution electron microscopy studies confirm that these crystals exhibit neither preferred cleavages nor satellites but that a large part of them show streaks of various intensities, along the c axis and a tendency to amorphization. Nevertheless, the regular stacking of the layers of heavy atoms along **c** is easily identified from the [010] HREM images, as shown in Fig.5.16, where the rows of white dots correspond to the sequence "Ba-Tl-Tl-Ba-Cu-Ca-Cu-Ba". The streaks that appear in this phase result from the existence of extended defects, which will be discussed in Chap.6.

Fig.5.16. [010] HREM images of a regular crystal with nominal composition $Tl_2Ba_2CaCu_2O_8$. Note the amorphization at the edges

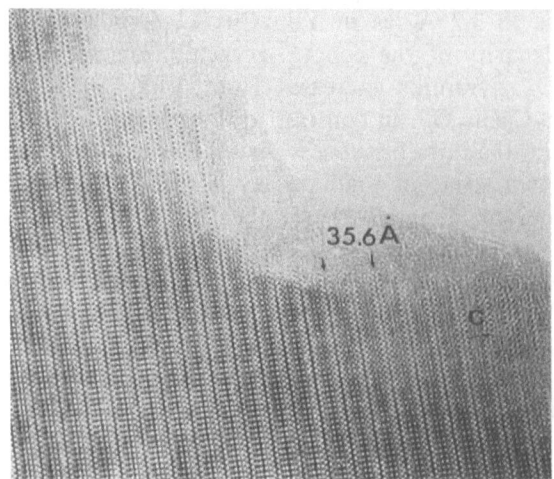

Fig.5.17. $Tl_2Ba_2Ca_2Cu_3O_{10}$. [100] HREM micrograph of a regular crystal showing a perfect stacking of the layers

As expected from the structural studies of the "2212" phase, the $Tl_2 \cdot Ba_2CuO_6$ superconductor and the 125 K superconductor $Tl_2Ba_2Ca_2Cu_3 \cdot O_{10}$ are characterized by closely related structures, which consist of intergrowths of multiple oxygen-deficient perovskite layers and triple distorted rock-salt-type layers $[BaTl_2O_3]_\infty$. The structure of $Tl_2Ba_2CuO_6$ [5.24] is essentially similar to $Bi_2Sr_2CuO_6$, so that it can be described (Fig.5.14) as an intergrowth of single octahedral perovskite layers $[BaCuO_3]_\infty$ with triple distorted rock-salt-type layers $[BaTl_2O_3]_\infty$. Again, unlike $Bi_2Sr_2CuO_6$, no satellite modulation was observed along the a axis and also no supercell reflections corresponding to $\sqrt{2}a_p$. This leads to the actual tetragonal cell (a=3.866Å and c=23.24Å). The phase $Tl_2Ba_2Ca_2Cu_3O_{10}$ [5.17,36], like other thallium oxocuprates of this series, exhibits a single tetragonal cell (a=3.85Å, and c=35.7Å), without any superstructure reflections or satellites with respect to the perovskite subcell in the (001) plane. The X-ray diffraction refinement and the high-resolution electron microscopy observation (Fig.5.17) allow identification of the following sequence of heavy atoms: Ba-Tl-Ba-Cu-Ca-Cu-Ca-Cu-Ba, showing that this structure corresponds to the model proposed for $Bi_2Sr_2Ca_2Cu_3O_{10}$ (Fig.5.13). Thus this 125 K superconductor corresponds to the intergrowth of triple $[Tl_2BaO_3]_\infty$ infinite layers with triple oxygen-deficient perovskite layers formed of two pyramidal $[CuO_{2.5}]_\infty$ sheets and one $[CuO_2]_\infty$ sheet of corner-sharing CuO_4 groups. Again, in this oxide refinement of the occupation factors of different sites suggested a substitution of Tl on the Ca sites accompanied by either Ca substitution on the Tl sites or thallium vacancies in the thallium bilayer. These observations lead, as for $Tl_2Ba_2CaCu_2O_8$, to different possibilities of nonstoichiometry, which can be represented by the formulae $Tl_{2-x}Ba_2Ca_{2-y}Tl_yCu_3O_{10}$ and $Tl_{2-x}Ca_xBa_2Ca_{2-y}Tl_yCu_3O_{10}$ with x and y smaller than 0.2. Such deviations from stoichiometry can be explained by the existence of extended defects, which correspond to a local variation of the composition, as can be shown by high-resolution electron microscopy (Chap.6). In both $Ba_2Tl_2CuO_6$ and $Tl_2Ba_2Ca_2Cu_3O_{10}$ one observes homogeneous Cu-O distances in the (001) plane ranging from 1.92 to 1.93 Å, whereas the Cu-O distances parallel to c are always much longer: 2.71 Å for the CuO_6 octahedra of $Tl_2Ba_2CuO_6$ and 2.48 Å for the CuO_5 pyramids of $Tl_2Ba_2Ca_2Cu_3O_{10}$. The TlO_6 octahedra of the thallium bilayers are, like those of $Tl_2BaCaCu_2O_8$, strongly distorted, but the difficulty of determining the position of the oxygen of the thallium bilayers does not allow a high precision to be obtained for these three compounds. The distances between copper planes (Cu-Cu = 3.2 Å) and between two successive thallium layers (2.2 Å) are remarkably constant.

Besides the three first-detected superconductors, three other superconductors were subsequently identified, with the ideal compositions $TlBa_2 \cdot Ca_2Cu_3O_9$ (called "1223"), $TlBa_2CaCu_2O_7$ (called "1212") and $Tl_2Ba_2Ca_3 \cdot Cu_4O_{12}$ (called "2234"). The "1223" phase was isolated for the first time by *Martin* et al. [5.37], while *Parkin* et al. [5.38] simultaneously obtained this oxide mixed with other phases and exhibiting numerous extended defects. This oxide exhibits a superconducting transition as high as 120 K for the fall in the resistivity (Fig.5.18), with a zero resistance at 105 K, or 110 K for the Meissner effect [5.39]. From the zero-field-cooled and field-cooled susceptibility of samples of nominal composition $Tl_{0.7}Ba_2Ca_2Cu_3O_{9-y}$ it was observed (Fig.5.19) that the Meissner effect corresponds to about 30% of the total volume, and the shielding effect in a 2 Oe field was found to correspond to a full diamagnetism. The powder X-ray and electron microscopy studies [5.37,38,40] allow the structural model of this 120 K superconductor to be established without any ambiguity (Fig.5.20). This oxide crystallizes in the tetragonal system (a=3.844 Å $\sim a_p$, c=15.88 Å) and does not exhibit any superstructure reflections with respect to this cell, as shown by the ED patterns (Fig.5.21). The structure of this phase is characterized, like that of the 125 K superconductor $Tl_2Ba_2Ca_2Cu_3O_{10}$, by the presence of triple layers $[BaCa_2Cu_3O_7]_\infty$ of oxygen-deficient perovskite type, i.e. formed of two pyramidal $[CuO_{2.5}]_\infty$ layers and one $[CuO_2]_\infty$ layer interleaved with calcium ions (Fig.5.20). Different from the perovskite-type layers are the rock-salt-type layers, which are only double

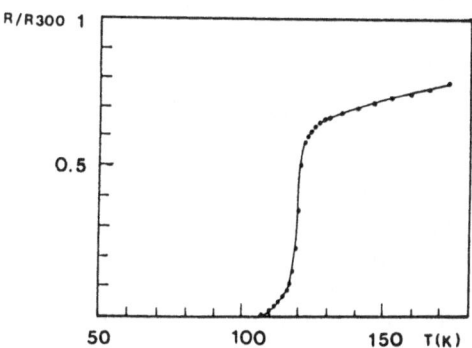

Fig.5.18. Relative resistance of the oxide $TlBa_2Ca_2Cu_3O_{10-y}$ as a function of temperature

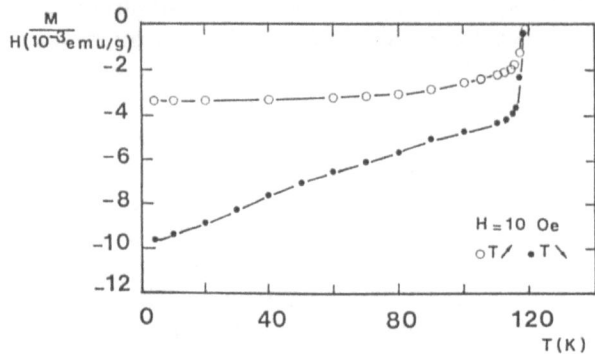

Fig.5.19. The zero-field-cooled (• T↓) and field-cooled susceptibilities (○ T↑) of the "1223" phase vs temperature

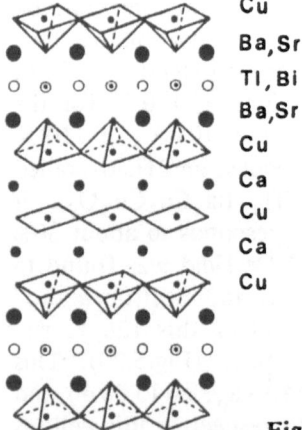

Cu
Ba,Sr
Tl,Bi
Ba,Sr
Cu
Ca
Cu
Ca
Cu

Fig.5.20. Schematized structure of $TlBa_2Ca_2Cu_3O_9$

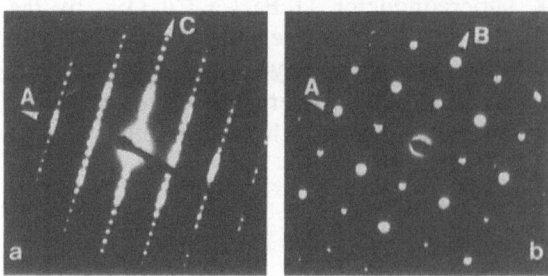

Fig.5.21. (a) (010) and (b) (001) electron diffraction patterns of $TlBa_{2-x}Ca_{2+x}Cu_3O_9$

$[TlBaO_2]_\infty$ slabs formed of a thallium monolayer sandwiched between two barium monolayers. The agreement between calculated and experimental HREM through-focus series (Fig.5.22) confirms the sequence "Ba-Tl-Ba-Cu-Ca-Cu-Ca-Cu-Ba". There is no doubt that this phase, like the other oxides, presents a possible thallium deficiency, but another interesting fea-

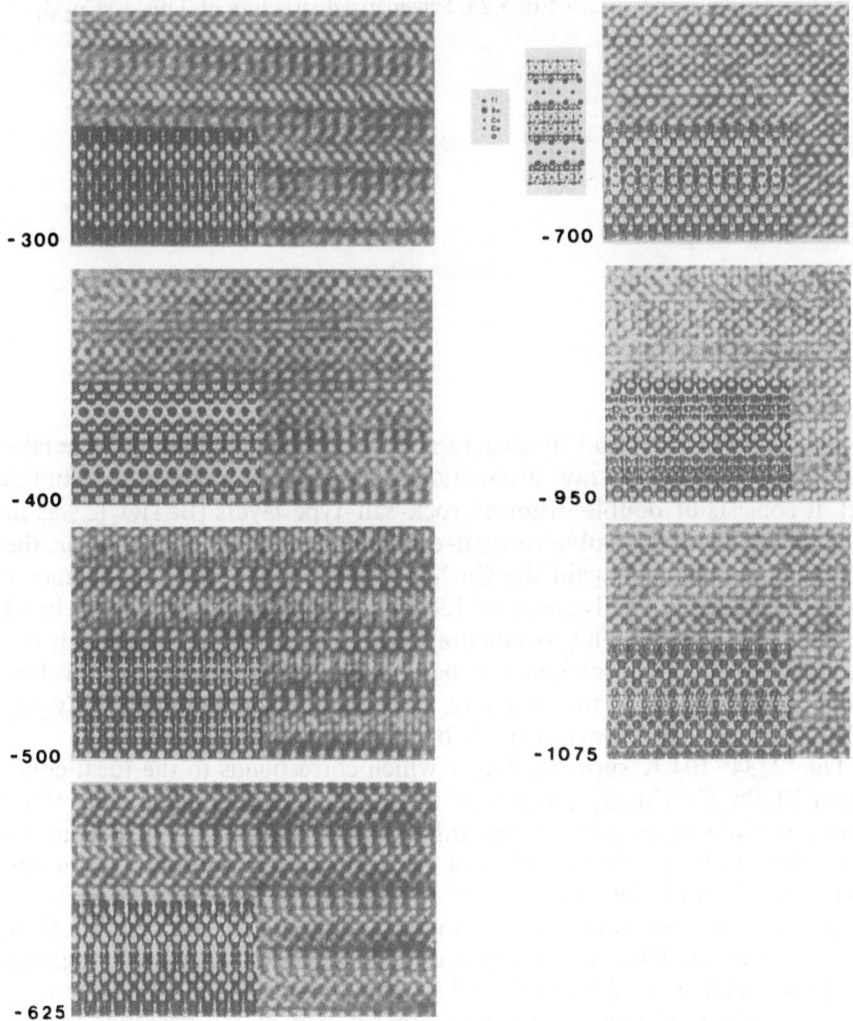

Fig.5.22. $TlBa_{2-x}Ca_{2+x}Cu_3O_9$: HREM images and through-focus series (calculated images in the bottom left-hand corners). Calculation parameters are: V = 200 kV, Cs = 0.8 mm, α = 0.8 mm, $\Delta/2$ = 75 Å, 155 beams, crystal thickness t = 30.7 Å, defocus values are in angstroms

ture deals with the possible substitution of calcium on the barium sites, leading to a rather large homogeneity range $TlBa_{2-x}Ca_{2+x}Cu_3O_9$ ($0 \leq x \leq 0.80$) characterized by a significant variation of the crystallographic parameters. It is also worth noting that superconductivity is observed over all the homogeneity range.

The "1212" superconductor $TlBa_2CaCu_2O_7$, discovered simultaneously [5.41, 42], can also be isolated by using an evacuated silica ampoule. Its critical temperature can be modified from 50 to 65 K by changing the thermal treatment. This oxide crystallizes in the tetragonal system

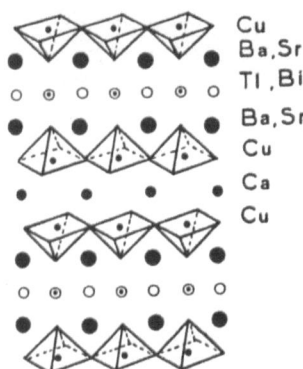 

Cu
Ba,Sr
Tl ,Bi
Ba,Sr
Cu
Ca
Cu

Fig.5.23. Schematized structure of $TlBa_2CaCu_2O_7$

(a=3.833Å, c=12.68Å) and its structural model (Fig.5.23) was easily established from powder X-ray diffraction and electron diffraction studies [5.41]. It consists of double distorted rock-salt-type layers $[BaTlO_2]_\infty$, as in the "1223" phase, and double oxygen-deficient perovskite layers, as in the "2212" superconductor. Again the CuO_5 pyramids have the usual symmetry (four short basal Cu-O distances of 1.92Å and one long apical Cu-O bond of 2.45Å), whereas the TlO_6 octahedra seem to be strongly distorted. It has been shown [5.43] that calcium can be completely replaced by a trivalent cation without changing the structure: $TlBa_2NdCu_2O_7$ and $TlBa_2YCu_2O_7$ are isostructural with $TlBa_2CaCu_2O_7$ but do not superconduct.

The "2234" 104 K superconductor which corresponds to the ideal composition $Tl_2Ba_2Ca_3Cu_4O_{12}$, was able to be prepared using the usual silica ampoule method described for the other oxides, starting from a nominal composition "1133" [5.44]. It could not be isolated as a pure phase: one obtained about 90% of this phase mixed with $BaCuO_2$. The electron diffraction patterns of this phase were studied for about 100 crystals. They showed that about 50% of the crystals correspond to the pure tetragonal phase (a = 3.84Å, c = 42.0Å) (Fig.5.24a) while in the other crystals streaks along c are observed (Fig.5.24b), which are a sign of the presence of extended defects in the matrix of the "2234" tetragonal phase. The high-resolution electron microscopy investigation of the crystals exhibiting a regular array (Fig.5.25) allowed the sequence of metallic atoms "Ba-Tl-Tl-

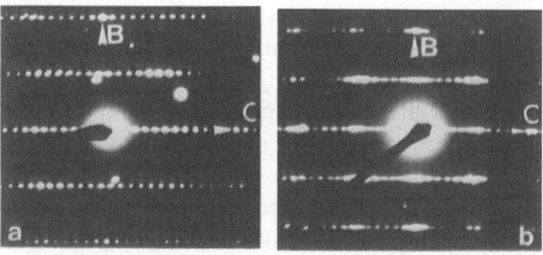

Fig.5.24a,b. [010] ED patterns of $Tl_2Ba_2Ca_3Cu_4O_{12}$. (a) Pattern for a regular crystal; (b) pattern with streaks along c

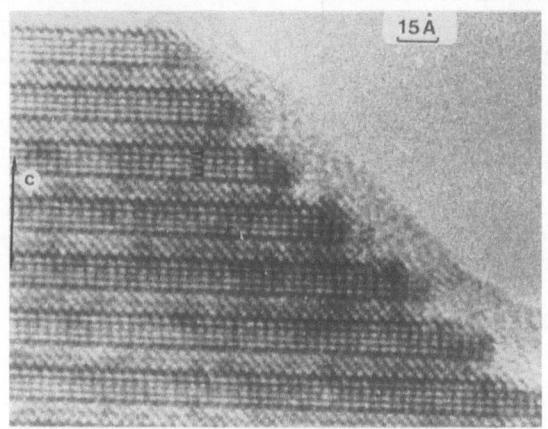

Fig.5.25. [010] high resolution image of the "2234" phase $Tl_2Ba_2Ca_3Cu_4O_{12}$

15 Å

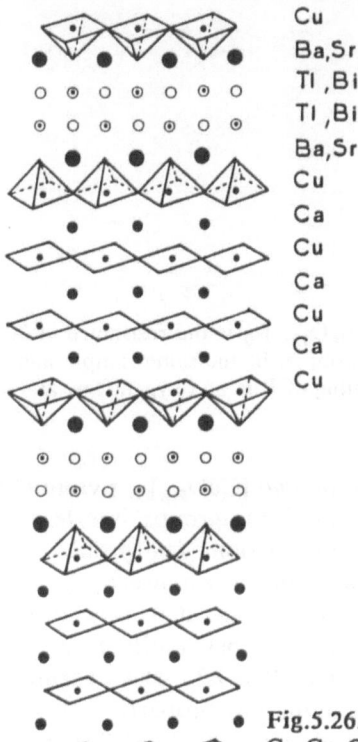

Cu
Ba,Sr
Tl,Bi
Tl,Bi
Ba,Sr
Cu
Ca
Cu
Ca
Cu
Ca
Cu

Fig.5.26. Idealized model of the "2234" phase $Tl_2Ba_2\cdot Ca_3Cu_4O_{12}$ built up from triple rock-salt-type layers and quadruple layers of oxygen-deficient perovskite

Ba-Ca-Cu-Ca-Cu-Ca-Cu-Ca-Cu-Ba" to be determined. Thus the structural model that corresponds to the oxide (Fig.5.26) consists of the intergrowth of triple distorted rock-salt-type layers $[Tl_2BaO_3]_\infty$ (already described for $Tl_2Ba_2CuO_6$, $Tl_2Ba_2CaCu_2O_8$ and $Tl_2Ba_2Ca_2Cu_3O_{10}$) and original quadruple oxygen-deficient perovskite layers $[Ba_2Ca_3Cu_4O_9]_\infty$.

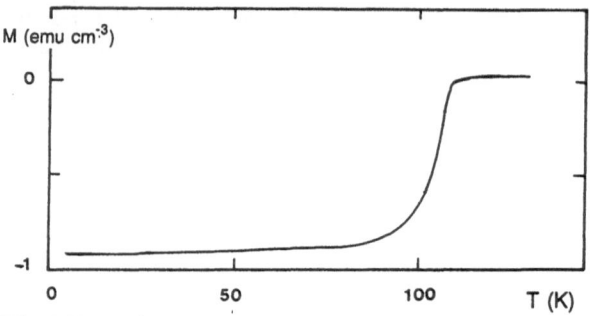

Fig.5.27. Meissner effect at B = 50 G in $Tl_2Ba_2Ca_3Cu_4O_{12}$

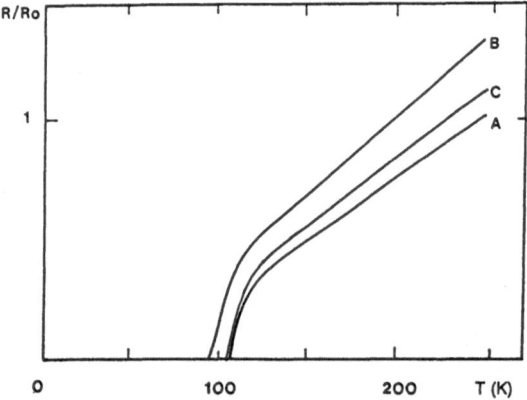

Fig.5.28. Evolution of R/R_0 vs T for $Tl_2Ba_2Ca_3Cu_4O_{12}$; R_0 is the resistance at 300 K of the as-synthesized sample. A: as-synthesized sample. B: the same sample annealed in Ar flow at T = 400°C for 1 h. C: after annealing of B in an oxygen flow at T = 400°C for 1 h

These latter layers can be described as formed of two $[CuO_{2.5}]_\infty$ pyramidal slabs and two $[CuO_2]_\infty$ slabs of corner-sharing CuO_4 groups interleaved with planes of calcium ions in pseudocubic coordination. The problem of extended defects will not be studied until Chap.6, but it can already be explained why numerous crystals are not absolutely regular. It is easy to understand that the tendency to disorder, i.e. to form perovskite layers of different thickness, will increase as the number of $[CuO_2]_\infty$ slabs forming such layers increases. Concerning the superconducting properties of this phase, one observes a beginning of diamagnetism at 114 K (Fig.5.27) and a diamagnetic volume of 42% at 4.2 K. The evolution of the ratio R/R_{300K} vs T shows that the T_c onset appears at 120 K, and zero resistance is obtained at 104 K (Fig.5.28A). Annealing of the samples in an argon flow at 400°C for 1 h leads to a decrease of T_c, zero resistance being reached at 93 K (Fig.5.28B). This phenomenon is reversible: annealing of the B sample in an oxygen flow at 400°C allows the zero resistance temperature to be restored to 104 K, exactly as in the as-synthesized oxide (Fig.5.28C). Attention should be drawn to the fact that the resistive transition observed here is rather broad compared to "2223" or "1223" for instance. This is easily explained by the fact that a great number of crystals exhibit defects corresponding to these "2223" or "1223" phases (Chap.6), which are characterized by T_c values of 125 K and 120 K, respectively. Thus, it appears that the T_c offset (i.e., T at which R = 0) of the oxide $Tl_2Ba_2Ca_3Cu_4O_{12}$, characterized by a quadruple perovskite layer, is smaller than the critical temperatures observed for the m = 3 members $Tl_2Ba_2Ca_2Cu_3O_{10}$ and $TlBa_2Ca_2Cu_3O_9$. Clearly, it is shown that T_c does not increase as the thickness of the oxygen-deficient perovskite layer increases, contrary to what was foreseen by several researchers. Moreover, these results are in agreement with the theory developed recently by *Bok* [5.45], which indicates a loss of two-dimensionality as the number of $[CuO_2]_\infty$ planes of each oxygen-deficient perovskite layer increases.

The 1234 superconductor, with the composition $TlBa_2Ca_3Cu_4O_{11}$ [5.46], is also obtained as a mixture and exhibits a T_c near 110 K. Its structure is built up from quadruple copper layers like in $Tl_2Ba_2Ca_3Cu_4O_{12}$ and from double rock salt containing thallium monolayers like $TlBa_2Ca_2 \cdot Cu_3O_9$. Thallium cuprates with perovskite slabs involving a greater number of copper layers are always obtained as mixtures or intergrowths with lower members, as shown for "1245" and "2245" oxides. They correspond to the ideal formula $TlBa_2Ca_4Cu_5O_{13}$ [5.47] and $Tl_2Ba_2Ca_4Cu_5O_{14}$ [5.48], respectively; T_c's of 105 K were reported for them.

Contrary to barium, strontium was not found to form such superconductive frameworks at first by many researchers, although *Sheng* et al. noticed superconductivity at 20 K, but with a weak Meissner effect (smaller than 1%) in the system Tl-Sr-Ca-Cu-O [5.49]. *Subramanian* et al. [5.50] did not succeed in synthesizing such oxides but they were able to prepare two new oxides, $Tl_{0.5}Pb_{0.5}Sr_2CaCu_2O_7$ and $Tl_{0.5}Pb_{0.5}Sr_2Ca_2Cu_3 \cdot O_9$, superconducting below 85 and 125 K, respectively, and isostructural with the thallium barium cuprates $TlBa_2CaCu_2O_7$ and $TlBa_2Ca_2Cu_3O_9$. However, *Martin* et al. [5.51] succeeded in synthesizing $TlSr_2CaCu_2O_7$ as an almost pure phase (at about 90%) by heating mixtures of SrO_2, PbO_2, CaO and CuO in evacuated quartz tubes at 800 °C. This "1212"-type oxide was found to superconduct below 47 K with a diamagetic volume fraction of 13% below 20 K.

The complete replacement of calcium by a trivalent rare earth element, such as yttrium or neodymium, is possible without changing the structure but it kills the superconducting properties, as shown for the oxides $TlBa_2LnCu_2O_7$ [5.43, 52]. On the other hand, the partial substitution of such a lanthanide for calcium enhances the critical temperature. For instance, an increase of T_c from 73 to 103 K was observed for the oxide $Tl \cdot Ba_2Ca_{1-x}Y_xCu_2O_7$ as x increases from 0 to 0.2 [5.53]; however, the Meissner fraction extrapolated to 0 K decreases drastically at the same time from 46% in the nondoped sample to 0.2% for x = 0.20. In the same way, superconductivity in the region 60-90K was found for the oxides $TlSr_2 \cdot Ca_{1-x}Ln_xCu_2O_7$ [5.54, 55]. Thallium can also be partly replaced by bismuth, leading to the 1212-type oxides $(Tl, Bi)Sr_2CaCu_2O_y$ [5.56] and

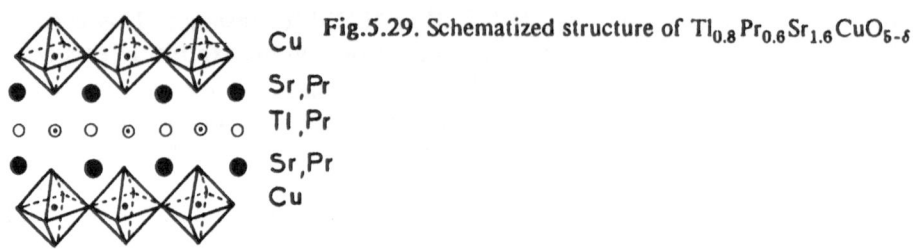

Fig.5.29. Schematized structure of $Tl_{0.8}Pr_{0.6}Sr_{1.6}CuO_{5-\delta}$

$Tl_{0.5}Bi_{0.5}Ca_{1-x}Y_xSr_2CuO_y$ [5.57], which exhibit critical temperatures ranging from 83 to 102 K.

$Tl_2Sr_2CaCu_2O_8$ prepared by *Greenblatt* et al. [5.58] is the only pure strontium cuprate which contains thallium bilayers. This phase which is isotypic with $Tl_2Ba_2CaCu_2O_8$ exhibits a T_c of 44 K.

Besides these thallium strontium cuprates, a new member of the family has been isolated by *Bourgault* et al. [5.59]. The structure of this oxide, $Tl_{1-x}Pr_xSr_{2-y}Pr_yCuO_{5-\delta}$ (Fig.5.29), corresponds to the intergrowth of single perovskite layers $[(Sr,Pr,CuO_3)]_\infty$ and double rock-salt-type layers $(Tl,Pr,Sr)_2O_2]$ involving thallium monolayers. This compound was the first found to exhibit superconducting properties in spite of the presence of praseodymium: $Tl_{0.8}Pr_{0.6}Sr_{1.6}CuO_{5-\delta}$ superconducts below 40 K with 2% diamagnetism at 4 K, while traces of superconductivity are obtained for $Tl_{0.7}Pr_{0.7}Sr_2CuO_{5-\delta}$. The isostructural phase $TlSr_2CuO_{5-\delta}$ could not be isolated, but was observed in the form of a mixture with an unknown phase. Traces of diamagnetism were observed for this mixture which can be attributed to $TlSr_2CuO_{5-\delta}$. Substitution of lead for thallium stabilizes this structural type: $Tl_{0.5}Pb_{0.5}Sr_2CuO_{5-\delta}$ can be synthesized as a pure phase [5.60] and has recently been demonstrated to exhibit a T_c of 40 K [5.61].

The nonstoichiometry phenomena in the thallium cuprates are in fact more complex than described above. Thallium and oxygen nonstoichiometry can drastically affect the superconducting properties of those materials. Thallium cuprates characterized with an excess thallium, corresponding to the formulation $Tl_{3-4x/3}Ba_{1+x}LnCu_2O_8$ [5.62], have been isolated. In these oxides, which are of "2212" type, a significant part of the thallium is located over the barium sites, suggesting a mixed valency Tl(I)/Tl(III); monovalent thallium would be located in the $[BaO]_\infty$ layers, whereas the thallium bilayers would contain Tl(III) only. Such compounds do not exhibit any superconducting properties.

The most striking feature deals with the spectacular effect of oxygen nonstoichiometry upon the critical temperature of the thallium cuprates. It was found that for the same composition the T_c' varies from one report to another. This is the case for the "1212" phase $TlBa_2CaCu_2O_7$, whose T_c ranges from 50 K [5.41] to 100-112 K [5.42,63]; the "2212" oxide $Tl_2Ba_2 \cdot CaCu_2O_8$, with $T_c's$ ranging from 97 to 108 K; and $Tl_2Ba_2CuO_6$, for which a wide range of behaviours, from nonsuperconducting to a T_c of 90K, have been reported [5.24,28,42,64]. Such results suggest that the ox-

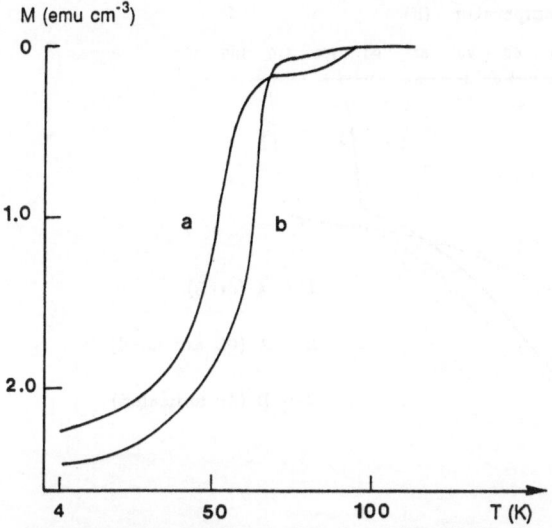

Fig.5.30. Magnetization vs T for $TlBa_2CaCu_2O_7$ (*a*) after synthesis and (*b*) after annealing in an innert-gas flow

ygen content can play a role in superconductivity. This was indeed shown first for the oxides $TlBa_2CaCu_2O_7$ and $Tl_2Ba_2CuO_6$ [5.41]. These compounds, prepared under an oxygen pressure of several atmospheres in sealed ampoules, were found to be superconducting (T_c = 50K) and nonsuperconducting, respectively. Annealing these samples at 450-500 °C under an argon flow allowed T_c to be increased up to 65 K for the first one (Fig.5.30) and superconductivity at 30 K to be obtained for the second one. The fact that the superconducting transition can be increased by decreasing the oxygen content was then confirmed later for $Tl_2Ba_2CuO_6$ [5.65,66] whereas the opposite was found for $Tl_2Ba_2CaCu_2O_8$ [5.67] and $Tl_2Ba_2Ca_2Cu_3O_{10}$ [5.68]. A study of the annealing of $Tl_2Ba_2CaCu_2O_8$ crystals under argon and vacuum [5.63] showed a more complex behaviour: for one of the crystals T_c decreased by vacuum annealing, whereas for the others T_c increased significantly. A recent study of $Tl_2Ba_2CaCu_2O_8$ clearly shows that in those cuprates the critical temperature is mainly governed by the oxygen stoichiometry [5.69]. Starting from a sample synthesized under an oxygen pressure of about 4-7 bars in sealed tubes three sorts of annealings were performed in a first step, corresponding to samples A, B and C respectively. For sample A, annealing was performed at 400 °C in an oxygen flow. For sample B, annealing was carried out in an argon flow at 400 °C. For sample C, the sealed silica ampoule containing the "as-synthesized" sample was cooled down to 400 °C, kept at this temperature for 12 hours and then slowly cooled down to room temperature. Figure 5.31a shows the real part χ' of the ac susceptibility of the A (or B) "as-synthesized" sample (A and B were synthesized in the same evacuated ampoule), the oxygen annealed sample A and the argon annealed sample B. The data are not corrected for the demagnetizing effects, as shown by the χ' values smaller

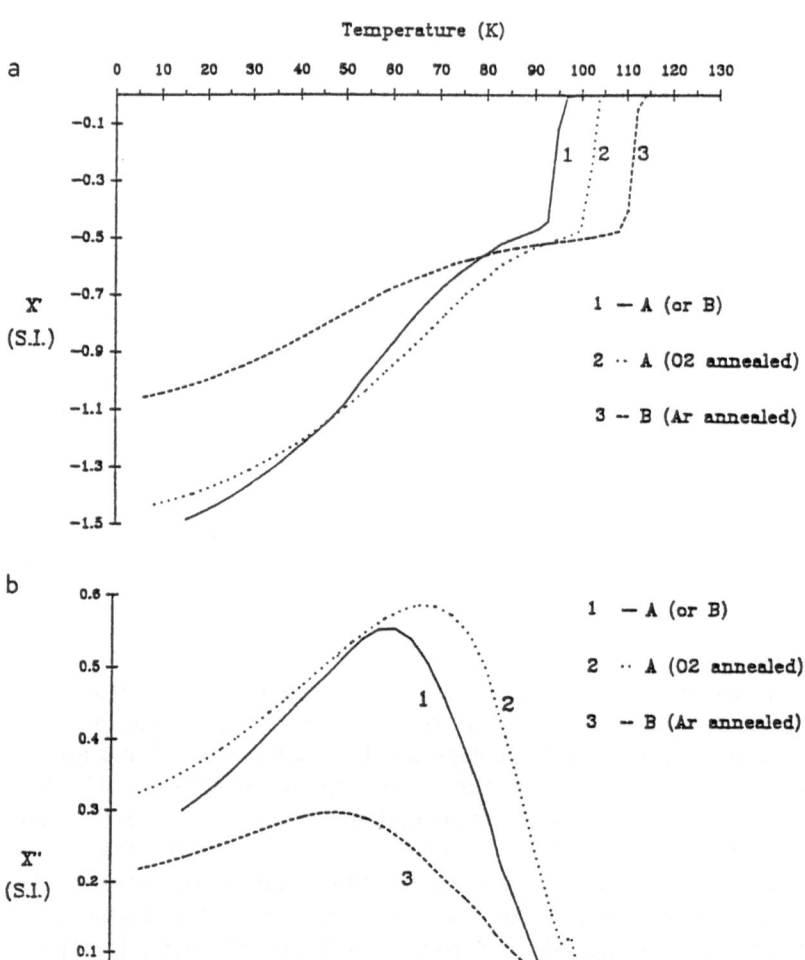

Fig.5.31. (a) Real part $\chi'(T)$ of the magnetic AC susceptibility of thallium cuprates $Tl_2Ba_2CaCu_2O_8$: (1) as-synthesized sample ($T_c \simeq 96K$): (2) sample annealed under oxygen flow at 400°C. (b) Imaginary part $\chi''(T)$ of the magnetic ac susceptibility corresponding to the curves in (a)

than -1. The first important result deals with the fact that the critical temperature is increased by argon annealing as well as by oxygen annealing. One indeed observes that T_c is about 97 K for the as-synthesized samples, whereas it is increased to 104 K by oxygen annealing (sample A) and to about 112 K by argon annealing (sample B). The ac magnetic field amplitude (5Oe) was the same for the three curves reported in Fig.5.31. The difference between the magnetization in the low temperature range comes from the grain decoupling, as can be seen from the imaginary part χ'' of

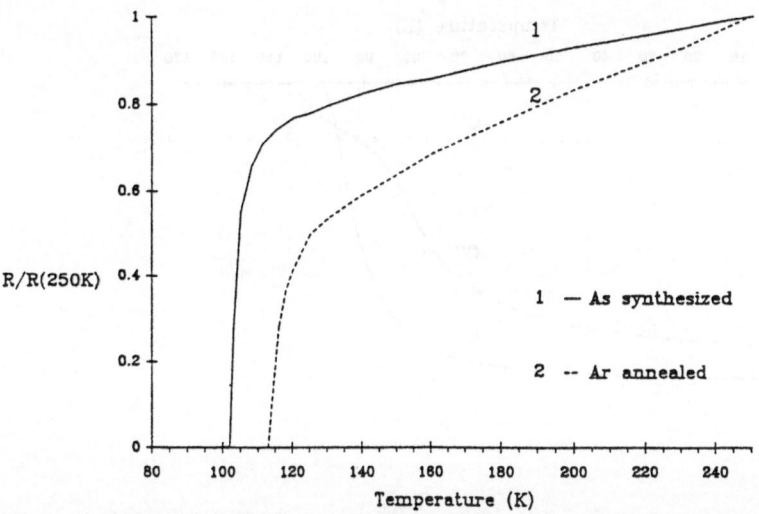

Fig.5.32. Electrical resistance vs T for a $Tl_2Ba_2CaCu_2O_8$ sample: (*1*) as-synthesized sample ($T_c \simeq 100 K$); (*2*) the same sample annealed under argon flow at 400° C

the magnetic susceptibility versus temperature (Fig.5.31b). These magnetization measurements are confirmed by electrical resistance measurements as can be seen in Fig.5.32 for the effect of argon annealing on the as-synthesized samples.

At this stage of the experiment it appears that oxygen loss during annealing under either argon or oxygen flow should be responsible for the increase of $T_c's$, suggesting that it allows the optimal hole carrier density to be reached by those thermal treatments. In order to confirm this hypothesis, sample C ($T_c \sim 98 K$) was heated in an evacuated ampoule in the presence of zirconium turnings for 12 h at 400 ° C. The $\chi'(T)$ curve after annealing of the C'' sample shows a spectacular increase of the critical temperature (Fig.5.33): $\Delta T_c \simeq 17$ K. However, the diamagnetic volume of the grains started to decrease, owing to the long annealing time in these conditions. The C' sample was then resealed in an evacuated silica ampoule, heated at 800 ° C for 2 h and finally rapidly cooled down to room temperature. The resulting C'' sample shows a decrease of T_c down to 105 K (Fig. 5.33) but the diamagnetic volume in the low temperature range is increased compared to C', showing clear evidence of grain boundary "reconstruction".

It is clear from these experiments that there exists an optimum hole carrier density leading to a maximum of T_c. Nevertheless, although the oxygen nonstoichiometry seems to play an important role in this phenomenon, the weight loss may also result from the thallium oxide volatilization. The volatility of thallium oxide at 400 ° C was indeed demonstrated for this compound by chemical analysis. In order to avoid this thallium loss the annealing of $Tl_2Ba_2CaCu_2O_8$ was then performed at 300 ° C in a reducing atmosphere (argon-hydrogen); in these conditions no thallium loss was observed. The spectacular effect of this method of annealing upon $T_c's$ is

137

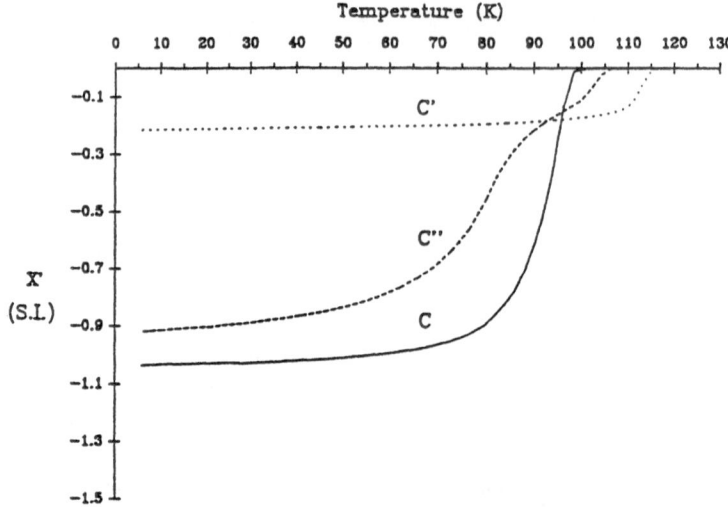

Fig.5.33. Real part $\chi'(T)$ of the AC susceptibility of a $Tl_2Ba_2CaCu_2O_8$ sample pre-pared with the same synthesis conditions as samples A and B (Fig.5.31): (C) sample annealed at 400° C under oxygen pressure in a sealed silica ampoule; (C') the same sample annealed in an evacuated silica ampoule with Zr turnings at 400° C for 12 h; (C'') the C' sample heated for 2 h at 800° C in an evacuated sealed ampoule

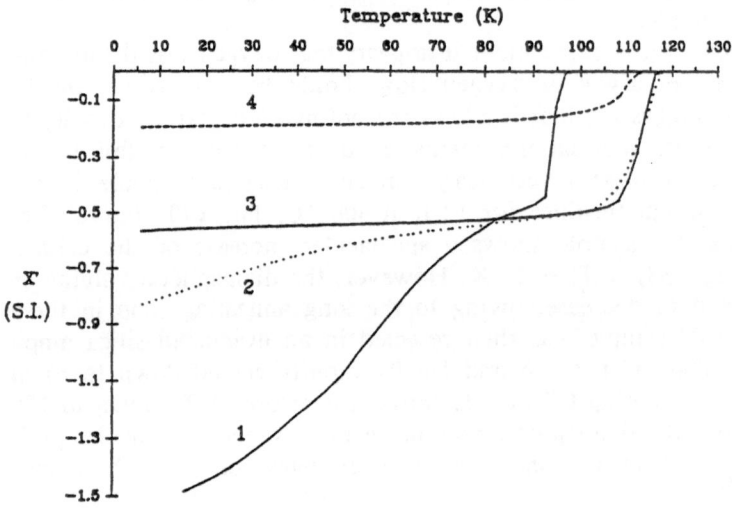

Fig.5.34. AC susceptibilities $\chi'(T)$ of the same sample Tl"2212" for different annealing times in a gas mixture (Ar90%+H$_2$10%): (1) unannealed (T_c = 96K); (2) annealed for 15 min (weight loss 0.07%, T_c = 118K); (3) annealed for 2 h (weight loss 0.4%, T_c = 117K); (4) annealed for 12 h (weight loss 2.2%, T_c = 115K)

shown in Fig.5.34. Indeed, T_c can be increased from 96 to 118 K for a very short annealing time, 15 min. It can also be seen that the variation of the annealing time does not influence the T_c's excessively but that a prolon-

138

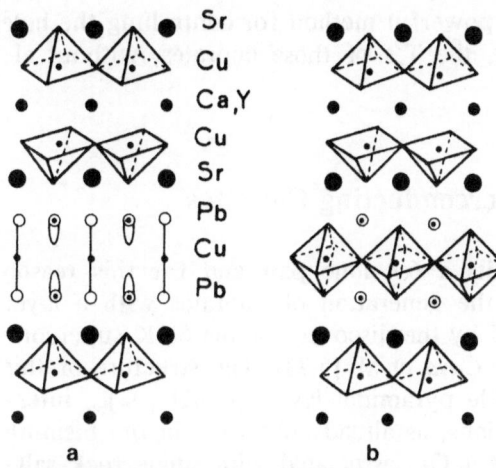

Fig.5.35. Schematized structure of $Pb_2Sr_2Y_{1-x}Ca_xCu_3O_8$ (a) derived from the hypothetical "double inter-growth" $Pb_2Sr_2YCu_3O_{10}$ (b)

Sr
Cu
Ca,Y
Cu
Sr
Pb
Cu
Pb

a   b

gated annealing tends to decrease the diamagnetic volume. It is worth pointing out that the weight losses are very small, and that the smallest weight loss of 0.07% leads to a dramatic increase of T_c: $\Delta T_c \simeq 22$ K.

Similar effects were observed for the "1212" phase $TlBa_2CaCu_2O_7$ and for $Tl_2Ba_2CuO_6$ [5.70], as shown for the latter phase (Fig.5.35) for which the T_c's can be increased from 10 to 90 K. On the opposite the influence of this annealing is considerably decreased as the number m of copper layers increases; the critical temperature of $TlBa_2Ca_2Cu_3O_9$ is only increased by 5 K, whereas for $Tl_2Ba_2Ca_2Cu_3O_{10}$ T_c remains unchanged (125K). It must also be emphasized, from the HREM observations, that the treatment under hydrogen at 300° C does not alter the materials, contrary to the annealings at 400° C-500° C under different atmospheres.

The oxygen nonstoichiometry is the key to the optimization of the superconducting properties and especially the critical temperature of the thallium cuprates, owing to its great influence upon the hole carrier density. This explains why the values of T_c observed for the "2212" phase could vary greatly, depending upon the oxygen pressure and temperature and time used for its preparation [5.42, 71, 72].

It is worth pointing out that the effect of oxygen nonstoichiometry upon T_c becomes more important as the number of copper layers forming the perovskite slabs becomes smaller, i.e. as m becomes smaller. One observes ΔT_c variations of 90K for $Tl_2Ba_2CuO_6$ (m = 1), of more than 20K for $TlBa_2CaCu_2O_7$ and $Tl_2Ba_2CaCu_2O_8$ (m = 2), and of 0-5K for $TlBa_2 \cdot Ca_2Cu_3O_9$ and $Tl_2Ba_2Ca_2Cu_3O_{10}$ (m = 3). Thus, the important factor which governs T_c in thallium cuprates is the hole carrier density per mole of copper in the structure [equivalent to the number of Cu(III) per mole Cu] and this factor seems to be predominant with respect to the effect of the number m of copper layers in the perovskite slabs.

The main issue which remains to be solved deals with the knowledge of the oxygen content and distribution in the structure, and consequently the actual carrier density. Nevertheless "annealing at 300 °C in a reducing atmosphere" appears to be a very powerful method for controlling the hole carrier density and, consequently, the T_c's of those cuprates, without altering the structure.

5.3 Lead Alkaline-Earth Superconducting Cuprates

Divalent lead, like Bi(III), exhibits a $6s^2$ lone pair and for this reason should be a good candidate for the generation of cuprates with a layer structure. This idea was confirmed by the discovery of the 68 K superconductor $Pb_2Sr_2Ln_{1-x}Ca_xCu_3O_8$ by *Cava* et al. [5.73]. The structure of this phase (Fig.5.35a) consists of double pyramidal layers $[(CuO_{2.5})_2]_\infty$ interleaved with calcium and yttrium ions, as already observed in the bismuth and thallium oxides and in $YBa_2Cu_3O_7$, associated with single rock-salt-type layers $[(Sr,Pb)_1O]_\infty$ similar to those observed in La_2CuO_4-type oxides and in $TlSr_2CuO_5$-type compounds. The rock-salt layers are themselves connected through layers of monovalent copper, with Cu(I) in twofold coordination, showing a great similarity with $YBa_2Cu_3O_6$. In fact this oxide belongs to the same structural family as the thallium and bismuth cuprates. Let us consider the hypothetical oxide "$Pb_2Sr_2YCu_3O_{10}$", whose structure (Fig.5.35b) consists of double oxygen-deficient perovskite layers and single perovskite layers intergrown with single rock-salt-type layers $[(Pb,Sr)_1O]_\infty$. Such an oxide corresponds to a double intergrowth, i.e. an intergrowth of the La_2CuO_4-type structure (single perovskite layers) with the $La_2CaCu_2O_6$-type structure (double perovskite layers). The elimination of the oxygen atoms of the basal planes of the CuO_6 octahedra, belonging to the single perovskite layers of "$Pb_2Sr_2YCu_3O_{10}$" leads to the 68 K superconductor. It results that oxygen can be intercalated in the structure of $Pb_2Sr_2Y_{1-x}Ca_xCu_3O_8$. It is worth pointing out that such an oxide can be obtained as a superconductor only by reaction in a reducing atmosphere. This leads one to consider a disproportionation of Cu(II) into Cu(III) and Cu(I) for the final formula. This latter point, which may be important for the explanation of superconductivity in these oxides, will be discussed in Chap.8. The best flux expulsion, obtained for x = 0.50, equal to 20%, confirms that superconductivity exists in the bulk. *Retoux* et al. [5.74] have shown that the substitution of bismuth for lead improves the superconducting properties of these oxides. The cuprates $Pb_{2-x}Bi_xSr_2Y_{1-y}Ca_yCu_3 \cdot O_8$ are isostructural for $0 \leq x \leq 0.50$ and $0 \leq y \leq 1$, and characterized by a variation of T_c. For instance, T_c = 60 K for x = 0.3 and y = 0.50, whereas a zero resistance is observed at 80 K for x = 0.60 and y = 0.50. Diamagnetic volumes ranging from 15% to 25% are obtained for those two samples.

The oxide $Pb_2Sr_{2-x}La_xCu_2O_{6+\delta}$ discovered by *Zandbergen* et al. [5.75] exhibits a great similarity with $Pb_2Sr_2Ca_{0.5}Y_{0.5}Cu_3O_8$. It is indeed also characterized by a mean oxidation state of copper smaller than 2 in spite of its superconducting properties ($T_c \simeq 30K$). Its structure (Fig.5.36a)

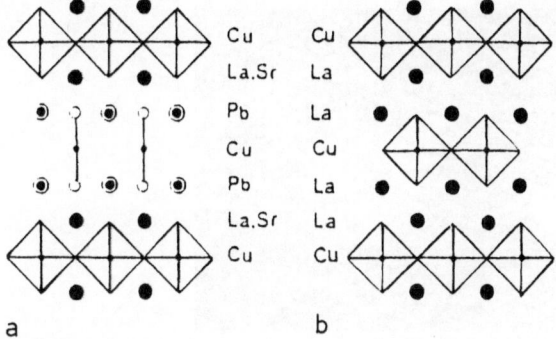

Fig.5.36. The idealized structure of $Pb_2Sr_{2-x}La_xCu_2O_{6+\delta}$ (a) compared to that of La_2CuO_4-type oxides (b)

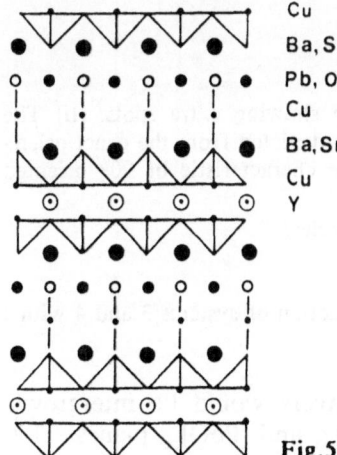

Fig.5.37. Schematized structural model for $PbBaYSrCu_3O_7$

can be described as the intergrowth of single rock salt layers [(Sr, La, Pb) O]$_\infty$ with single oxygen-deficient perovskite layers. Two sorts of copper layers must be distinguished: octahedral layers, i.e. nondeficient perovskite layers as in La_2CuO_4, and layers of CuO_2 sticks as in $Pb_2Sr_2Ca_{0.5}Y_{0.5}$· Cu_3O_8. Thus, this structure is derived from that of La_2CuO_4 (Fig.5.36b) by removing rows of oxygen atoms in alternate perovskite layers. Again such a distribution of the oxygen atoms implies a copper disproportionation with monovalent copper in the CuO_2 sticks and mixed valence copper, Cu(II)-Cu(III), in the superconductive octahedral layers.

$PbBaYSrCu_3O_{8-\delta}$, recently isolated by *Rouillon* et al. [5.76a] represents a new member of the intergrowth family. The structure of this phase has not been completely elucidated. Although its structure consists of triple copper layers, the geometry of the layer seems to be different from that of the other members. A structual model deduced from neutron-diffraction studies [5.76b] is proposed in Fig.5.37. The limit oxide $PbBaYSrCu_3O_7$ would consist of triple copper layers built up from two pyramidal copper layers between which yttrium ions are interleaved and one layer of Cu^IO_2

141

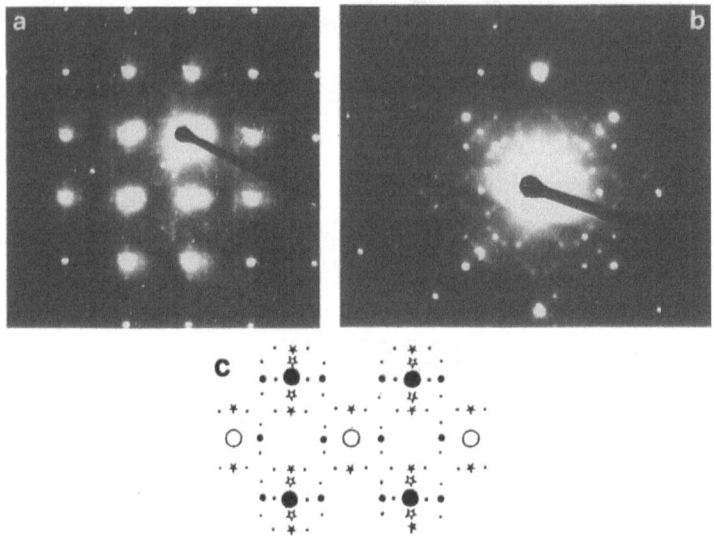

Fig.5.38a-c. PbBaYSrCu$_3$O$_{8-\delta}$. (a) [001] ED pattern showing extra spots. (b) The extra spots are clearly visible when the zone is slightly deviated from the exact orientation, (c) Two sets of satellites are visible; they are characteristic of 90° oriented domains.

o : basic spots at level 1/2 are represented as open circles.

★ : systems 1 and 2 (~ 4 d$_{110}$)

☆ : systems 3 and 4 (~ 8 d$_{110}$)

• : small spots: extra spots arising from double diffraction of systems 3 and 4 with 1 and 2 as secondary origins

sticks. These triple oxygen-deficient copper layers would be intergrown with single rock salt layers involving [(Ba,Sr)O]$_{\infty}$ and [PbO]$_{\infty}$ planes. This phase does not superconduct but the partial replacement of yttrium by calcium, leading to the phase PbBaSrY$_{1-x}$Ca$_x$Cu$_3$O$_{8-\delta}$, allows 1%-2% diamagnetism to be detected below 50 K. Bulk superconductivity was recently confirmed for x = 0.3 with T$_c$ = 37 K [5.77].

The electron microscopy study of those oxides shows that PbBa-YSrCu$_3$O$_{8-\delta}$ is characterized, like the bismuth cuprates, by an incommensurate structure. All the ED patterns exhibit extra spots whose intensity varies from one crystal to another. Such satellites are particularly visible along a zone slightly deviated from the exact [001] orientation (Fig.5.38). The satellites, which look like a two-dimensional modulation, correspond to a complex superposition of two systems characterized by variants oriented at 90°. Both systems are set up along [110] and [1$\bar{1}$0]. The more intense satellites exhibit a wavelength of modulation slightly variable but close to 4 times d$_{110}$ (2.8Å). In some crystals, with a longer exposure time, satellites in the position (1/8)(hh0) are visible (Fig.5.38b), and also extra spots arise from double diffraction phenomena. The idealized section [001] is reproduced in Fig.5.38c. The [1$\bar{1}$0] ED pattern is shown in Fig.5.39, where the incommensurate nature of the modulation is clearly visible. On the other hand, most of the crystals of the superconductive phase Pb$_2$Sr$_2$·

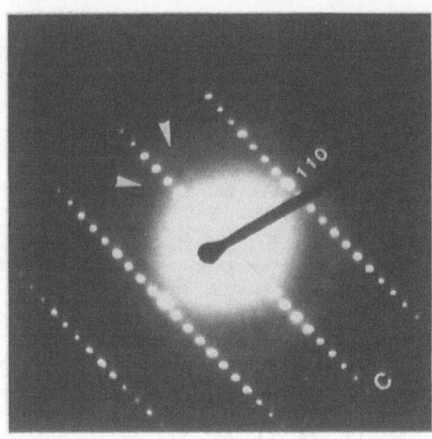

Fig.5.39. $PbBaYSrCu_3O_8$-δ. $[1\bar{1}0]$ ED pattern with extra spots in incommensurate positions ($\frac{1}{2}\ell$ and $4.27d_{110}$)

$Y_{1-x}Ca_xCu_3O_8$ do not exhibit such satellites characteristic of an incommensurate structure. These results can be explained by the stereoactivity of the $6s^2$ lone pair of Pb(II). In the Bi(III) intergrowths, as well as in BaPbSrYCu$_3$O$_{8-\delta}$, Pb(II) and Bi(III) are contained in a rock-salt layer, so that their $6s^2$ lone pair has no room for an extension. This results in a distortion of the BiO_6 or PbO_6 octahedra, and a modulated displacement of the Bi(III) or Pb(II) cations and oxygen atoms in these layers. In contrast, in $Pb_2Sr_2Y_{1-x}Ca_xCu_3O_8$, the $6s^2$ lone pair of Pb(II) can spread in the Cu(I) layers (Fig.5.35), so that no distortion of the rock-salt layer will be involved.

The oxide $Pb_{0.5}Sr_{2.5}Y_{1-x}Ca_xCu_2O_{7-\delta}$ [5.78, 79] is the only pure lead cuprate which exhibits double copper layers. Its structure (Fig.5.40) belongs to the "1212" type already described for $TlA_2CaCu_2O_7$ (A = Ba, Sr). Mixed lead-strontium monolayers $[Pb_{0.5}Sr_{0.5}O]_\infty$ replace the $[TlO]_\infty$ monolayers in the double rock-salt slabs. However, an important difference with the thallium cuprates concerns the existence of oxygen vacancies at the boundary between the perovskite and rock-salt-type layers. It results in the existence of several superstructures which can be interpreted in terms of ordering of lead, strontium and anionic vacancies. An isostructural solid solution has been isolated by completely replacing strontium by calcium in the mixed monolayers, leading to the formulation $Pb_{0.50}Ca_{0.50}Sr_2Y_{1-x}Ca_x \cdot Cu_2O_{7-\delta}$ [5.80]. Both oxides exhibit a wide homogeneity range, $0 \leq x \leq 0.60$, but curiously superconductivity is only observed for $0.50 \leq x \leq 0.60$.

In all cases, one again observes very broad resistive transitions, as shown for instance for $Pb_{0.50}Sr_{2.50}Y_{0.50}Ca_{0.5}Cu_2O_{7-\delta}$ (Fig.5.41). Moreover, it can be seen that the experimental conditions, especially the temperature, influence the superconducting properties of the materials dramatically. This behaviour is confirmed by the magnetic study performed with a SQUID magnetometer (Fig.5.42). For instance, one observes that annealing under an oxygen flow at about 500 °C increases T_c (onset) from 60 to 70 K, and that, in the same way, synthesis performed at 825 °C leads to much better results than at 870 °C. An interesting feature deals with the

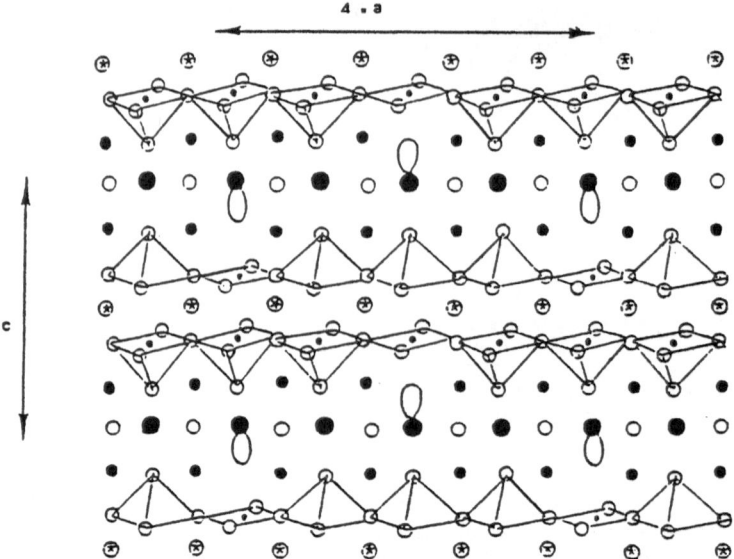

Fig.5.40. $Pb_{0.5}Sr_{2.5}Y_{1-x}Ca_xCu_2O_{7-\delta}$: model of a perfect ordering of the Pb(II) and Sr in the intermediate rock salt layers. This idealized drawing shows that 4axc superstructure can be built up. Other models can be obtained by translation of the adjacent layers

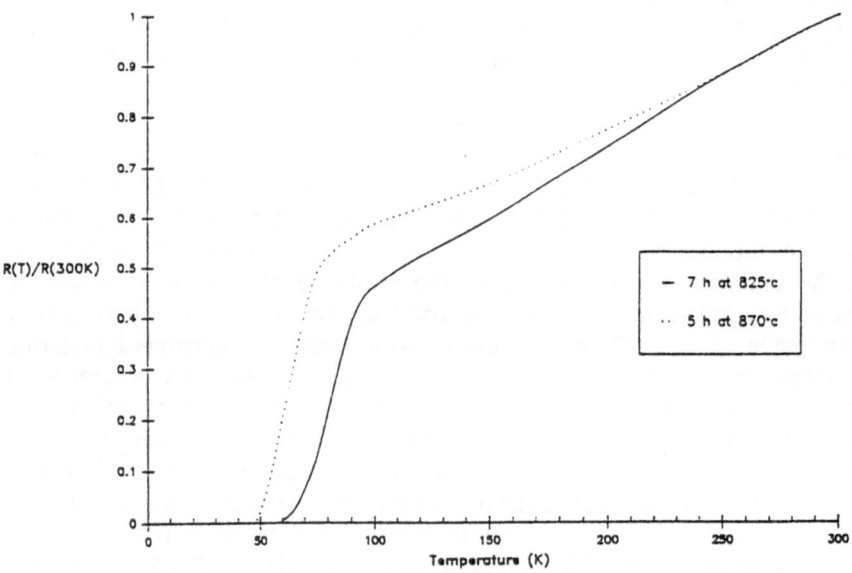

Fig.5.41. Temperature dependence of the ratio R(T)/R (300K) for different synthesis conditions (unannealed samples) for $Pb_{0.5}Sr_{2.5}Y_{1-x}Ca_xCu_2O_{7-\delta}$

compositions corresponding to x greater than 0.60 which are not single phased, but which present superconductivity at high temperatures. This is indeed the case of the composition "x = 0.90" which presents a diamagnetic signal corresponding to 3%-4% of the sample volume and a T_c of 104 K

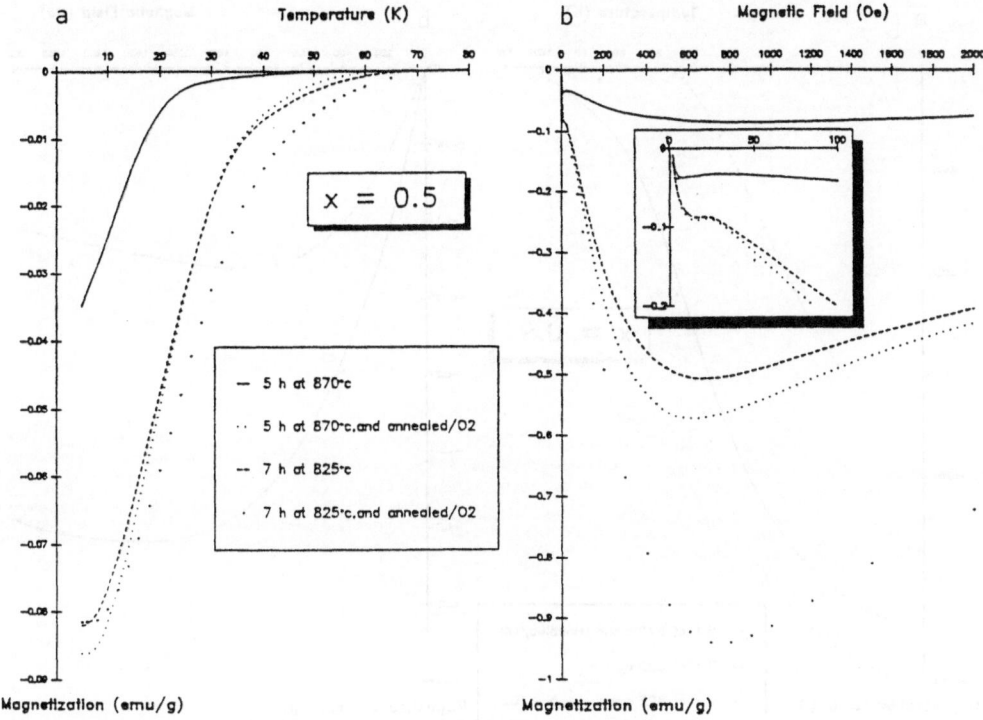

Fig.5.42. (a) Magnetization vs T for $Pb_{0.5}Sr_{2.5}Y_{1-x}Ca_xCu_2O_{7-\delta}$ (y = 0.5) at 10 Oe for different thermal treatments; (b) magnetization vs H (H < 2000 Oe) at 5 K for the same samples. Insert shows the low field part of the curves

(Fig.5.43); of course, for this latter composition, zero resistance cannot be reached, owing to the large volume of the insulating phase belonging to the Sr_2PbO_4 type which prevents the percolation. Nevertheless, this confirms the possibility of superconductivity up to 100K in these systems, and suggests the possible existence of another superconductive lead cuprate.

Taking into account the redox property of the system Pb(II)/Pb(IV), and the role of the $6s^2$ lone pair of Pb(II), it is clear that other members of this series should be synthesized in the near future.

5.4 Layered Cuprates Involving Double Fluorite-Type Layers

Consideration of the space between the pyramidal copper layers in 2212- or 1212-type oxides shows that calcium or yttrium cations and the surrounding oxygens of the basal planes of the CuO_5 pyramids form a layer of edge-sharing distorted CuO_8 cubes similar to those observed in the fluorite structure (Fig.5.44a). This suggests the possibility of replacing these single fluorite-type layers by double layers (Fig.5.44b). Two series of oxides with double fluorite layers [5.81-83] were synthesized simultaneously for the first time. The oxides $Tl_{1-x}A_{2-y}Ln_2Cu_2O_9$ [5.81] with Ln = Pr,

145

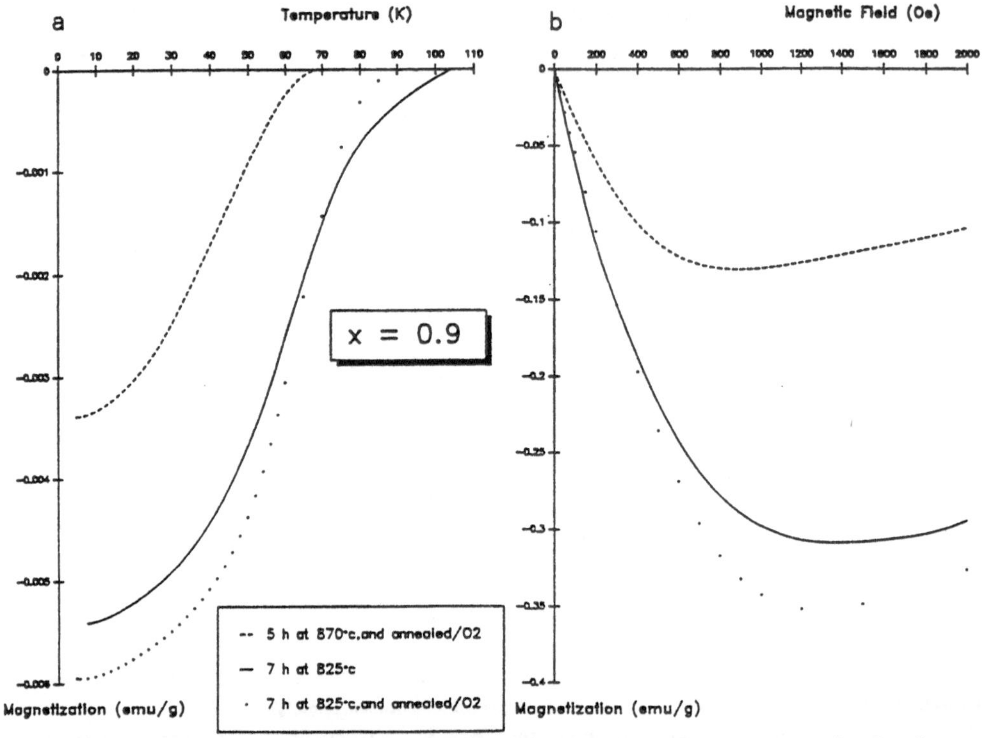

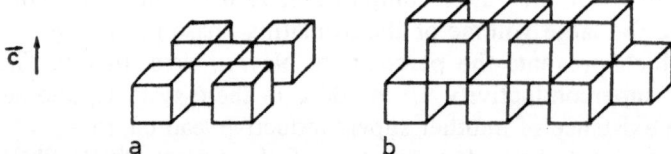

Fig.5.43a,b. Multiphase sample of nominal composition $Pb_{0.5}Sr_{2.5}$ $Y_{1-x}Ca_xCu_2O_{7-\delta}$ (x = 0.9): (a) magnetization vs T at 10 Oe for different thermal treatments; (b) magnetization vs H (H < 2000 Oe) at 5K

Fig.5.44. Schematic drawing of single fluorite layers (a) and of double fluorite layers (b) (edge sharing AO_8 cubes)

Nd, Ce and A = Sr, Ba crystallize in the tetragonal system (a ≃ 3.9Å, c ≃ 30Å). As can be seen from the structure of two of them, $Tl_{1.30}Sr_{1.6}$·$Pr_2Cu_2O_9$ and $Tl_{1.3}Ba_{1.7}Pr_2Cu_2O_9$ (Fig.5.45), one observes a triple intergrowth of double pyramidal copper layers (oxygen deficient perovskite layers) with double fluorite-type layers and double rock-salt-type layers. The latter are built up from one thallium monolayer sandwiched between two "BaO" or "SrO" layers. In these structures some of the barium (or strontium) sites are occupied by the excess thallium. In fact, this structure is directly derived from the 1212-type oxides $TlBa_2CaCu_2O_7$ by just replacing the $[CaO_2]_\infty$ single fluorite layer by a $[Pr_2O_4]_\infty$ double fluorite layer, leading to the idealized formula "$TlBa_2Pr_2Cu_2O_9$". Contrary to the 1212 thallium cuprates, this oxide does not superconduct.

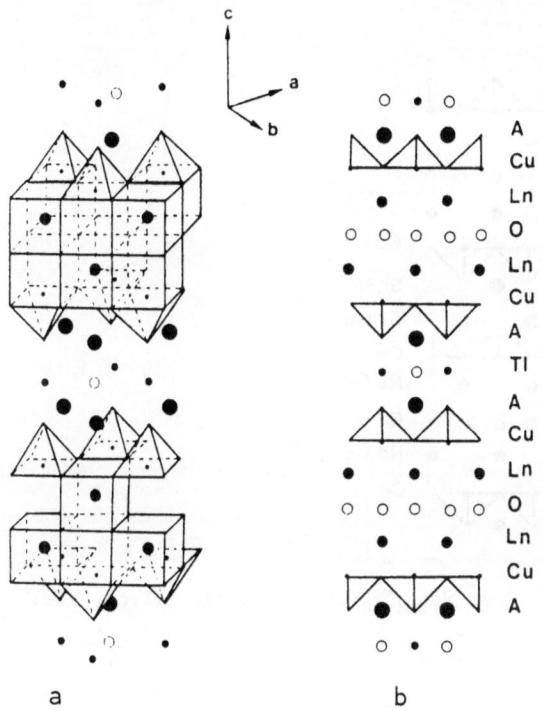

a   b

Fig.5.45. Crystal structure of the oxides $Tl_{1+x}A_{2-y}Ln_2Cu_2O_9$ (A = Sr, Ba; Ln = Pr, Nd, Ce)

The oxide $Nd_{2.64}Sr_{0.82}Ce_{0.54}Cu_2O_{8-y}$ [5.82, 83] called T^* is also tetragonal (a $\simeq$ 3.9Å, c $\simeq$ 12.5Å). Its structure (Fig.5.46) is also built up from double pyramidal copper layers, double $[(Nd, Ce)_2O_4]_\infty$ layers and single rock salt layers $[(Nd, Sr)O]_\infty$. In fact, this structure can be derived from the $La_2CaCu_2O_6$-type structure (Fig.2.22) by replacing the $[CaO_2]_\infty$ single layer by a double fluorite-type layer $[(Nd, Ce)_2O_4]_\infty$.

The oxides $Bi_2Sr_2Ln_{2-x}Ce_xCu_2O_{10}$ and $Tl_2Ba_2Ln_{2-x}Ce_xCu_2O_{10}$ [5. 84] which were synthesized for Ln = Sm, Eu or Gd correspond to the replacement of thallium monolayers by thallium or bismuth bilayers in the idealized "$TlA_2Ln_2Cu_2O_9$"structure". They crystallize in the tetragonal system (a $\simeq$ 3.88, c $\simeq$ 17.28Å) for thallium. In contrast to the Tl derivatives, the Bi compounds are slightly orthorhombic. Their structure (Fig.5.47) can be described as a triple intergrowth of double copper layers, with double fluorite type layers and triple rock-salt-type layers. Thus this structure can be derived from the 2212 oxides, $Bi_2Sr_2CaCu_2O_8$ and $Tl_2Ba_2CaCu_2O_8$ (Fig.5.6a), by just replacing $[CaO_2]_\infty$ single fluorite layers by $[Ln_2O_4]_\infty$ double fluorite layers. Thermal treatment at 823-873K of Bi oxides under an oxygen pressure of 80 bars allowed superconducting properties to be obtained with onset T_c around 25 K.

The oxides $Pb_2Sr_2LnCeCu_3O_{10+\delta}$ [5.85] with Ln = Pr,Nd,Sm,La are orthorhombic (b $\simeq$ $a_p\sqrt{2}$ ~ 5.4Å, c $\simeq$ 36.9Å). They represent the most complex intergrowth in this family. Their structure (Fig.5.48) consists of dou-

147

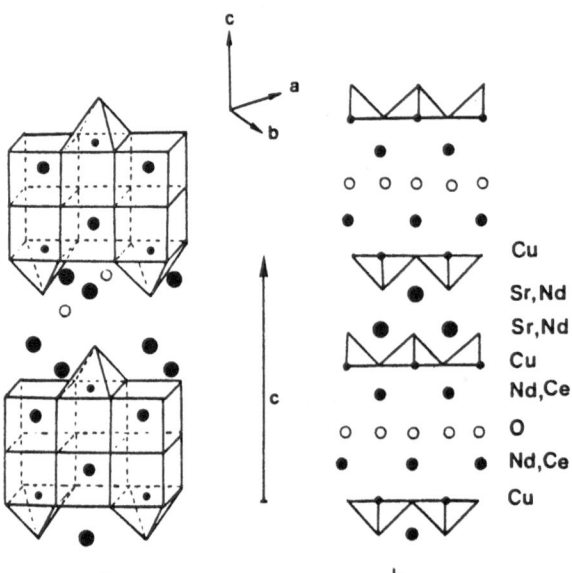

a   b

Fig.5.46. Crystal structure of the T^* phase $Nd_{2.64}Sr_{0.82}Ce_{0.54}Cu_2O_{8-y}$. (From [5.82])

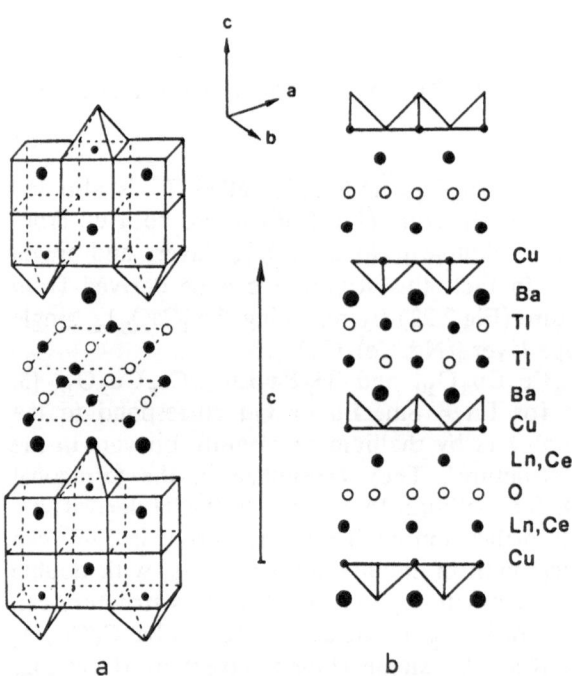

a   b

Fig.5.47a,b. Crystal structure of the oxides $Tl_2Ba_2Ln_{2-x}Ce_xCu_2O_{10}$. (a) Perspective view. (b) Projected view onto (010). (From [5.84])

ble pyramidal copper layers, $[LnCeO_4]_\infty$ double fluorite-type layers, single layers of CuO_2 sticks and single rock-salt-type layers $[(Pb,Sr)O]_\infty$. In fact these oxides are derived from the $Pb_2Sr_2Ca_{0.5}Y_{0.5}Cu_3O_8$ structure [5.73]

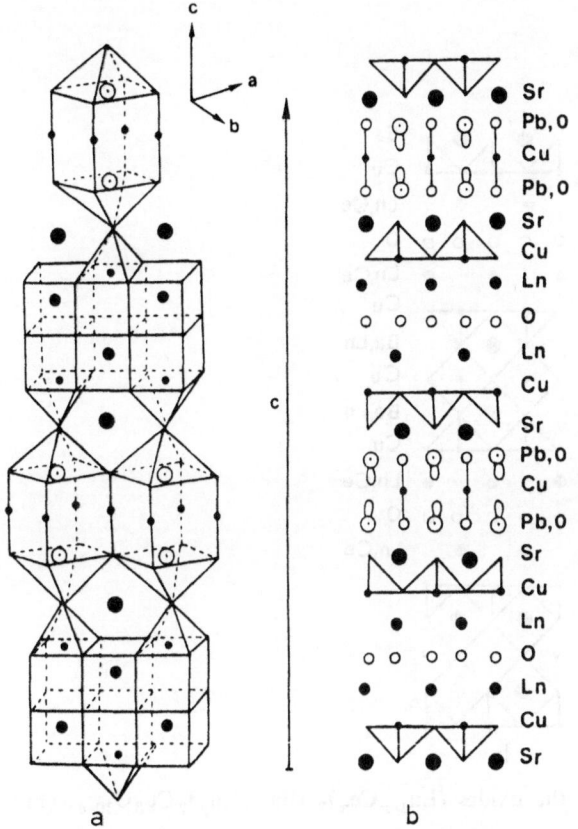

Fig.5.48a,b. Crystal structure of the oxides $Pb_2A_2LnCeCu_3O_{10+\delta}$ (A = Sr, Ba; Ln = Pr, Nd, Sm, La). (a) Perspective view. (b) Projected view onto (010). (From [5.85])

(Fig.5.35a) with the single $[Ca_{0.5}Y_{0.5}O_2]_\infty$ fluorite layers replaced by double fluorite type layers $[LnCeO_4]_\infty$. Thus, they can also be described as an intergrowth of the $(Nd, Sr)_2(Nd, Ce)$ Cu_2O_8-type structure (Fig.5.46) with the oxygen deficient K_2NiF_4-type structure $A_2CuO_2\blacksquare_2$. The disproportionation of copper, leading to the mixed valency Cu(II)-Cu(III) in the pyramidal layers while monovalent copper forms the CuO_2 sticks, appears as most likely. In spite of that, no superconductivity has been detected in this oxide until now. This may be due to the excess oxygen with respect to the "O_{10}" ideal structure, which may be distributed over the O(1) sites at the same level as Cu(1). One must indeed bear in mind that the introduction of an excess oxygen in $Pb_2Sr_2Ca_{0.5}Y_{0.5}Cu_3O_8$ [5.73] kills the superconducting properties of this material.

The oxides $(Ln_{1-x}Ce_x)_2(Ba_{1-y}Ln_y)_2Cu_3O_{10-\delta}$ [5.86, 87] do not exhibit any rock salt layers. Those tetragonal oxides (a ~ 3.9Å, c ~ 28.6Å) can be described as an intergrowth of oxygen-deficient triple perovskite copper layers with double fluorite-type layers. In fact the idealized structure $(Ln_{1-x}Ce_x)_2(Ba_{1-y}Ln_y)_2Cu_3O_9$ (Fig.5.49) corresponds to the replacement of the $[YO_2]_\infty$ single fluorite layers by $[Ln_{2-2x}Ce_{2x}O_4]_\infty$ double fluorite

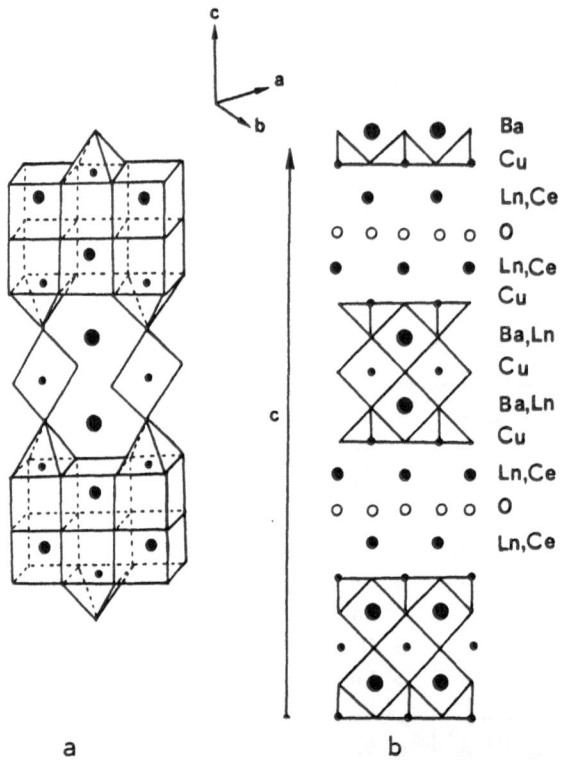

Fig.5.49. Crystal structure of the oxides $(Ln_{1-x}Ce_x)_2 (Ba_{1-y}Ln_y)_2 Cu_3 O_{10-\delta}$. (From [5.86])

type layers in the "123"-type $Ba_2 YCu_3 O_7$ structure. The synthesis of those oxides under high oxygen pressures of several hundred bars allows critical temperatures ranging from 36 to 43 K to be reached.

5.5 Structural Relationships

A feature common to all the high-T_c cuprate superconductors is the high covalency of the Cu-O bonds and two-dimensional character of the copper-oxygen framework closely related to that of the perovskites. Thus a large family of layered cuprates with the general formula $[ACuO_{3-x}]_m \cdot [AO]_n$ can be considered. Such a class of oxides corresponds to the intergrowth of multiple oxygen-deficient peroskite layers $ACuO_{3-x}$, with rocksalt-type layers AO. They can be represented by the symbol [m, n], where m and n represent the number of copper layers and $[AO]_\infty$ layers in the perovskite and rock-salt slabs, respectively. All these different members are summarized in Table 5.1. $La_{2-x}M_xCuO_4$, $Bi_2Sr_2CuO_6$, $Tl_2Ba_2CuO_6$ and $TlSr_2CuO_{5-\delta}$ represent the m = 1 members formed of single perovskite layers according to the formulae $[La_{1-x}M_xCuO_{3-\delta}] [La_{1-y}M_yO]$, $[SrCuO_3]$

Table 5.1. The [m, n] layered cuprates: P/RS-intergrowths $(ACuO_{3-x})_m(AO)_n$

n m	1	2	3
1	[1,1] $La_{2-x}A_xCuO_4$ (A=Ca, Sr, Ba) T_c = 20–40K	[1,2] $TlBa_2CuO_{5-\delta}$ Nonsuperconducting $TlSr_2CuO_{5-\delta}$ traces	[1,3] $Tl_2Ba_2CuO_6$ Nonsuperconducting to $T_c \simeq 92$K
	La_2CuO_4 $T_c \simeq 38$K $Pb_2Sr_{2-x}La_xCu_2O_{6+\delta}$ T_c = 30K	$Tl_{1-x}Pr_xSr_{2-y}Pr_yCuO_{5-\delta}$ $T_c \simeq 40$K (x=0.2, y=0.4) $Tl_{0.5}Pb_{0.5}Sr_2CuO_{5-\delta}$ Nonsuperconducting to $T_c \simeq 40$K $TlBa_{1+x}La_{1-x}CuO_{5-\delta}$ $T_c \simeq 50$K (x=0.2)	$Bi_2Sr_2CuO_6$ $T_c \simeq 10$–22K
1.5	[1.5, 1] $Pb_2Sr_2Y_{1-x}Ca_xCu_3O_8$ T_c = 50K (x=0.5) to 80K $Pb_{2-x}Bi_2Sr_2Y_{1-y}Ca_yCu_3O_8$ T_c = 79K (x=0.6; y=0.5, 1)		
2	[2, 1] $La_{2-x}A_{1+x}Cu_2O_6$ A = Ca, Sr Nonsuperconducting to $T_c \simeq 60$K	[2, 2] $TlBa_2Ca_2Cu_2O_7$ $T_c \simeq 50$ to 112K $TlSr_2CaCu_2O_7$ $T_c \simeq 50$K $Tl_{0.5}Pb_{0.5}Sr_2CaCu_2O_7$ $T_c \simeq 85$K $TlBa_2LnCu_2O_7$ Ln = Pr, Y, Nd Nonsuperconducting $Pb_{0.5}Sr_{2.5}Y_{0.5}$ $Ca_{0.5}Cu_2O_{7-\delta}$ $T_c \simeq 59$K	[2, 3] $Tl_2Ba_2CaCu_2O_8$ $T_c \simeq 97$ to 108K $Tl_{3-4x/3}Ba_{1+x}LnCu_2O_8$ (x=0.25, Ln=Pr, Nd, Sm) Nonsuperconducting $Bi_2Sr_2CaCu_2O_8$ $T_c \simeq 85$K $Tl_2Sr_2CaCu_2O_8$ $T_c \simeq 44$K $Bi_{2-x}Pb_xSr_2Ca_{1-x}Y_xCu_2O_8$ $T_c \simeq 85$K to nonsuperconducting
3	[3, 1] $PbBaYSrCu_3O_{8-\delta}$ Nonsuperconducting $PbBaSrY_{0.7}Ca_{0.3}Cu_3O_7$ T_c = 37K	[3, 2] $TlBa_2Ca_2Cu_3O_9$ $T_c \simeq 120$K $Tl_{0.5}Pb_{0.5}Sr_2Ca_2Cu_3O_9$ $T_c \simeq 120$K	[3, 3] $Tl_2Ba_2Ca_2Cu_3O_{10}$ $T_c \simeq 120$K $Bi_{2-x}Pb_xSr_2Ca_2Cu_3O_{10}$ $T_c \simeq 110$K
4	[4, 1]	[4, 2] $TlBa_2Ca_3Cu_4O_{11}$ $T_c \simeq 110$K	[4, 3] $Tl_2Ba_2Ca_3Cu_4O_{12}$ $T_c \simeq 104$–108K
5	[5, 1]	[5, 2] $TlBa_2Ca_4Cu_5O_{13}$ T_c = 105K	[5, 3] $Tl_2Ba_2Ca_4Cu_5O_{14}$ T_c = 105K

$[(BiO)_2 (SrO)]$, $[BaCuO_3]$ $[(TlO)_2(BaO)]$ and $[SrCuO_3]$ $[(TlO)(SrO)]$. They differ from one another in the thickness of the rock-salt-type layer. One observes single $[La_{1-y}M_yO]_\infty$ layers in La_2CuO_4-related oxides (n = 1) and triple rock-salt layers (n = 3) in the other two oxides formed by a monolayer of [SrO] or [BaO] and bilayers of $[(BiO)_2]$ or $[(TlO)_2]$. The m = 2 members are represented by the five oxides $TlBa_2CaCu_2O_7$, $Bi_2Sr_2Ca\cdot Cu_2O_8$, $Tl_2Ba_2CaCu_2O_8$, $La_{2-x}A_{1+x}Cu_2O_6$ and $Pb_{0.5}Sr_{2.5}Y_{1-x}Ca_xCu_2\cdot O_{7-\delta}$. Here, the double oxygen-deficient perovskite layers (m=2) are formed by two pyramidal $[CuO_{2.5}]_\infty$ sheets, leading to formulae$[BaCa\cdot (CuO_{2.5}\blacksquare_{0.5})_2]$ $[(SrO) (BiO)_2]$ and $[BaCa(CuO_{2.5}\blacksquare_{0.5})_2]$ $[(BaO) (TlO)_2]$, where $\blacksquare$ represents an oxygen vacancy. Thus, $TlBa_2CaCu_2O_7$ and $Pb_{0.5}\cdot Sr_{2.5}Y_{1-x}Ca_xCu_2O_{7-\delta}$ exhibit a double rock-salt layer (n = 2) $[(BaO) (TlO)]_\infty$ or $[(SrO)(Pb_{0.5}Sr_{0.5})]_\infty$ whereas triple rock-salt layers (n = 3) formed of double layers $[(BiO)_2]$ or $[(TlO)_2]$ and a monolayer of [SrO] or [BaO] are present in $Bi_2Sr_2CaCu_2O_8$ and $Tl_2Ba_2CaCu_2O_8$. Triple oxygen-deficient perovskite layers (m = 3) are obtained in $TlBa_2Ca_2Cu_3O_9$, $Tl_2Ba_2Ca_2Cu_3O_{10}$ and $Bi_2Sr_2Ca_2Cu_3O_{10}$. In these structures, the triple perovskite layer is formed of two $[CuO_{2.5}]_\infty$ pyramidal layers and one $[CuO_2]_\infty$ slab of CuO_4 groups. There are two rock-salt slabs (n = 2) in $TlBa_2Ca_2Cu_3O_9$ and three (n = 3) in the other two oxides. Thus, the general formula of these three oxides can be written as $[A'Ca_2(CuO_{2.5}\blacksquare_{0.5})_2]$ $[(AO)_1$ or $(A'O)_2]$ where A' = Sr or Ba and A = Bi or Tl. $Tl_2Ba_2Ca_3\cdot Cu_4O_{12}$ corresponds to the member m = 4 of the series: its quadruple perovskite layers are formed of two $[CuO_2]_\infty$ slabs sandwiched between two pyramidal $[CuO_{2.5}]_\infty$ slabs and its rock-salt-type layers correspond to n = 3, i.e., one thallium bilayer sandwiched between barium monolayers. Thus the formula of this oxide is $[BaCa_3(CuO_{2.5}\blacksquare_{0.5})_2 (CuO_2(\blacksquare)_2]$ $[(BaO) (TlO)_2]$. The oxide $Pb_2Sr_2Y_{0.5}Ca_{0.5}Cu_3O_8$ corresponds to the simultaneous presence of double (m = 2) and single (m = 1) copper layers leading to the mean value m = 1.5, intergrown with single (n = 1) rock-salt layers; the latter are built up from one SrO layer and one PbO layer.

One can also subdivide this large family of intergrowths into different series which differ in the multiplicity of the rock-salt-type layers:

• The oxides $Tl_2Ba_2Ca_{m-1}Cu_mO_{2m+4}$ and $Bi_2Sr_2Ca_{m-1}Cu_mO_{2m+4}$, whose triple rock-salt layers are formed of thallium or bismuth bilayers.

• The oxides $TlBa_2Ca_{m-1}Cu_mO_{2m+3}$ and $TlSr_2Ca_{m-1}Cu_mO_{2m+3}$, whose double rock-salt-type layers are formed of thallium monolayers.

• The oxides $A_2(Ca, Y, Sr)_{m-1}Cu_mO_{2m+2}$, which are characterized by single rock-salt layers and are up to now only represented by four oxides: La_2CuO_4-type (m = 1), $La_2SrCu_2O_6$ (m = 2), $PbBaSrYCu_3O_{8-\delta}$ (m = 3) and $Pb_2Sr_2YCu_3O_8$ (m = 1.5: double intergrowth).

$YBa_2Cu_3O_7$ and related "123" oxides can be considered to constitute $[\infty, 0]$ members of the general structural class of oxygen-deficient perovskites described above. The ordering of oxygen and anionic vacancies in the perovskite framework leads to a layer structure which bears considerable similarity to the other cuprates. We can recognize similar layers of corner-sharing CuO_5 pyramids interleaved with planes of yttrium ions. The main difference in the "123" oxides is that two pyramidal layers are connected through rows of corner-sharing CuO_4 square-planar groups, resulting in the formula $[BaY(CuO_{2.5} \blacksquare_{0.5})_2] [BaCuO_2 \blacksquare]$.

The double fluorite-type layered cuprates can be deduced from the above intergrowths by intercalation of one LnO_2 layer within the $[ACu_2O_5]_\infty$ double copper layers so that the double intergrowths $(AO)_n$ $(ACuO_{2.5})_2$ lead to the triple intergrowths $(AO)_n LnO_2(ACuO_{2.5})_2$. In the same way the interpolation of one LnO_2 layer between the double pyramidal copper layer of $YBa_2Cu_3O_7$ leads to the ideal structure $Y(LnO_2)\cdot Ba_2Cu_3O_7$.

Another way of describing all these layered cuprates consists of considering that the principles which govern the crystal chemistry of these compounds are founded on the easy association of three structures: perovskite, rock salt and fluorite. Starting from a single fluorite-type layer built up from corner-sharing AO_8 cubic groups (Fig.5.50a) one can generate $[CuO_2]_\infty$ layers of corner-sharing CuO_4 square planar groups by adding copper on both sides of the layer (Fig.5.50b), leading to the $Sr_{0.15}\cdot$

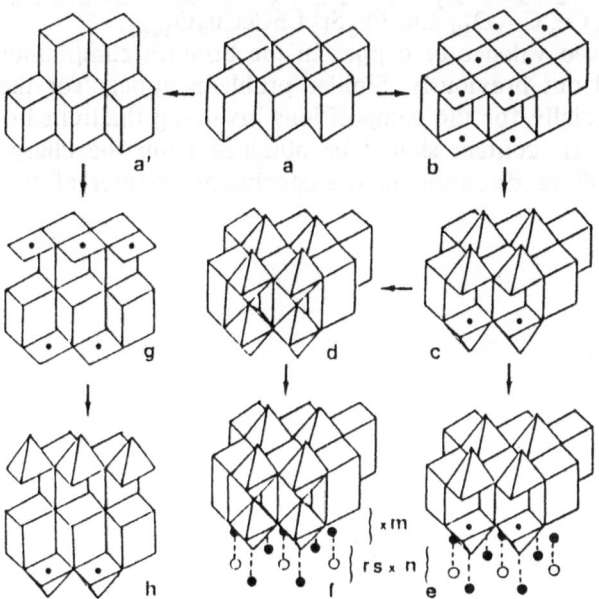

Fig.5.50a–h. Structural filiation in layered copper oxides. (a) Single fluorite layers (edge sharing AO_8 cubes); (a') double fluorite layer; (b) formation of $[CuO_2]_\infty$ layers (corner-sharing CuO_4 square planar groups; copper is symbolized by small filled circles) leading to $Sr_{0.15}Ca_{0.85} CuO_2$ structure; (c) formation of pyramidal layers (corner-sharing CuO_5 pyramids) leading to $YBaFeCuO_5$ structure; (d) formation of the octahedral layers characteristic of the stoichiometric perovskite $LaCuO_3$ (note the tetragonal distortion of the AO_8 groups); (e), (f) intergrowths of (c) and (d) with rock-salt-type layers (oxygen and A cation are symbolized by empty and large filled circles, respectively); (g) formation of the Nd_2CuO_4-type structure by adding copper to a double fluorite; (h) formation of pyramidal layers from the Nd_2CuO_4-type structure

153

$Ca_{0.85}CuO_2$ -type structure [5.88]. The addition of oxygen to this latter structure allows $[Cu_2O_5]_\infty$ pyramidal layers to be created (Fig.5.50c), leading to the $YBaFeCuO_5$-type structure [5.89]. Further introduction of oxygen into the $YBaFeCuO_5$-type structure leads to the stoichiometric perovskite ABO_3 in which the fluorite cage exhibits a tetragonal distortion (Fig.5.50d), A being in 12-fold coordination and copper in octahedral coordination as shown for $LaCuO_3$. The juxtaposition of AO layers to the $YBaFeCuO_5$-type layers and to the ABO_3 stoichiometric perovskite leads to the formation of an intergrowth between rock-salt-type layers and the perovskite structure (Fig.5.50e,f). Thus, all those oxides which can be represented by the general formula $[ACuO_{3-x}]_m^P[AO]_n^{RS}$ (P:perovskite; RS:rock salt; m,n are integers) form a very large family. They correspond to the La_2CuO_4-type oxides, the bismuth cuprates, the $YBa_2Cu_3O_7$ family, the thallium cuprates and the lead cuprates. In the same way, starting from a double fluorite-type layer (Fig.5.50a), one obtains the Nd_2CuO_4-type structure just by adding copper in square planar coordination on both sides of the layer (Fig.5.50g). Addition of oxygen to this latter layer leads to the formation of pyramidal layers (Fig.5.50k) which are characteristic of $Tl_{1+x}A_{2-y}Ln_2Cu_2O_9$, $Ba_{2-y}Ln_yLn_{2-x}Ce_xCu_3O_{10}$, $Bi_2Sr_2Ln_{2-x}Ce_xCu_2O_{10}$, $Tl_2Ba_2Ln_{2-x}Ce_xCu_2O_{10}$ and $Pb_2Sr_2LnCeCu_3O_{10+\delta}$.

The issue of the mixed valence of copper in the bismuth compounds has already been pointed out previously. Similar problems appear for the thallium oxides, and especially for the compositions involving thallium bilayers, for which no Cu(III) content should be obtained from the charge balance. This question will be discussed in the concluding chapter of this book.

6. Extended Defects in Superconducting Oxides: High-Resolution Electron Microscopy

Oxygen-deficient perovskite-type structure, ion ordering and intergrowth of perovskite with rock-salt-type layers are keywords for the mechanisms, described in Chaps.2 and 5 that generate the high-T_c superconductor structures. They are known by solid-state chemists to favor nonstoichiometric features. The phenomena thus generated often play a prominent role in the material properties and there are numerous examples that illustrate their influence on the physical (electron transport, magnetism) as well as the chemical properties. An understanding of the relationships between properties and structure thus needs a good knowledge of the crystalline state. In such an investigation, High-Resolution Electron Microscopy (HREM) is obviously an essential tool, since it achieves direct imaging of the complex features. It should be emphasized, however, that this attractive technique needs to be used carefully, both experimentally (where the required conditions are perfectly defined [6.1]) and in the interpretation.

In the case of all the perovskite-related compounds, and so especially in the superconducting copper oxides, the wide range of potential phenomena implies that interpretation will be done from both microscopy and chemistry viewpoints. Such various and essential issues as atom size, possible coordinations, mixed valence, electronegativity and covalency have indeed to be considered. Moreover, although some images, for example those related to layer stacking sequences, can be interpreted in an intuitive way, the large majority need to be compared with computed images calculated for different experimental parameters (electron microscopy) and structural hypotheses (crystal chemistry). Considering the perovskite structure, and more especially oxygen-deficient frameworks, one finds that the contribution of the oxygen atoms to the potential is indeed very weak. Variations in the contrast more often arise from the cation displacements related to the existence of vacancies rather than from the actual absence of the oxygen atoms. The aim is to find the real origin of such contrast variations, and the only possible solution lies in calculated images.

A comprehensive study of the relationships between structure and properties in the high-T_c superconductors, with the help of high-resolution electron microscopy, is illustrated here in sections related firstly to the two building principles (ordering in the perovskite framework, Chap.2 and intergrowth of two structural types, Chap.5) and to the origin of extra reflections in the ED patterns. With this aim in mind, the images of "perfect" crystal surfaces are often omitted here, although they were recorded and used for the usual contrast interpretation.

6.1 YBa$_2$Cu$_3$O$_7$-Type Superconductors: Ordering in the Perovskite Framework

The ordering of the oxygen atoms and oxygen vacancies in the plane of the median copper layer between the barium atoms is the essential structural feature of the "123" phases. As shown in Chap.3, the variations of the electron transport properties with the oxygen nonstoichiometry are particularly spectacular in these systems, since both T_c and J_c values are strongly dependent on the thermal history of the samples. However, a straightforward correlation with the oxygen content, $O_{7-\delta}$, or with the symmetry is not sufficient to explain the behavior of the phases $ABa_2Cu_3O_{7-\delta}$, where the A cation may be yttrium or lanthanum. Previous HREM investigations of various oxygen-deficient perovskites showed that the "homogeneity" of the matrix must be considered as an important factor to be integrated into the interpretation of the physical properties. Its role is prominent in the superconducting behavior of the copper oxides.

6.1.1 YBa$_2$Cu$_3$O$_{7-\delta}$

a) HREM Images

As previously mentioned, the topic of contrast interpretation in the oxygen-deficient perovskite images is highly delicate and the main question is: is it possible to differentiate [100] and [010] images in orthorhombic YBa$_2$·Cu$_3$O$_7$? These two projections differ only in the orientation of the square planar groups CuO$_4$. Thus in the [100] projection we observe, in the median copper plane (arrowed in Fig.6.1), a sequence of one oxygen and one copper along the b axis (Fig.6.1a), whereas, in the [010] projection, the sequence along a is one oxygen vacancy and one copper superimposed on an oxygen (Fig.6.1b). From the calculated images, produced with the positions obtained from the neutron diffraction study [6.2], it appears that these orientations exhibit similar contrasts for most of the images and that few values of focus allow the two orientations to be distinguished. However, the researchers agree that, in well-defined experimental conditions, a distinc-

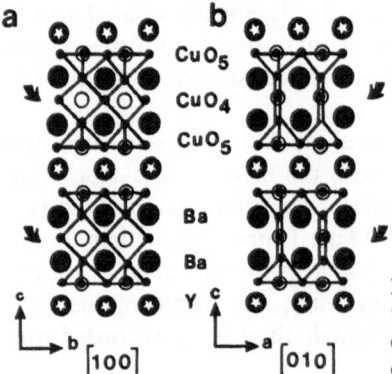

Fig.6.1. (a) [100] and (b) [010] projections of the YBa$_2$Cu$_3$O$_7$ idealized structure. The nature of the copper polyhedra is marked. The median copper plane (z = 0) is indicated by arrows

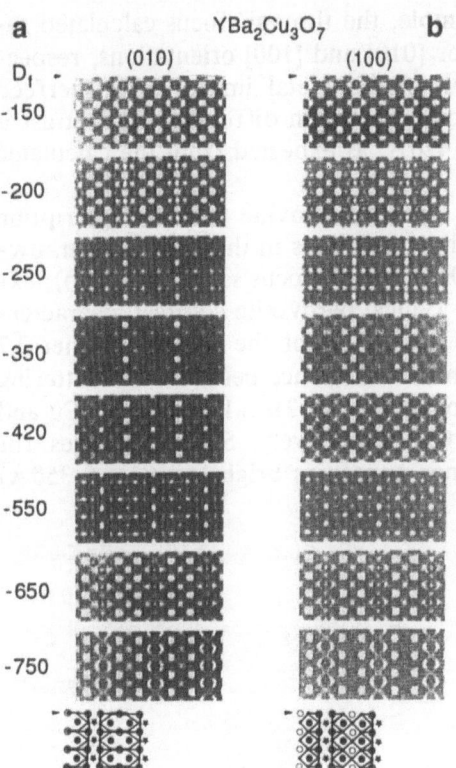

a YBa$_2$Cu$_3$O$_7$ b

D$_f$ (010) (100)

-150

-200

-250

-350

-420

-550

-650

-750

Fig.6.2a–d. Calculated through-focus series and corresponding projections of the structure: (a) [010], (b) [100]. (c, d) Experimental images obtained with three different values of focus: (c) [010], (d) [100]. (Defocus specified in Ångstroms)

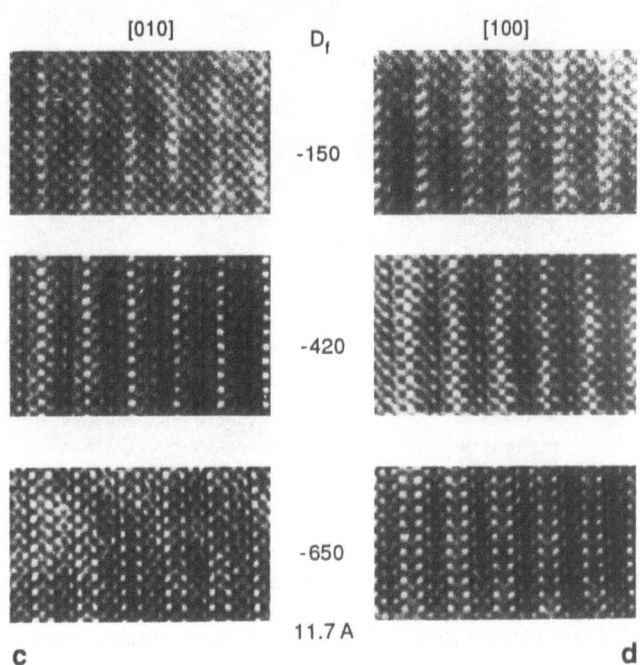

[010] D$_f$ [100]

-150

-420

-650

11.7 Å

c d

tion can be made [6.3-8]. As an example, the through-focus calculated series are shown in Figs.6.2a and b for [010] and [100] orientations, respectively, while Figs.6.2c and d exhibit experimental images of a "perfect" crystal taken for these two orientations. The main difference in contrast is observed for a focus value close to -420Å, as expected from the calculated images.

The [110] orientation (Fig.6.3a) does not provide the best description of the oxygen distribution or of possible variations in that distribution, owing to their mean projection. The [001] through-focus series (Fig.6.3b), except for the -250Å image, shows the typical perovskite contrast, characterized by two main features. First, the periodicity of the images is either 2.7 or 3.8 Å, the latter value arising from the difference between the scattering powers in the Ba and Y columns (proportional to $2Ba+1Y$) and the Cu and O columns (proportional to $3Cu+1O$), respectively. Second, images for which the low electronic density zones appear as bright contrast (-350Å)

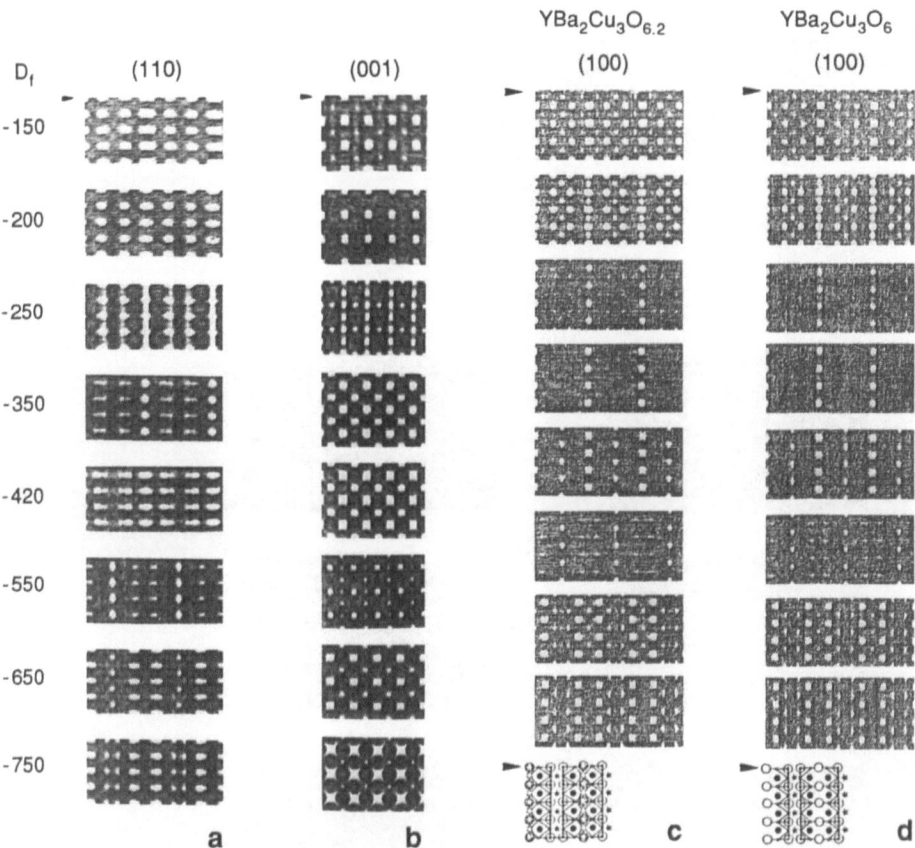

Fig.6.3a-d. $YBa_2Cu_3O_7$: (a) [110], (b) [001]; (c) $YBa_2Cu_3O_{6.2}$, (d) $YBa_2Cu_3O_6$. Through-focus simulated images: voltage $V = 200$kV, spherical aberration constant C_s = 0.8mm, semi-angle of beam convergence $\alpha/2 = 0.2$mrad, focus spread: 150Å, 111 beams, thickness $t = 30$Å

differ only by a 2.8 Å translation from those where high-electronic-density zones are brightest (~650 Å); this implies that the characterization of the experimental images definitely requires the experimental through-focus series to be carried out.

The images in Figs.6.2 and 3a,b are those of "perfect" crystals with a formulation close to $YBa_2Cu_3O_7$, and a perfect ordering of both cations and anions. The point is then to detect any deviation from that ideal ordering and to understand it. Systematic simulations of the through-focus series were performed for variations of the oxygen content, taking into account the displacements of the cations implied by such a change in the oxygen distribution. Some examples are shown in Figs.6.3c and d. Other structural hypotheses were considered, such as oxygen in the yttrium planes and Y-Ba disorder. They are not presented here but are needed in any contrast variation interpretation.

b) 92 K Orthorhombic Superconductor ($0 \leq \delta \leq 0.1$)

Deviations from the perfect ion ordering are common features, even in the 92 K superconductor. Their complexity and intensity are strongly dependent on the thermal treatments, which therefore need to be systematically specified in any microstructural or physical study. These deviations are classified here in three sections related to oriented domains, oxygen-content variations and cationic disorder.

Twins and oriented domains. Early on, the existence of systematic twinning domains for the orthorhombic superconductor was reported [6.3, 9- 20]. This phenomenon results from the phase transition between the tetragonal high-temperature phase and the orthorhombic low-temperature form (Chap.2). Twins are evidence of the orthorhombic distortion resulting from the oxygen-vacancy ordering and thus large twinning domains can be considered as a sign of superconducting behavior of the bulk. However, it should be mentioned that this feature is only significant in the pure (undoped) oxides $YBa_2Cu_3O_{7-\delta}$. Twinned crystals are characterized by electron diffraction patterns with split reflections; only hh0 reflections are unsplit as they are common to both components (Fig.6.4a). The dark-field image, obtained by selecting a 220 reflection with the objective aperture, allows the domains to be displayed (Fig.6.4b). The domain widths are variable and the mean value ranges from 500 to 1000 Å. By a careful electron beam heating [6.11, 17, 21], we can observe the reversible transformation of both orthorhombic and tetragonal forms (Fig.6.5).

The twinning domains can be readily described by considering vacancy ordering along [010]. The change in direction implies that the CuO_4 groups in one domain are oriented perpendicular to those in the next one (Fig.6.6). This change in CuO_4 group orientation raises two questions: how is the junction ensured, and does it imply stress or defects, which can modify the electron transport properties?

Several models of a boundary have been proposed [6.3, 9, 10, 13, 20]. They are schematically represented in Fig.6.7. In order to simplify the

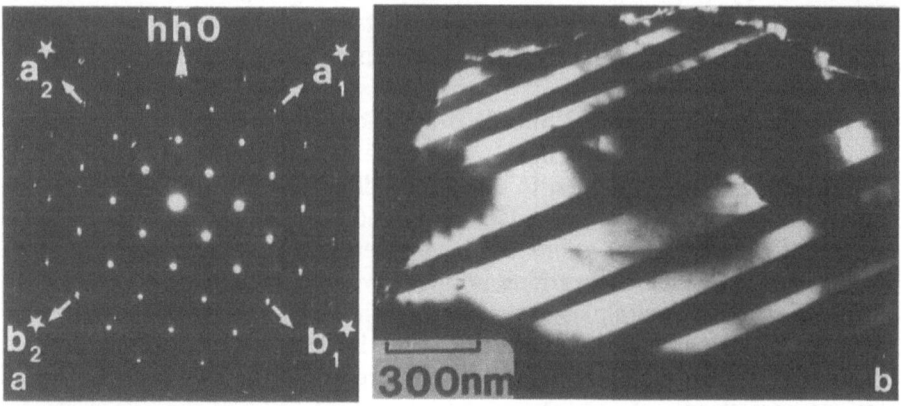

Fig.6.4. (a) Typical electron diffraction pattern and (b) dark-field [001] image of an orthorhombic crystal $YBa_2Cu_3O_{7-\delta}$

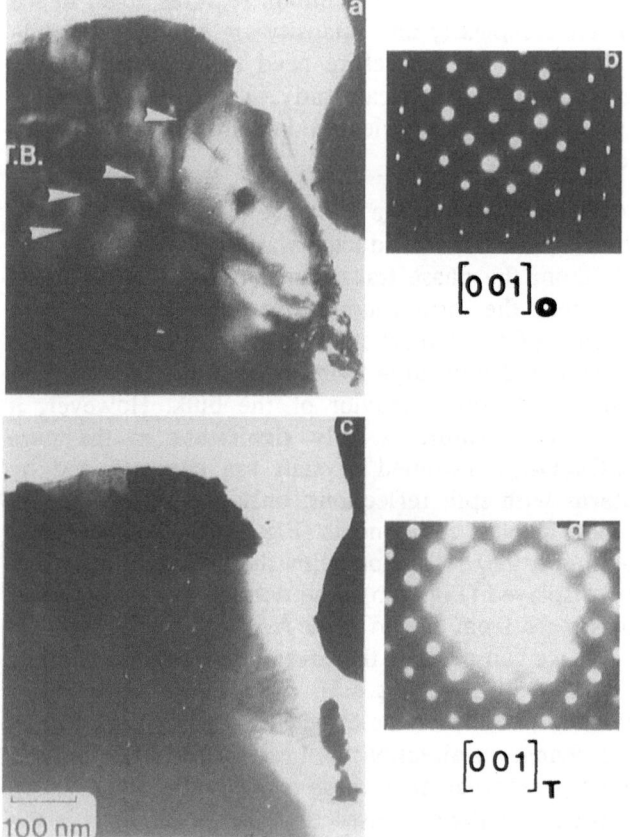

Fig.6.5. (a) [001] bright-field image of an orthorhombic crystal. Twin boundaries (T.B.), which appear as black lines, are arrowed. (b) Corresponding ED pattern. (c) Bright-field image and (d) ED pattern of the same crystal under beam heating. Note that the spots are now unsplit and that the twinning domains have disappeared

160

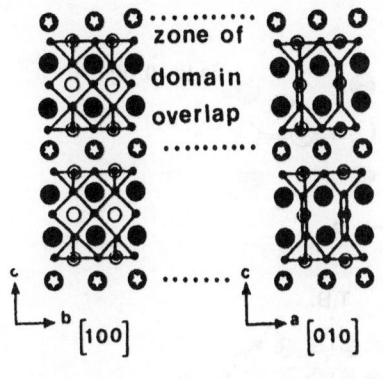

zone of domain overlap

c ⟶ b [100]

c ⟶ a [010]

◀— domain I   domain II —▶

Fig.6.6. Projection of twinning domains in a crystal. The change in the orientation of the CuO_4 groups is observed in one perovskite triple layer

drawing, only three types of atoms are presented in the projection of the structure along [001]; the others are omitted. The oxygen and copper atoms ($z = 0$) of the median plane are presented (Fig.6.7a) so that the CuO_4 groups appear as straight lines of circles (oxygen atoms) and small black dots (copper atoms). The large black dots represent the barium positions at the upper level ($z = 0.18$), in order to show that the cationic lattice is unchanged from one domain to the next. Two models for the junction (Figs.6.7b and c) imply the presence of additional oxygen at the boundary, through the existence of CuO_6 octahedra and CuO_5 pyramids, but no drastic displacements in the cation framework. A third model (Fig.6.7d), the simplest because the junctions are achieved solely through tetrahedra, can be proposed: it does not lead to a variation of the oxygen content, but the existence of such CuO_4 tetrahedra would involve a displacement of the copper atoms, and thus this appears to be less favorable. The last model (Fig.6.7e) implies a substoichiometry with a twofold coordination of copper [Cu(I)] and a threefold coordination of the adjacent coppers, and so seems unrealistic. However, it should be noted that this latter model, implying Cu(I) in a twofold coordination, was chosen assuming diffuse boundaries [6.22] and detwinning mechanisms [6.23]. The equivalence of the [110] and [1$\overline{1}$0] directions is illustrated in Fig.6.8a by the simultaneous existence of quasi-perpendicular twinning domains in the thicker crystals. The idealized model (Fig.6.8b) shows the arrangement of the $[CuO_2]_\infty$ rows at the quasi-perpendicular boundaries (the junction through octahedra of the model in Fig.6.7b was chosen for the drawing, but the other models are valid, too). Such a phenomenon can easily occur because the domains result from a single change in vacancy ordering in the core of the matrix without any large displacement of the cationic host lattice.

Four crystal orientations can be chosen to observe the twin boundaries by high-resolution electron microscopy. [010] and [100] are, theoretically, favorable for the observation of these changes in the orientation of the CuO_4 groups, since they can be distinguished through characteristic variations of the contrast for some particular values of the focus. However, the orthorhombic distortion of the cell implies a slight misorientation of a do-

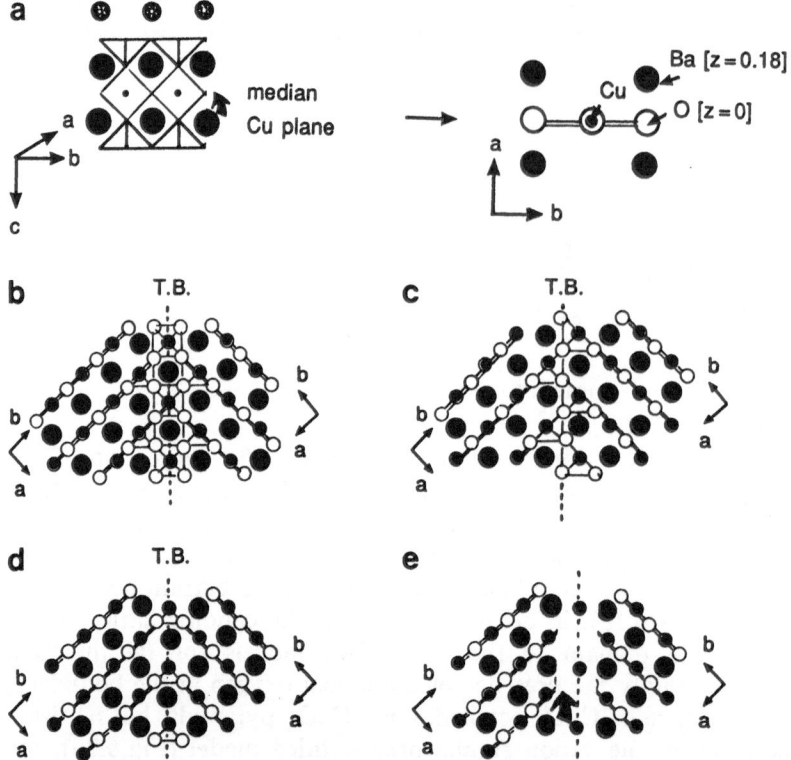

Fig.6.7. (a) Principle of schematization of the drawing of the twin boundaries (T.B.s). (b-e) Four idealized models of junctions between the twinning domains in the $[CuO_2]_\infty$ layers: (b) junction in a mirror plane through CuO_6 octahedra and CuO_5 pyramids. (c) Junction through CuO_5 pyramids. (d) Junction in a mirror plane through CuO_4 tetrahedra. (e) Junction through twofold coordinated copper, according to *Bordet* et al. [6.10]. This implies a threefold coordination for the adjacent copper atoms (*arrowed*). Note that, unlike the third and fourth models, the first two imply additional oxygens at the boundary; such arrangements between octahedra and pyramids are frequently observed in oxygen-deficient perovskites.

main with respect to the adjacent one and, additionally, the 45° orientation of the domain boundary with respect to the axes **a** and **b** implies an overlapping of the two adjacent domains whose width is equal to the crystal thickness. Such an overlapping obviously affects the contrast and hinders the detection of local defects in the immediate neighborhood of the boundary.

Observations of the twinned crystals along [110] do not provide any important structural information, owing to the mean projection of the oxygen atoms, in agreement with the calculated images. However, these images clearly show that no particular defect areas are coupled with the domain boundaries.

The best orientation for observing the twin boundaries is [001]. The micrographs recorded with an aperture of 0.85 Å^{-1} (V = 200kV, C_s = 0.8

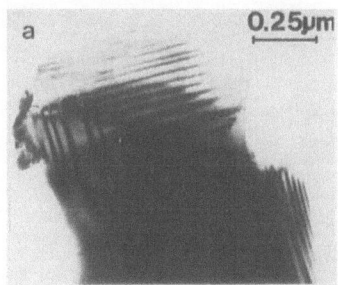

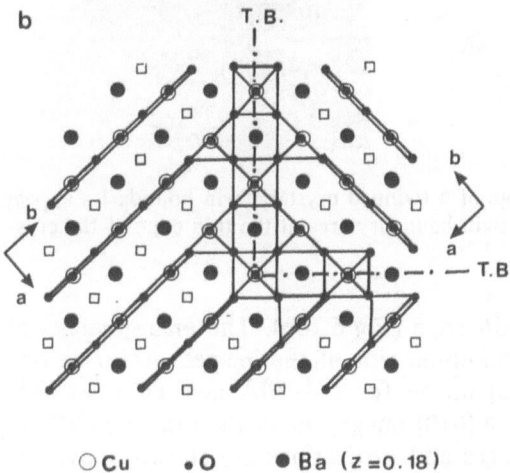

Fig.6.8. (a) [001] image of quasi-perpendicular twinning domains. (b) Idealized model of the boundaries

○ Cu   • O   ● Ba $(z = 0.18)$

mm) from a well-oriented crystal show a similar contrast from one domain to another, making them hardly distinguishable. It is necessary to use a smaller objective aperture (0.48Å^{-1}) to make the twin boundaries appear as black lines in the thicker part of the bulk, as shown in Fig.6.9a. Enlargements of the thin part of the crystal close to the Twin Boundary (T.B.), as in Fig.6.9b, allow two points to be noted. 1) Confirmation of the observations along [110]. The crystals exhibit a lot of defects (detailed in the next sections), which are not localized specifically at the boundary areas but extend throughout the whole matrix. 2) The nature of the junction. It is not possible to draw any conclusions about either any particular model of the junction or the existence of a single model for such a junction.

The actual role played by the twins in the superconducting behavior is still an enigma. The existence of "tetragonal" superconducting oxides $YBa_2 \cdot Cu_{3-x}Fe_xO_7$ (Chap.4) shows that they are not essential for the appearance of such behavior. However, the possible presence of additional oxygens (Fig.6.7b, c) at the twin boundaries can be a favorable factor for formation of Cu(III) and, thereby, for superconducting properties.

Observations of the $YBa_2Cu_3O_7$ crystals along [100] or [010] show a second type of domain [6.1-9], parallel to the (001) plane, which can slice the crystal perpendicularly to the c axis, as shown in Fig.6.10a. The nature of such domains can be more easily interpreted from images of crystals

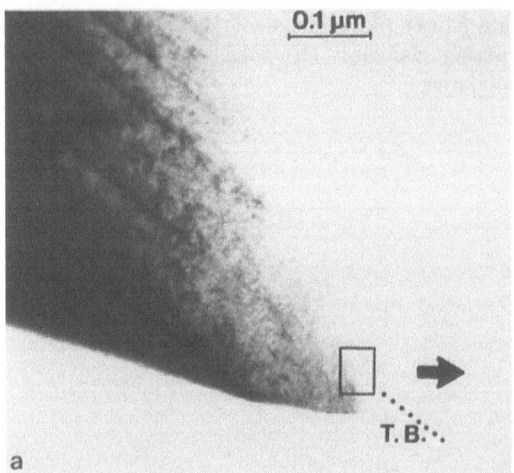

Fig.6.9. (a) [001] high-resolution image of a twinned crystal. Twin boundaries appear as black lines. (b) Enlargement of the twin boundary area at the thin edge of the crystal

whose c axis is parallel to the thin edge (Fig.6.10b). The enlargements of areas labelled *1* and *2* are shown on either side of the images: area *1* exhibits the typical contrast of a [100] image for t ~ 30Å and Df = -430Å, whereas area *2* is characteristic of a [010] image, under the same conditions. The corresponding calculated images and projections are shown above the experimental images. It appears clearly that a change in the orientation of the CuO_4 square planar groups has still to be invoked. Contrary to the twinning domains, this phenomenon occurs from one triple $[Cu_3O_7]_\infty$ layer to the adjacent one, forming "oriented" slices perpendicular to the c axis. The idealized model is illustrated in Fig.6.10c. It can be seen that from one domain to the next along c, the $[CuO_2]_\infty$ rows are turned through 90°. The nature of the domain boundaries is easy to understand with such a theoretical model, since one yttrium plane appears to be common to two adjacent domains. However, it is worth pointing out that, in contrast to the twinning domains, the "oriented" domains involve a junction that imposes the juxtaposition of two different pairs of a and b parameters. The mismatch of the a and b parameters in the orthorhombic cell is large enough (~2%) to entail strains and the domain interfaces are then particularly disturbed, as observed in the thicker part of the bulk (Fig.6.10a). It should be noted that such variations in the orientation of CuO_4 groups from one slice to the adjacent ones are assumed to occur in the pseudotetragonal T_3 (Chap.2) [6.24]. However, the equivalence of the a and b parameters of the latter allows this phenomenon to occur very frequently since it does not imply any mismatch.

Another typical structural feature corresponds to a misorientation of crystal zones. A characteristic Electron Diffraction (ED) pattern of such a crystal is shown in Fig.6.11. We clearly observe the contribution of every

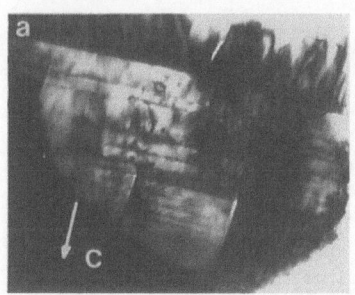

Fig.6.10. (a) Low-resolution image of oriented domains. (b) High-resolution image of "oriented" domains; areas labelled 1 and 2 show contrast corresponding to [100] and [010] images respectively; calculated images included in the enlargements were calculated for a defocus of -430Å and a thickness ~30Å. (c) Idealized model showing the orientations of the CuO$_4$ groups.

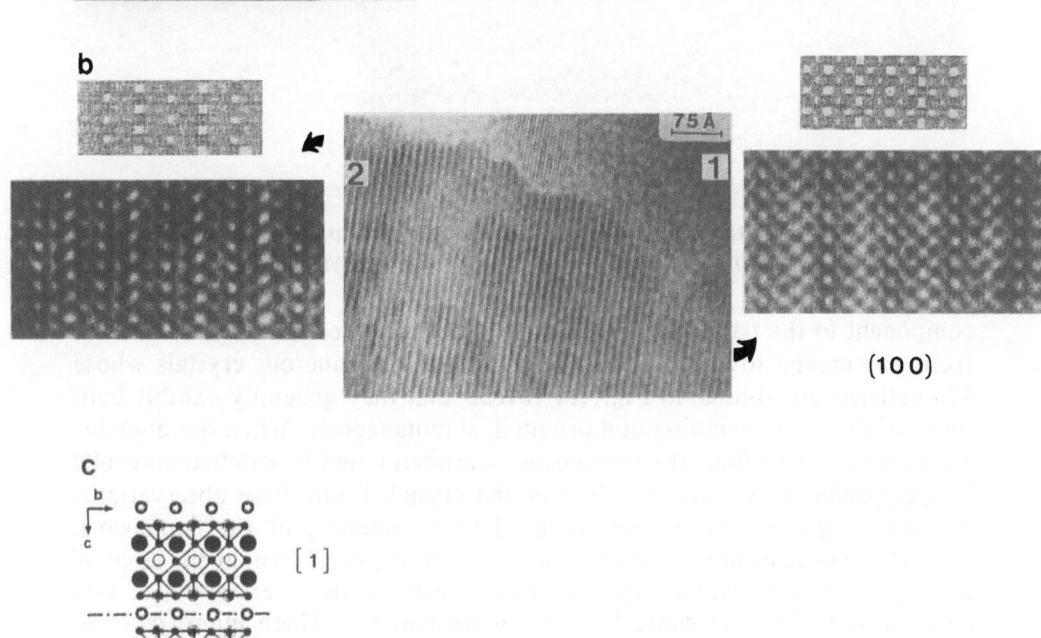

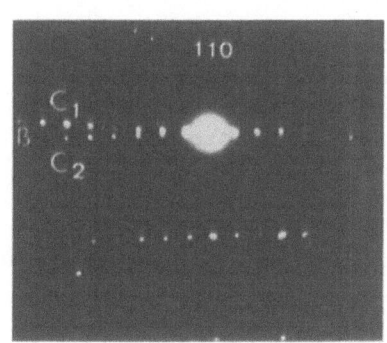

Fig.6.11. Typical diffraction pattern of a crystal with misoriented areas; the two c axes differ by an angle β

165

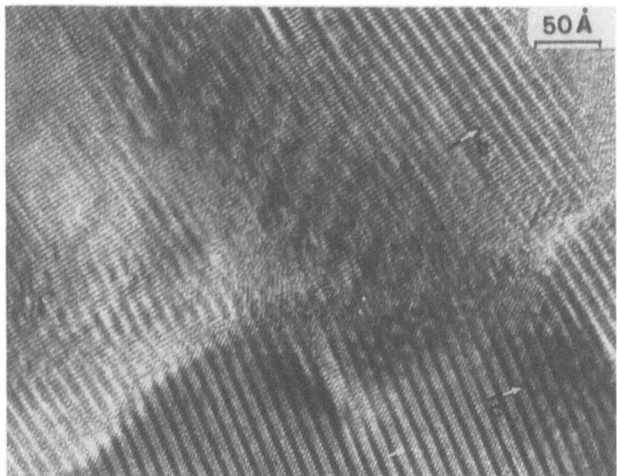

Fig.6.12. High-resolution image of a crystal showing overlap of misoriented domains. Strains related to microfractures and moiré patterns are observed

component to the pattern. However, the β angle between the domains varies from one crystal to another. The investigation of numerous crystals whose ED patterns are similar to Fig.6.11 reveals that they generally exhibit both types of domains, twinning and oriented, simultaneously. When the domains are numerous and thin, the discrepancy between a and b parameters results in spectacular strains and bending of the crystal. From these observations, we assume that the strains are reduced by a loosening of the framework through a mechanism of microfractures, which begin to form at the edge of the crystals but are not always completely realized; the overlapping of two adjacent domains sometimes leads to moiré patterns. These microfractures explain the variations in the β angle values. A HREM image of such a misorientation is shown in Fig.6.12, where the angle β between two adjacent domains and their overlap are clearly visible. Such features would undoubtedly affect the electron transport properties by layer and O-Cu-O chain breaking.

Oxygen over- and substoichiometry. The HREM investigation of numerous samples obtained from different thermal treatments shows that the distribution of oxygen is not as homogeneous as expected from the X-ray and neutron diffraction results, in spite of a 93 K resistive transition observed in all cases.

The orthorhombic $YBa_2Cu_3O_{7-\delta}$ synthesized in the now "classical" way, i.e. sintered at 950°C with a slow decrease of the temperature under oxygen, exhibits an oxygen content ranging from 6.85 to 7. The [100], [010] and [001] orientations of the crystals are best for the observation of oxygen nonstoichiometry features. The real difficulty in the study of the local oxygen variations must be emphasized again: besides the oxygen contribution to the potential and the cationic displacement consequences mentioned previously, it is essential to pay attention to small misorientations of crystal areas and other similar artefacts, which can drastically affect the contrast. Moreover, in order to be imaged, the variations of the oxygen content must correspond either to an extended defect, in the form of a domain, or to a column of atoms along the direction of observation. They should be observed on the thin edges of the crystals in order to avoid average images or features such as the overlapping of oriented domains [6.25].

As depicted in Fig.6.9, [001] images of the ceramics show numerous defects localized all over the matrix, and the investigation of several microcrystals reveals the formation of domains, whose size varies from about 30Å^2 to some hundreds of Å^2, characterized by a change of the contrast. Another typical example is given in Fig.6.13. From the enlargement of the HREM image (Fig.6.14), it is clear that the contrast corresponding to big and small white spots, correlated to the projected positions of (2 oxygens + 1 vacancy) and (3 oxygens), respectively, has disappeared: it is replaced by quite equivalent white spots. Such a variation is interpreted in terms of the

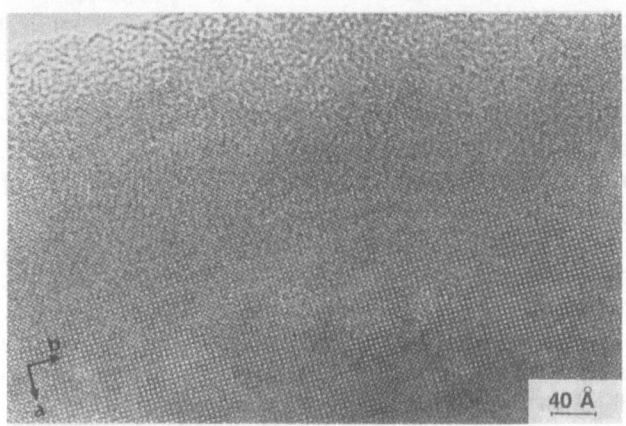

Fig.6.13. Typical [001] HREM image of a ceramic synthesized in the "classical" way. Note the contrast variations in the form of larger or smaller domains

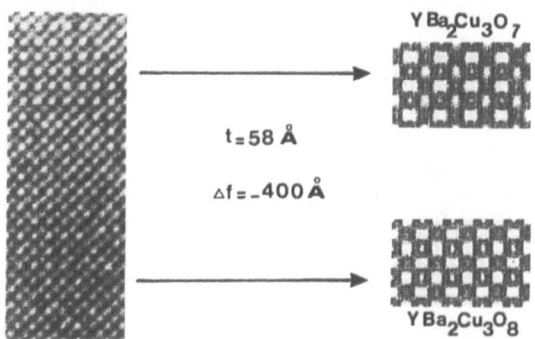

Fig.6.14. Enlargement of a defect area. At the top of the micrograph the bright spots are spaced at 3.8Å, in agreement with the calculated image for $YBa_2Cu_3O_7$. At the bottom, all the spots are equivalent. The images were calculated for a focus value close to -400Å and a crystal thickness of ~85Å

167

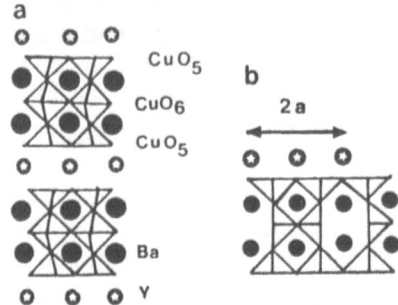

a

CuO_5
CuO_6
CuO_5

b

2a

Ba

Y

Fig.6.15. The presence of additional oxygens involves the formation of (**a**) octahedra (one oxygen per cell) or (**b**) pyramids (for 0.5 oxygen per cell). A regular model built up from pyramids involves a doubling of the a parameter

presence of additional oxygen atoms in the $[CuO_2]_\infty$ layer, in agreement with the calculated images (Fig.6.14). It is obvious that such a contrast variation is evidence for additional oxygens but it does not allow an exact occupancy ratio of the corresponding sites to be determined. Thus the stoichiometry of these domains can be characterized by $YBa_2Cu_3O_{7+\delta}$. The local structure is then described as a succession of one CuO_6 octahedron and two CuO_5 pyramids along the c axis. The idealized model in Fig.6.15a corresponds to the hypothetical limit $YBa_2Cu_3O_8$ [6.3]. It is worth noting that, owing to their size, such domains can only be observed and interpreted at the thin edges (<200Å) of the crystals. As previously discussed, the copper coordination must be taken account of in the interpretation of the extended defects. In that case, the presence of a single additional oxygen in the median copper plane transforms, in a simple way, two CuO_4 groups to two CuO_5 pyramids, keeping a coordination usual for copper. The idealized model in Fig.6.15b corresponds to $YBa_2Cu_3O_{7.5}$. The only consequence then is a change in the Cu-O distances, which would involve slight ion displacements in the neighborhood of the defect, for the formation of octahedra as well as for the pyramids.

Observations of the crystals along [100] and [010] confirm the existence of domains whose shape is in agreement with those observed along [001]. Moreover, along these directions, some very localized additional oxygens

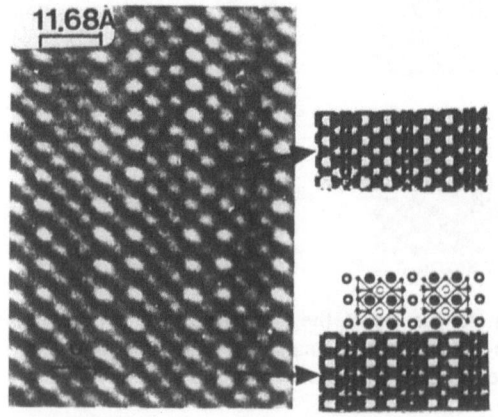

11.68Å

Fig.6.16. Experimental and calculated images [100] of a local variation of the oxygen content (t~30Å and Δf~-650Å): $YBa_2Cu_3O_8$ (*top*) and $YBa_2Cu_3O_7$ (*bottom*)

are sometimes observed, as shown in Fig.6.16. For a focus value of about -650Å the essential feature of the contrast of $YBa_2Cu_3O_7$ along [100] is two rows of strong white spots with a row of smaller ones between then. The contrast in a $YBa_2Cu_3O_8$ area, for the same focus value and indeed the same thickness, consists of three rows of white spots of equal intensity, in agreement with the experimental image.

This oxygen overstoichiometry, involving $YBa_2Cu_3O_{7+\delta}$ regions, is in agreement with the ability of Cu(III) to take pyramidal and octahedral coordinations. However, this local excess of oxygen with respect to the apparent oxygen content of the samples (ranging from 6.85 to 7) implies that other regions can be expected to show oxygen deficiency. The nonstoichiometry mechanisms that can generate such domains are of two types: a) phases that are lacking in oxygen but exhibit the same cationic framework as the perovskites (and therefore numerous vacancies) and b) new phases whose formulation differs from that of the perovskite in the cation content as well as oxygen content.

1) For the existence of perovskite-type areas with an oxygen content less than 7, the question of the copper coordination must be considered. Chapter 2 deals with this subject, and the hypothesis of a disproportionation of Cu(II) into Cu(III) and Cu(I) allows this point to be understood. Unlike the additional oxygen atoms, whose presence is in agreement with the usual coordinations of Cu(III) and/or Cu(II), the existence of isolated Cu(I) atoms and the model of their linkage with the adjacent oxygen and copper atoms in the perovskite matrix are difficult to reconcile with the usual coordinations of copper. Considering the threefold coordination of Cu(II) as unlikely and the ability of Cu(I) to take the twofold coordination, it appears that the oxygen substoichiometry is ensured either through isolated $[Cu^IO]_\infty$ chains or through local superstructures corresponding to an ordering of the oxygens and vacancies [6.26]. No evidence of the existence of isolated Cu(I)-rich domains was observed in the experimental images of $YBa_2 \cdot Cu_3O_{7-\delta}$ for $0 \le \delta \le 0.1$. We can assume that the existence of $[Cu(I)O]_\infty$ chains, isolated and single, cannot be detected. Superstructures in the form of domains or microcrystals are observed, but only for samples corresponding to an oxygen content less than $O_{6.85}$. They will be considered in the following sections.

2) The existence of extended defects involving both oxygen and copper atoms is often observed [6.27-29]. The HREM image of such a defect is displayed in Fig.6.17a, where we clearly observe a variation of the spacing of the rows of spots. The zone axis is [100] and the focus value about -650Å. The correlation between the image and the projected potential shows that the large white spots, in double rows, correspond to the barium atoms, and the smaller ones, in the middle of the double rows, correspond to the copper atoms of the CuO_4 groups. It can be observed that the rows of white spots move apart, implying a bending of the adjacent double row and that an extra row of small spots appears in the middle. Such a feature can be explained by the formation of a double row of edge-sharing CuO_4

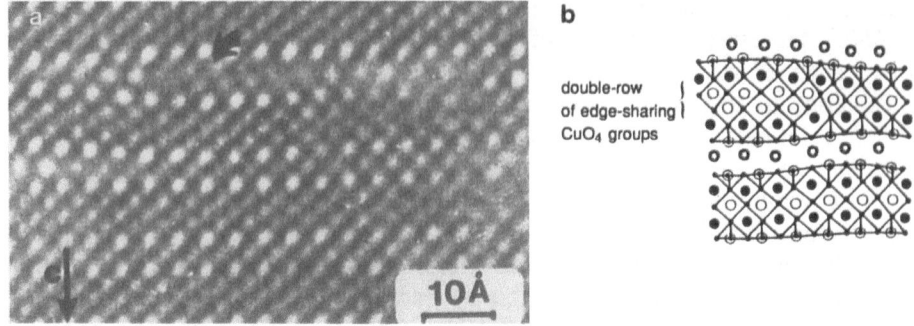

Fig.6.17. (a) HREM image of a defect corresponding to the insertion of an extra row of copper. (b) Idealized model corresponding to a local $YBa_2Cu_4O_8$ stoichiometry

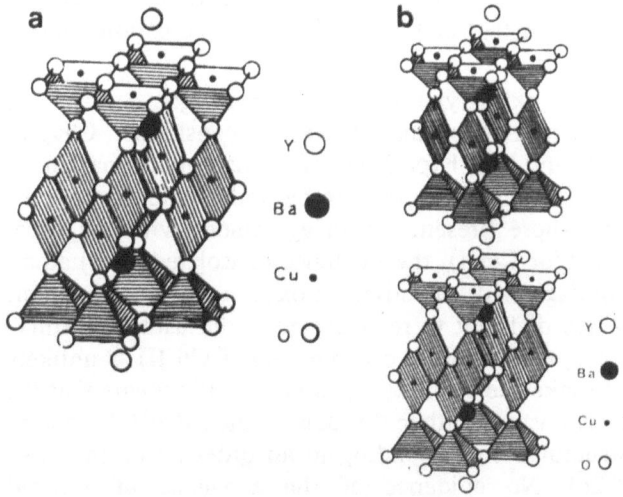

Y ○

Ba ●

Cu ·

o ○

Y ○

Ba ●

Cu ·

o ○

Fig.6.18. Idealized structures of (a) $YBa_2Cu_4O_8$ and (b) $Y_2Ba_4Cu_7O_{15}$

groups, similar to those observed in the $SrCuO_2$ framework. An idealized model of the defect is illustrated in Fig.6.17b. The occurrence of such defects involves, locally, an oxygen substoichiometry. The local composition of such a slice can be formulated $YBa_2Cu_4O_8$. It corresponds to the hypothetical structure $YBa_2Cu_4O_8$ given in Fig.6.18a, which consists of quadruple $[Cu_4O_8]_\infty$ layers formed of two pyramidal $[CuO_{2.5}]_\infty$ layers connected through double $[Cu_2O_3]_\infty$ rows of edge-sharing CuO_4 square planar groups running along b. The existence of such defects suggests the possibility of generating intergrowths of the two structures corresponding to the general formula $[YBa_2Cu_3O_7]_n[YBa_2Cu_4O_8]_{n'}$. The first member of such a series (Fig.6.18b), which for $n = n' = 1$ corresponds to an intergrowth $Y_2Ba_4Cu_7O_{15}$, has recently been synthesized [6.30], and its structure has been confirmed by *Chaillout* et al. [6.31].

It is interesting to compare, from the oxygen homogeneity point of view, these crystals with others collected in differently synthesized samples.

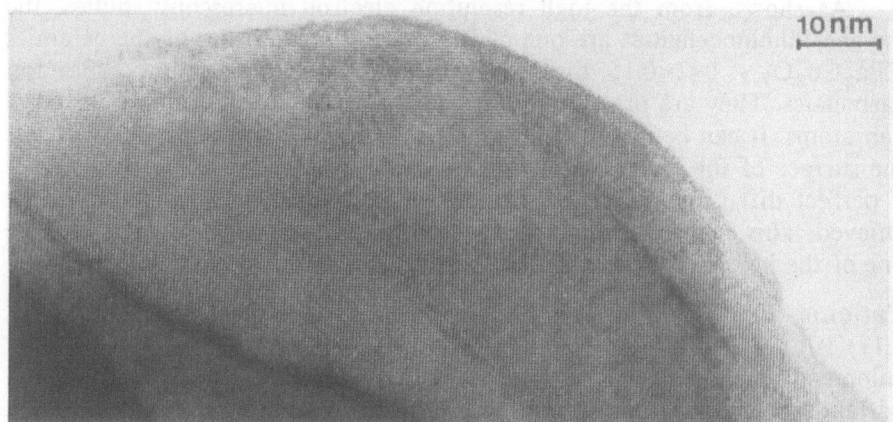

Fig.6.19. [001] HREM image of $YBa_2Cu_3O_{6.93}$ obtained by controlled reoxidation of a reduced precursor. Note the even contrast and clear edges, contrasting with Fig.6.13

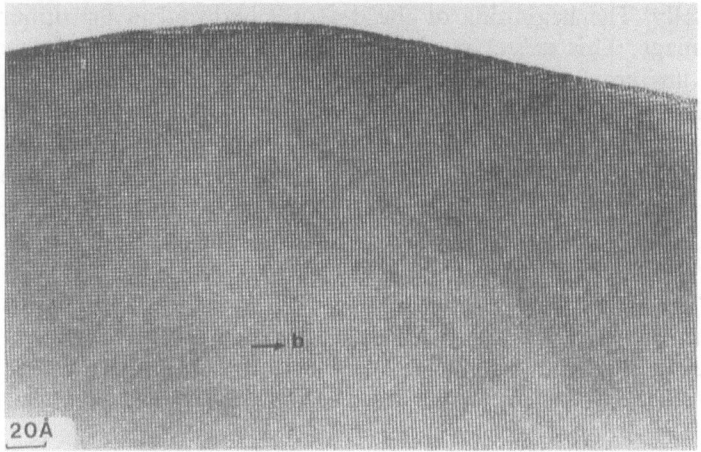

Fig.6.20. [001] HREM image of a microcrystal observed in a thin layer deposited by sputtering on a silicon substrate

An orthorhombic superconducting phase with a formula $YBa_2Cu_3O_{6.93}$ can be obtained from the controlled re-oxidation of the reduced $YBa_2Cu_3O_{6.1}$ [6.32]. The precursor $YBa_2Cu_3O_{6.1}$ exhibits a lamellar morphology, which is retained to a large extent by the oxidized sample. The main characteristic feature of this as-synthesized sample is that the anionic disorders encountered in the ceramics obtained from direct solid-state reactions are almost entirely absent: the [001] images (Fig.6.19) exhibit an even contrast and clear edges. This is to be compared to the microcrystals observed in poly-crystalline thin layers [6.33, 34]: the very thin twinning crystals exhibit the same quite perfect contrast (Fig.6.20). This feature is assumed to be related to the thickness of the particles: the lamellar morphology allows an almost perfect diffusion of the oxygen to be established during annealing.

As shown from the high-resolution electron microscopy studies, the oxygen inhomogeneities are one of the main characteristics of the ceramics $YBa_2Cu_3O_{7-\delta}$, $0 \leq \delta \leq 0.15$, directly synthesized from the starting oxides and carbonates. They are probably related to the difficult diffusion of the oxygen atoms. It can be assumed that the oxygen-rich domains are localized at the surface of the grains: the usual annealing time is not sufficient to allow a perfect diffusion of the oxygen throughout the whole matrix to be achieved. This suggests that the formation of $YBa_2Cu_3O_{7+\delta}$ clusters may be one of the keys to the structural mechanisms of these phases.

Cationic disorder. The "$YBa_2Cu_4O_8$"-type defect described above (Fig. 6.17) is an example of cationic disorder involving a variation of the stoichiometry, on the anionic as well as on the cationic lattice. A second important type of defect involving the Y-Ba cation ordering is observed in the superconducting ceramics $YBa_2Cu_3O_{7-\delta}$ ($0 \leq \delta \leq 0.15$) [6.27]. A typical example of such a defect is shown in Fig.6.21. Looking at the fringes resulting from the triple layers $[Cu_3O_7]_\infty$, we observe a c/3 shift, which crosses the image diagonally. The beginning of the defect is arrowed in the upper left part of the image. This defect is easily explained by a reversal of one barium and one yttrium layer from one part of the defect to the other, while the second barium layer remains unchanged through the defect, as shown by an idealized model (Fig.6.22a). In that example, the defect extends over twelve triple $YBa_2Cu_3O_7$ layers with a clear contrast and the junction of the reverse layers is regularly translated, perpendicular to the c axis. Notice that such a defect keeps the ratios Y/Ba and Y+Ba/Cu constant, but it can involve a local variation of the oxygen content at the domain boundary. Thus, the model drawn in Fig.6.22a corresponds to a substoichiometry with only six oxygens per YBa_2Cu_3 group, but an examination of this structure shows that additional oxygen atoms can easily be incorporated, since they involve the formation of one CuO_5 pyramid and one

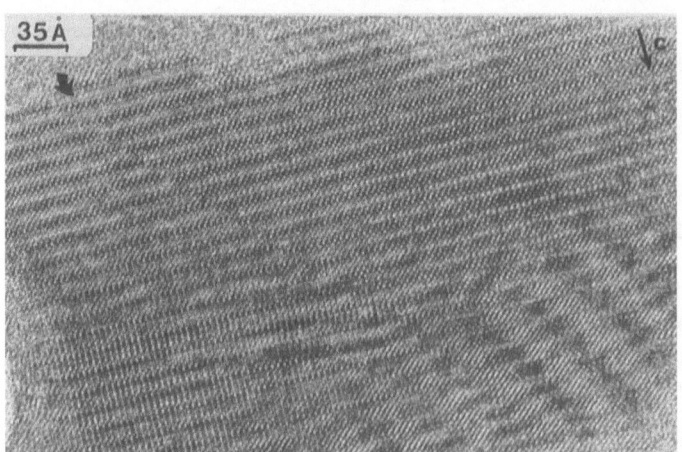

Fig.6.21. HREM image of cation disorder corresponding to a reversal of one barium layer and one yttrium layer

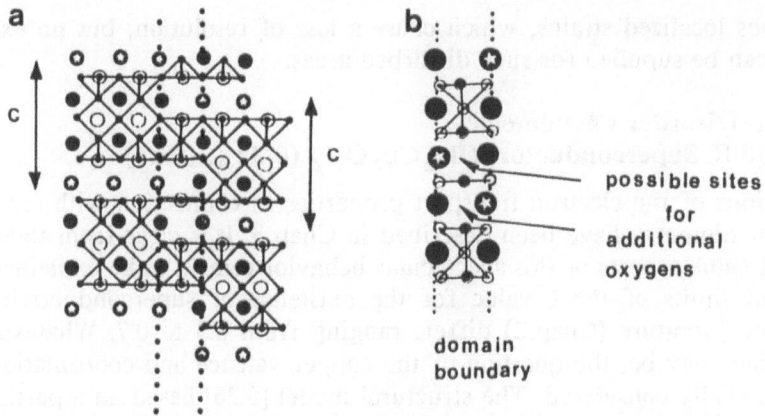

a

c

b

possible sites
for
additional
oxygens

domain
boundary

Fig.6.22. (a) Idealized model of the translation of the triple perovskite layers along the c axis. (b) Example of possible sites for additional oxygen at the domain boundary

CuO_6 octahedron (Fig.6.22b). In the lower part of the micrograph Fig.6.21, a moiré pattern is observed in the thicker part of the crystal. These patterns are common features of HREM images. One origin is the overlapping of misorientated domains (see near the beginning of Sect.6. 1.1b) and a second one, observed in Fig.6.21, is related to a multiple shifting of the Y and Ba layers and the overlapping of the small domains. Note that such a type of defect does not affect the O-Cu-O chains but implies a break in the two-dimensionality of the structure.

Other extended defects are observed in the microcrystals. Some of them are well localized in regions of about 10 to 20 Å diameter where the contrast is highly disturbed. In the through-focus series, for a focus value close to -50Å, where metallic ions appear as white spots, it can be seen that cations are strongly displaced and sometimes missing, so that the perovskite framework can no longer be characterized (Fig.6.23). Other defects appear

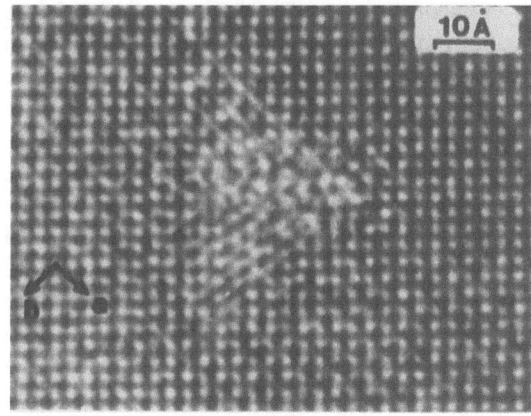

10 Å

Fig.6.23. Example of strongly displaced cation. [001] image: the perovskite framework can no longer be characterized

as numerous localized strains, which cause a loss of resolution, but no explanation can be supplied for such disturbed areas.

c) Order-Disorder Phenomena:
The 60 K Superconductor $YBa_2Cu_3O_{7-\delta}$ ($0.37 \leq \delta \leq 0.45$)

The variations of the electron transport properties in connection with oxygen nonstoichiometry have been described in Chap.3. It is clear from those results that some aspects of this spectacular behavior remain to be explained and current limits of the δ value for the existence of superconductivity given in the literature (Chap.2) differ, ranging from 0.3 to 0.7. Whatever the true limit may be, the question of the copper valence and coordination must be carefully considered. The structural model [6.26] based on a partial disproportionation of Cu(II) into Cu(III) and Cu(I) leads to the coexistence in the same structure of superconducting domains (or chains) $YBa_2 \cdot (Cu_2^{II}Cu^{III})O_7$ and of insulating domains (or chains) $YBa_2(Cu_2^{II}Cu^{I})O_6$. This can be achieved in either an ordered or a disordered fashion; only high-resolution electron microscopy studies can give information about the way these domains are arranged. The reported features [6.11, 12, 28, 29, 35-43] are really very complex and, once more, it appears essential to specify synthesis conditions. Moreover, such an investigation needs the observation of numerous crystals and the real crystalline state of the particles has to be reported; indeed most of the time the perfectly ordered crystals cannot be considered representative of the ceramic sample. In the same way, it is obvious that, in such inhomogeneous crystals, calculated images cannot be produced, because the ordering is established neither in a sufficiently wide area nor in a perfectly periodic way.

A 60 K superconductor with a sharp transition is obtained by the controlled reoxidation of a reduced precursor with the formula $YBa_2Cu_3O_{6.1}$ [6.35]. The chemical and thermal analyses show that this phase is obtained for $0.30 \leq \delta \leq 0.50$. This range corresponds to the compositions which lead to a plateau in the curve T_c vs δ observed by several researchers (Chap.3).

The striking features of the electron-microscopy study performed in δ = 0.45 and δ = 0.37 samples for about two hundred crystals are related, on the one hand, to the numerous microstructures and modulations, and on the other, to a highly disordered aspect of the microcrystals, which attests that the equilibrium has not been (or cannot be) reached.

The structural inhomogeneities of the crystals decrease, in both volume and intensity, when the δ value decreases. Two main directions of strains and contrast modulations can be drawn, which are, generally, $[100]_p$ and $[110]_p$; "o" and "p" subscripts refer to the orthorhombic cell and perovskite subcell, respectively. The latter (p) is used when contrast is too disturbed to allow the a_o and b_o axes to be distinguished. This is illustrated in Fig.6.24a, corresponding to a [001] HREM image. Strains can be correlated to the oxygen atoms beginning to order, which may induce local variations of the cell parameters and displacements of the cations. The contrast modulations probably result from a more advanced ordering.

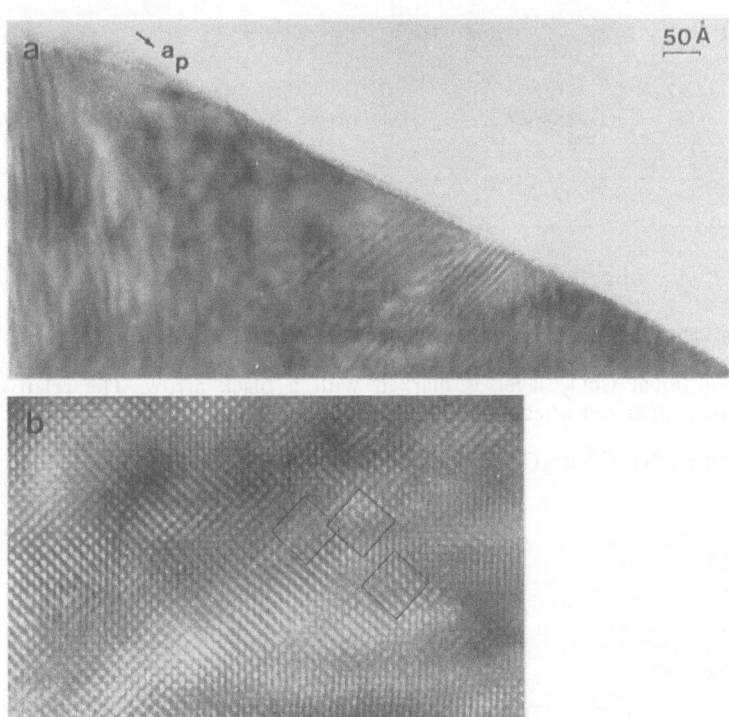

Fig.6.24. (a) [001] HREM image of $YBa_2Cu_3O_{6.66}$, characterized by strains and contrast modulations. (b) Enlargement showing variations in the spacing of the rows and their undulation

In the middle of the micrograph, a disturbed area without any preferential direction is observed. The enlargements (Fig.6.24b) show variations in the spacing of the rows of white spots, their undulation, and defects which look like "superdislocations", i.e. which appear as the termination of rows of atoms, but these defects move under the electron beam. "Superdislocations" as encountered in ReO_3-type structures [6.44] cannot be involved here, owing to the presence of the A cations. Moreover, the easy rearrangements observed in the microscope suggest rather metastable areas where the local ordering of oxygen is not achieved.

Microstructures are observed in regular areas of the crystals and are then directly interpretable in terms of projected structure. They show that the Y and Ba atoms and oxygen vacancies are ordered but, concomitantly, numerous contrast modulations attest to complex structural features set up in the bulk. The microstructures can be classified as two types, related to either cation or oxygen ordering. A typical defect in cation ordering is indicated by arrows in Fig.6.25. For this focus value (-420Å), the low-electron-density zones are highlighted and the brighter spots can be assigned to the oxygen atoms corresponding to the central $[CuO_2]_\infty$ row of corner-

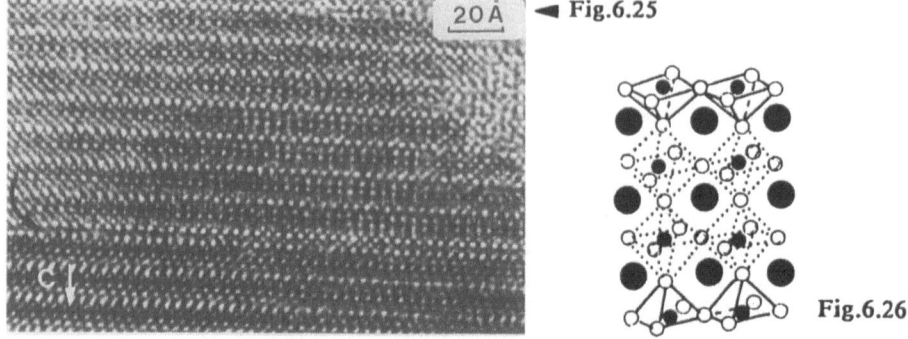

Fig.6.26

Fig.6.25. The additional CuO_2 layer is marked with a black arrow. The relative arrangement of the copper row attests to a local stoichiometry $YBa_3Cu_4O_{11-x}$

Fig.6.26. Idealized model of $YBa_3Cu_4O_{11-x}$

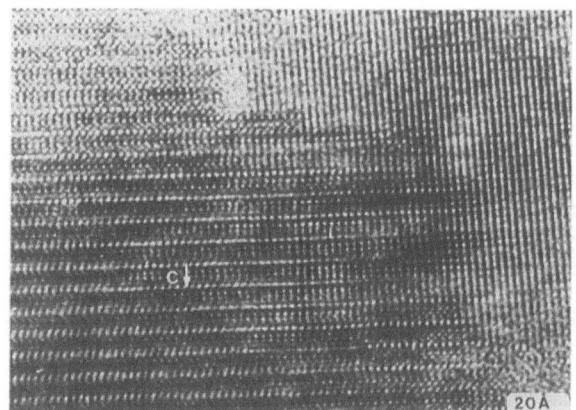

Fig.6.27. Local loss of the $3a_p$ superstructure along the c axis

sharing CuO_4 square planar groups in $YBa_2Cu_3O_7$. The three lines of smaller spots are associated with the oxygen atoms of the BaO layer, the vacancies of the Y plane and the BaO layer again, respectively. Next to one row of bright spots, we observe an identical additional row separated by 3.9Å (black arrow), a row of smaller spots intercalated between the two. This suggests that in the $[Ba_2Cu_3O_7]_\infty$ layer intercalation of additional BaO_y and CuO_y has occurred (Fig.6.26), leading to a formulation YBa_3Cu_4·O_{11-x}. Of course, the oxygen content y cannot be known with accuracy, so that CuO_6 octahedra as well as CuO_5 pyramids and CuO_4 groups may be involved between the pyramidal layers. Partial or complete loss of periodicity is observed locally. The first example corresponds to a loss of the $3a_p$ superstructure along the c_o axis (Fig.6.27) and is probably related to Y, Ba cation disorder. The second type appears as small domains (20–40Å wide) of crystallized material (Fig.6.28), whose contrast and distances suggest the existence of BaO microdomains.

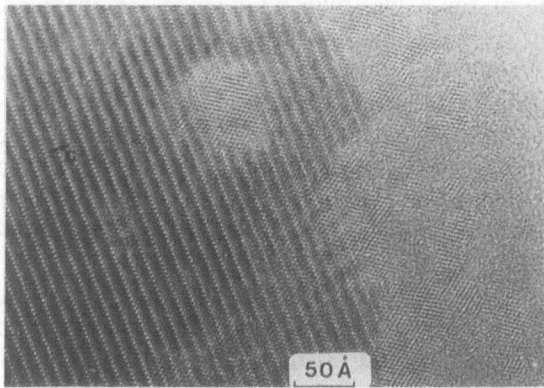

Fig.6.28. Example of microdomains embedded in a crystal. The contrast and separation of spots suggest a "BaO" nature for these domains

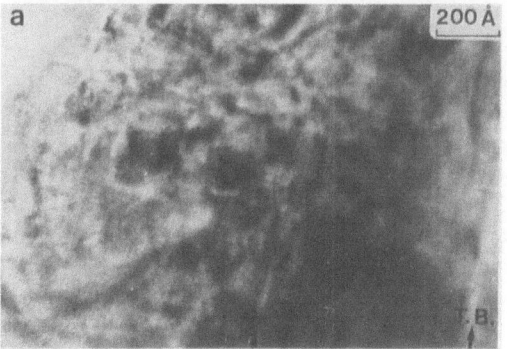

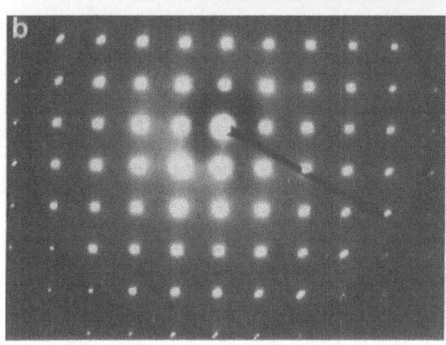

Fig.6.29a,b. Speckled appearance of the crystals $YBa_2Cu_3O_{7-\delta}$, $\delta = 0.37$ and 0.45. (a) [001] image and (b) corresponding ED pattern characterized by diffuse streaks along the a^* and b^* axes

Several types of microstructural features related to oxygen non-stoichiometry have been observed. The first feature is the speckled appearance of some crystals (Fig.6.29) encountered in samples with $\delta = 0.33$ and $\delta = 0.45$. These speckles are observed along $[110]_o$ and $[011]_o$ directions. The enlargements of such areas show that they correspond to variations of contrast resulting from strains and small local misorientations. These phenomena suggest the existence of numerous small defects or local variations of oxygen content. These disordered areas do not always extend to a whole crystal. At the right-hand side of Fig.6.29a, twin boundaries are observed. This means that the ordering of oxygens and oxygen vacancies is sufficiently established to have domains of aligned square planar groups, allowing the twinning mechanism as in $YBa_2Cu_3O_7$. The corresponding electron diffraction pattern attests, by the splitting of the spots, to the orthorhombic crystal symmetry, as for all the crystals showing a speckled contrast. The diffuse streaks along the $[100]^*$ and $[010]^*$ axes strengthen the hypothesis of local variations in the oxygen arrangements, which in this case can be considered nearly entirely disordered.

When the disorder is less complete, microstructures are observed in the crystals. The mean principal directions of the oxygen ordering leading to microstructures are $[100]_p$, $[110]_p$, $[210]_p$ and $[310]_p$. An example of ordering along the a axis is depicted in Fig.6.30; the resulting parameter is $3a_o$. Another, more complex, example is shown in Fig.6.31: an area of an orthorhombic crystal where superstructures are set up throughout the bulk with

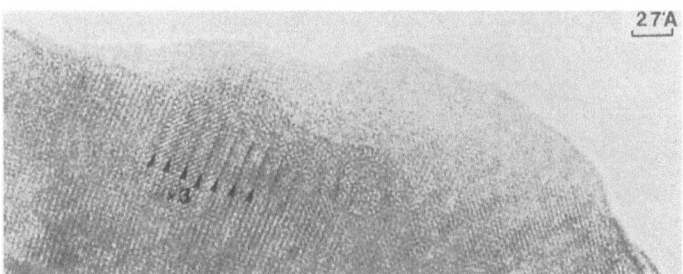

Fig.6.30. [001] HREM image of a local superstructure corresponding to a new periodicity a' = 3a

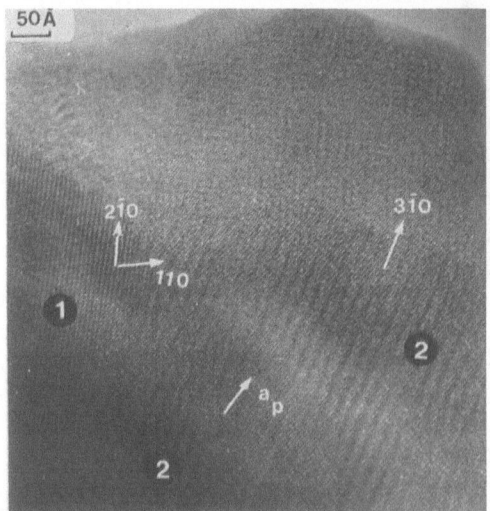

Fig.6.31. Complex modulations observed in an orthorhombic crystal $YBa_2Cu_3O_{6.66}$. New periodicities appear along the $[210]^*$ (in the areas "1") $[110]^*$ and $[310]^*$ (in the areas "2") directions of the perovskite subcell

shifted and modulated features. In fact, the superstructures are not well established, strictly speaking, over large domains, but, in very tiny areas, there are new periodicities. For example, in the zone labelled 1, two directions are enhanced, $[110]_p$, and $[210]_p$, so that contrast periodicities of $\sqrt{2}a_p \times \sqrt{5}a_p$ appear locally. In other zones, labelled 2, the $[310]_p$ direction is combined with local $2a_p$ superstructure.

Hypotheses about new oxygen orderings can be proposed to explain such phenomena. They are based on the copper disproportionation $2Cu(II) \rightleftharpoons Cu(III)+Cu(I)$ (Chap.7), assuming that threefold coordination for Cu is unlikely, and considering thus a twofold coordination for Cu(I) and the coordinations 4, 5 and 6 for Cu(II) and Cu(III) [6.19]. Some idealized models, in Fig.6.32, illustrate the way the oxygen ordering can settle along favored directions, these illustrations being far from unique. For instance, the doubling of "a" may correspond to the composition "$O_{6.5}$" (Fig.6.32a) in which $[CuO_2]_\infty$ rows of CuO_4 groups alternate with $[Cu^IO]_\infty$ rows [Cu(I) in two-fold coordination]. Tripling of "a" is explained either by the composition "$O_{6.66}$" (Fig.6.32c), in which double $[CuO_2]_\infty$ rows of CuO_4 groups alternate with $[Cu^IO]_\infty$ rows or by the composition $O_{6.33}$, in which double $[Cu^IO]_\infty$ rows of CuO_2 sticks alternate with $[CuO_2]_\infty$ rows (Fig.6.32b). Frameworks of interconnected CuO_5 pyramids and CuO_4 square planar groups can be proposed for the superstructures $2a_p \times \sqrt{5}a_p$ (Fig.6.32d), $2a_p \times \sqrt{10}a_p$ (Fig. 6.32e), and $2\sqrt{2}a_p \times 2\sqrt{2}a_p$ (Fig.6.32f) corresponding to "O_7", "$O_{6.83}$", and "$O_{6.75}$", respectively. Modulations suggest that these local variations of oxygen stoichiometry and ordering appear more or less simultaneously in the same crystal. It should be noted that the existence of a $2\sqrt{2}a_p \times 2\sqrt{2}a_p$ supercell was reported for two ranges of oxygen content: close to O_7 and close to O_6 [6.32,42,45]. For the latter composition, a model was proposed by *Krekels* et al. [6.46] based on a Jahn-Teller-type deformation of the CuO_5 pyramid deduced from electron-microscopy observations.

These complex orderings of oxygen vacancies are more clearly visible along the [001] direction, owing to the orientation of the crystal. For $[100]_p$ oriented crystals, only the superstructure along $\mathbf{a}_o$ (or $\mathbf{b}_o$) can actually be observed. Figure 6.33a exhibits a $2a_p$ superstructure, and Fig.6.33b displays a more complex one, existing along the $\mathbf{c}_o$ axis of the "123" cell. On the micrograph, two planes of atoms are implied: Y and Cu(1). The superstructure is easy to interpret as a double alternation of $[CuO_2]_\infty$ chains of corner-sharing CuO_4 square planar groups with $[Cu^IO]_\infty$ rows in which Cu(I) is in twofold coordination (Fig.6.34). The variation of the contrast in the "yttrium" layer may be correlated to cation displacements induced by oxygen disorder.

Besides all these complex structural phenomena which take place in the bulk in a more or less random way, systematic modulations of the contrast along $\mathbf{a}_o$ are observed, corresponding to the appearance of alternate rows of bright dots. They always appear in orthorhombic crystals and their direction varies from one twinning domain to the next (Fig.6.35a). The model proposed in Fig.6.32a, which corresponds to the alternation of one $[CuO_2]_\infty$ chain and one $[CuO]_\infty$ row, explains such a $2a_o$ periodicity; however, these

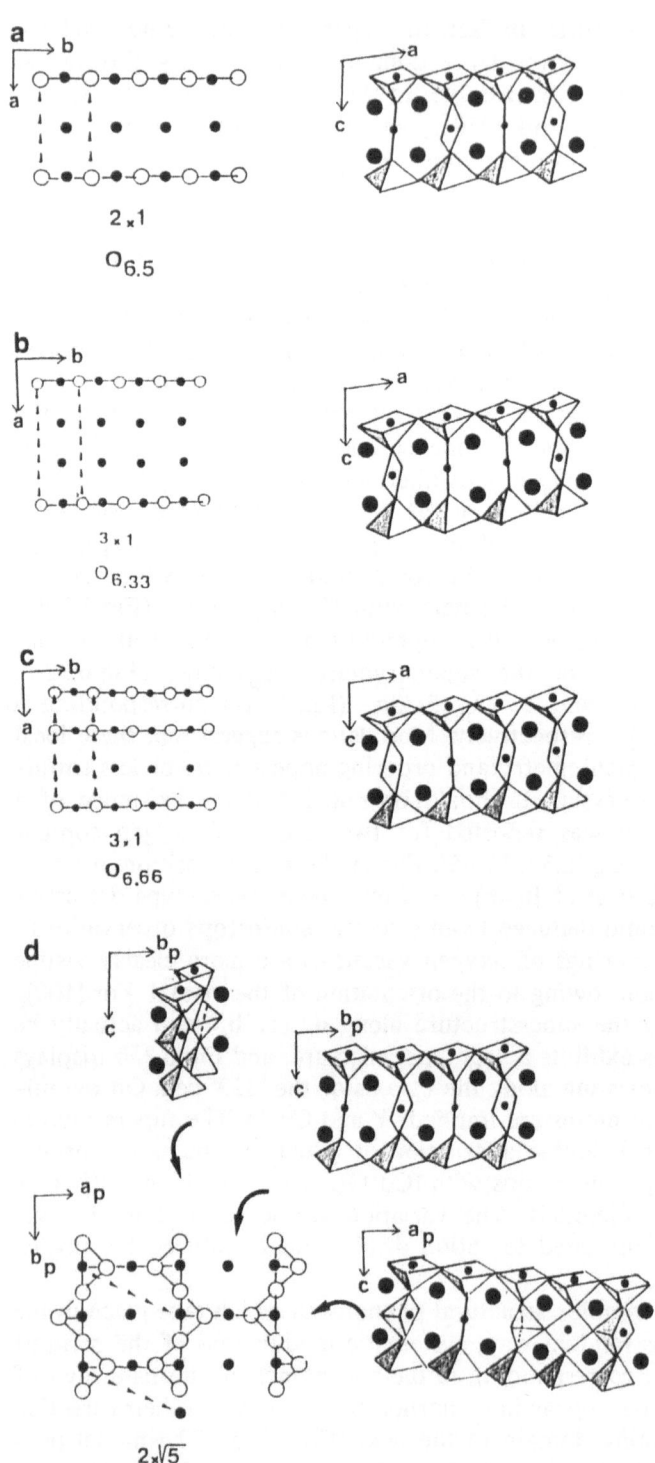

a

2 × 1

O$_{6.5}$

b

3 × 1

O$_{6.33}$

c

3 , 1

O$_{6.66}$

d

2 × √5

O$_7$

Fig.6.32a–d. The caption is given on the opposite page

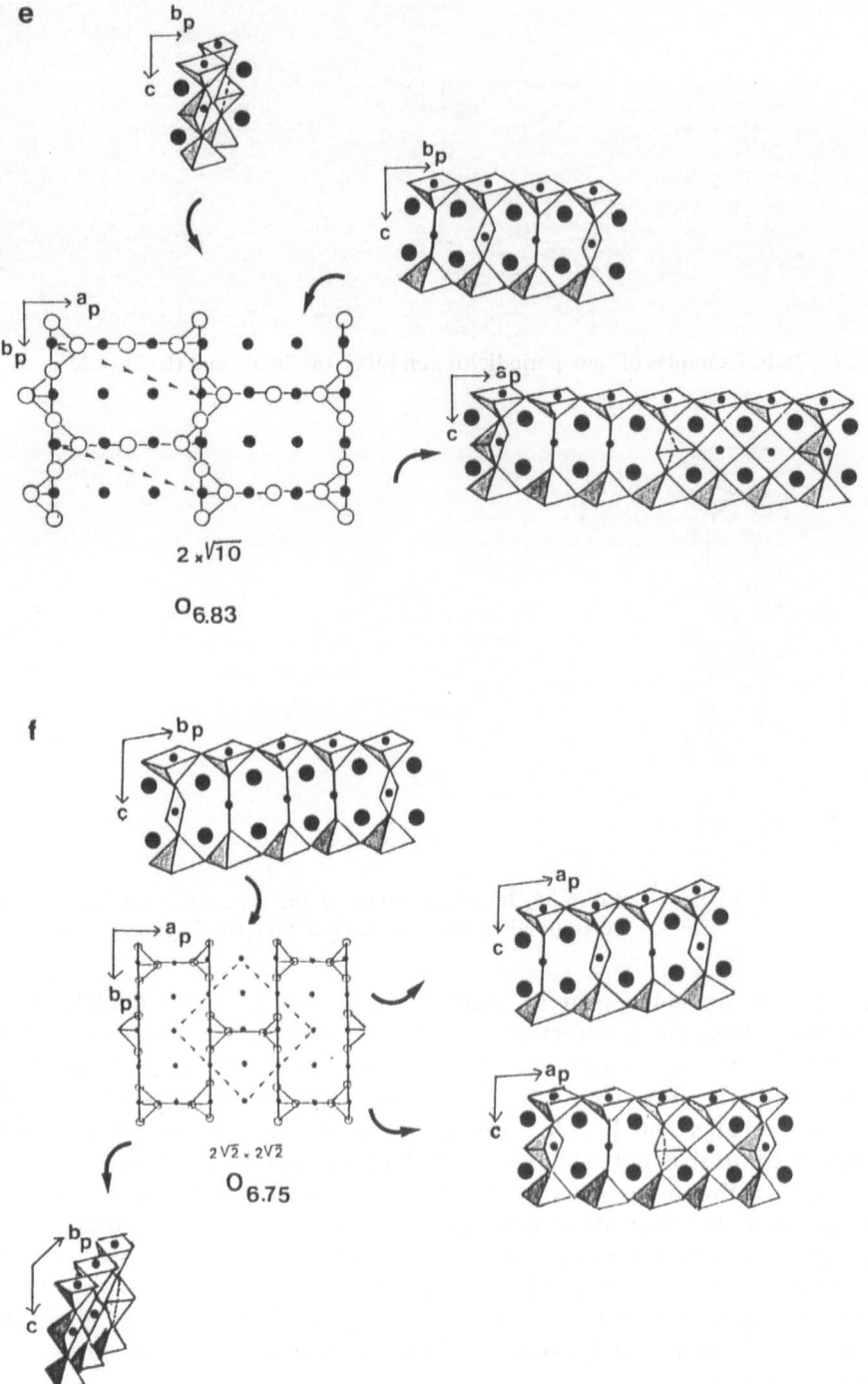

Fig.6.32a–f. Some idealized models of oxygen and vacancy ordering in the median copper plane. The new periodicities and oxygen contents are specified. The projected and perspective views along the different chains are drawn

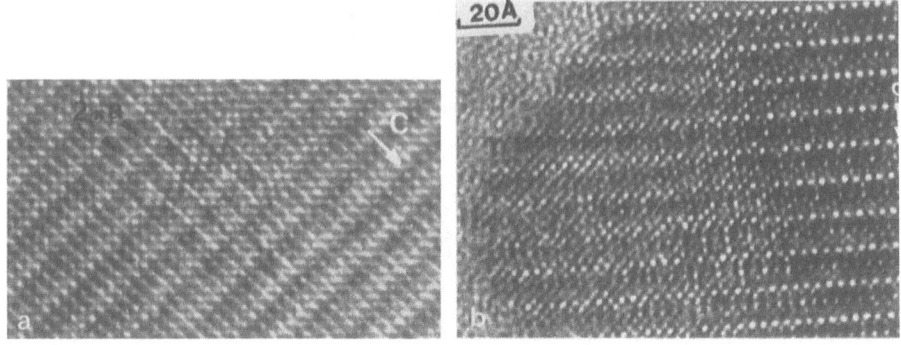

Fig.6.33a,b. Examples of new periodicities on [010]: (a) $2a_0 \times c$ and (b) $2a_0 \times 2c$

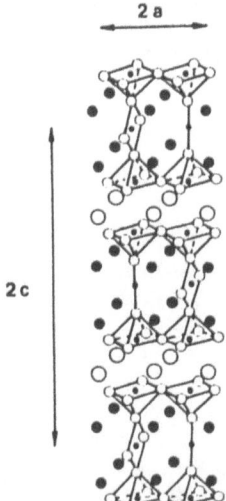

Fig.6.34. Idealized model of the superstructure $2a_0 \times 2c$. The corresponding stoichiometry is $YBa_2Cu_3O_{6.5}$

modulations do not appear as regularly as in this idealized model. These variations along the a_0 direction can easily be regarded as a way of accommodating a particular oxygen content. $YBa_2Cu_3O_{7-\delta}$ could, in this way, correspond to $[YBa_2Cu_3O_7]_{1-\delta}[YBa_2Cu_3O_6]_\delta$, i.e. a ratio $(1-\delta)/\delta$ of $[CuO_2]_\infty$ chains and $[CuO]_\infty$ rows, distributed all over the crystal. The brighter contrast is not established all along the row of white dots, and assuming that the three-fold coordination of copper is not realistic in oxides, the contrast variation along one row is more difficult to interpret. Three hypotheses can be proposed to explain this phenomenon. The first two correspond to real changes in the oxygen distribution along the row, i.e. a $[CuO_2]_\infty$ chain is transformed into a $[CuO]_\infty$ row. Two models allow the disruption to be realized, through either two additional oxygen atoms or a copper vacancy.

a) The additional oxygens imply the local formation of CuO_5 pyramids; a model is shown in Fig.6.35b, with the same schematical representation of the copper environment as in Fig.6.32.

182

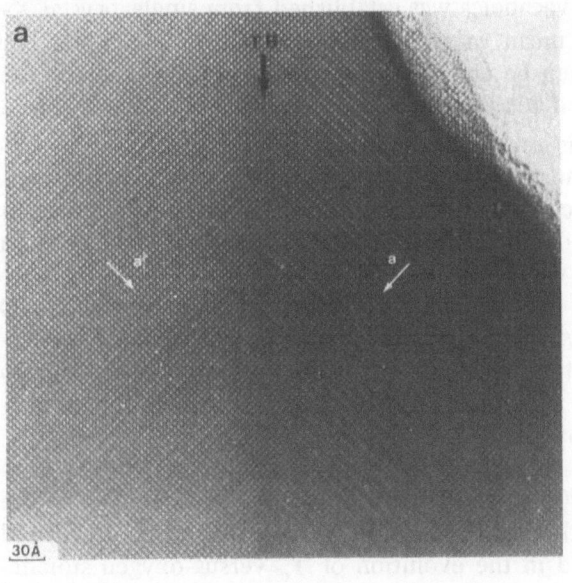

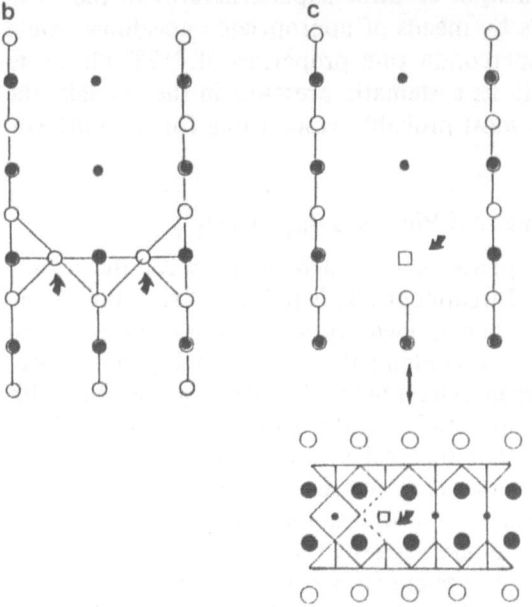

Fig.6.35. (a) Typical [001] HREM image of an orthorhombic crystal, $YBa_2Cu_3O_{6.66}$. Contrast modulations are observed corresponding to a doubling of the a parameter, whose orientation varies across the twin boundaries. (b) Model of disruption through additional oxygens (*arrowed*) that involves the formation of CuO_5 pyramids. (c) Model of disruption through a copper vacancy (*arrowed*). The corresponding projection along a is shown

b) The existence of copper vacancies was established from single-crystal X-ray diffraction studies; the mean vacancy rate on the Cu(1) site is 5%. The model in Fig.6.35c must then be considered as one possibility of chain interruption but not as a systematic feature.

The third hypothesis is related to contrast variations induced by cationic displacements. We must, indeed, keep in mind in such oxygen-deficient perovskites that the contrast is more sensitive to cation displacements than to a large variation of the oxygen content. However, the contrast variations reflect indirectly variations in oxygen content, as they induce cation displacements. In fact the variations of lattice parameters and Cu-O distances due to the changes in the oxidation state of copper [from Cu(III)-Cu(II)-Cu(I)] and the resulting changes in coordination are large enough to induce strong distortions in such mixed crystals and can explain why a uniform contrast all over the matrix is never observed.

These modulations and superstructures are observed in differently synthesized samples [6.12, 35, 40, 41] as phenomena of various intensities. The partial ordering of oxygen that is at the origin of such superstructures could explain the plateau observed in the evolution of T_c versus oxygen stoichiometry in this material. The isolation of these superstructures in the form of pure phases or single crystals by means of appropriate annealings would allow their influence on the superconducting properties of "123" phases to be explained. However, owing to its systematic presence in the crystals, the 2a superstructure appears to be most probably responsible for the 60K superconductivity.

d) Nonsuperconducting Tetragonal Phases $YBa_2Cu_3O_{7-\delta}$

The synthesis of the tetragonal phases can be achieved, as for the orthorhombic ones, by various thermal treatments. This point and structural considerations have been treated in Chap.2. Two types of tetragonal phases can be considered, related to the oxygen content: the oxygen-rich phase, labelled T_3, with $\delta \sim 0.2$ and the oxygen-poor phases, labelled T_2 and T_1, with $0.75 \leq \delta \leq 1$. The electron microscopy study of the first one (Chap.2) [6.24] showed without any ambiguity that the real symmetry of this metastable phase is orthorhombic, with a very dense microtwinning. For the oxygen poor phases, whatever the synthesis conditions, the high background observed in the neutron-diffraction study suggests the presence of amorphous or highly disordered phases, which was confirmed by the electron-diffraction studies. Two examples are considered here to illustrate the crystalline state of the tetragonal phase. The first one is a sample obtained by annealing a superconducting $YBa_2Cu_3O_7$ oxide under vacuum. The chemical and X-ray powder analyses give evidence of a "pure" phase $YBa_2Cu_3O_{6.1}$. In fact, the electron diffraction investigation reveals that few particles can be considered as representative of a $YBa_2Cu_3O_6$ phase, i.e. of an ordered oxygen-deficient perovskite structure [6.32]. The majority exhibit ED patterns made up of fine to broad rings or partial rings (Fig.6.36), which indicate a polycrystalline state, more or less well crystallized. The HREM images re-

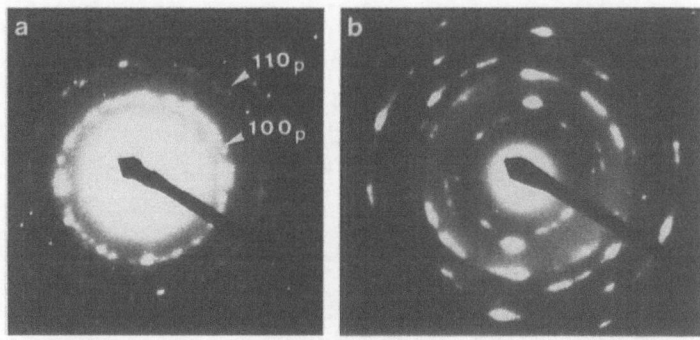

Fig.6.36a,b. ED patterns of $YBa_2Cu_3O_{6.1}$ sample. **(a)** Polycrystalline sample; rings can be indexed in the perovskite cell. **(b)** Dots cannot be indexed; the framework has been destroyed

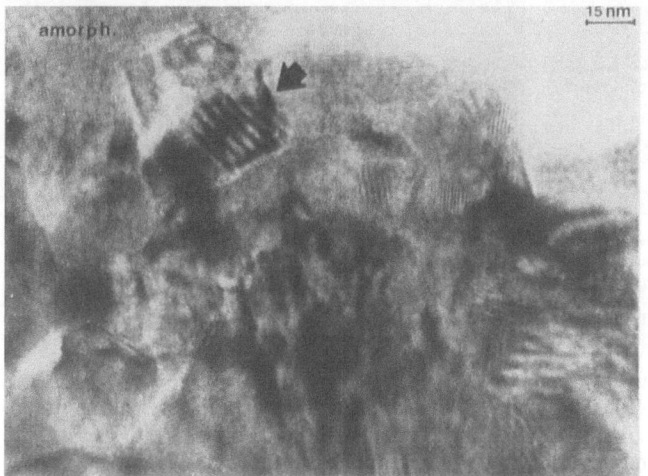

Fig.6.37. High-resolution image of $YBa_2Cu_3O_{6.1}$ sample showing the overlapping of domains (crystallized or not). Moiré patterns (arrowed) and strain contrast are observed

corded for this sample are in agreement with such an analysis (Fig.6.37). Crystallized domains, whose size ranges from some tens to some thousands of Angstroms, are embedded in an amorphous matrix; their coexistence and overlap give rise to numerous strains and moiré patterns.

The second example corresponds to a sample obtained by quenching a superconducting oxide from 980 °C to room temperature, in air. The average oxygen content, deduced from thermal and chemical analyses, is 6.55 but the neutron-diffraction results attest that the crystalline part of the sample is poorer in oxygen with an oxygen content of $O_{6.25}$ [6.37].

High-resolution electron microscopy investigation confirms that some particles and crystal areas are strongly disturbed (Fig.6.38), and that most of the crystals are coated with an amorphous layer some tens of Ångstroms thick. Such layers can influence the electron transport properties of the material, but they are not responsible for the absence of superconductivity: no

Fig.6.38. (100) high-resolution image of a disturbed crystal area

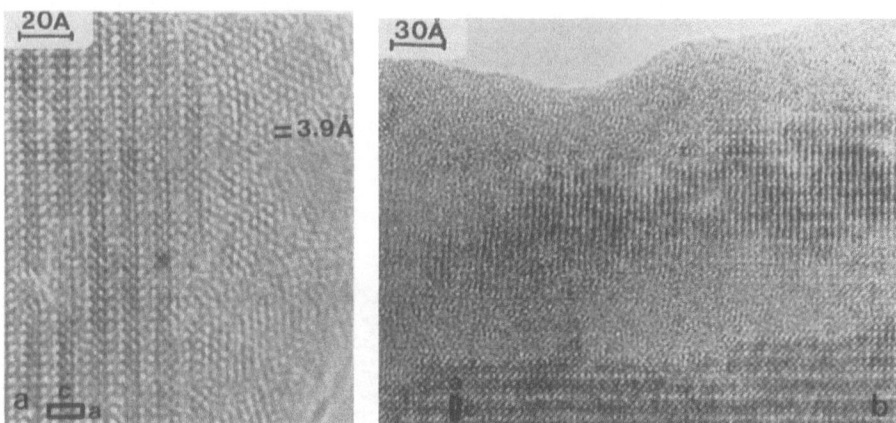

Fig.6.39. (a) High-resolution image of the pseudo-amorphous layer coating a crystal. The hexagonal array of white dots spaced at 3.9Å probably indicates BaO domains. (b) High-resolution image showing an amorphous area surrounded by a disordered atomic layer on the upper side and a $YBa_2Cu_3O_7$-like contrast on the lower side

diamagnetism was observed in the bulk. In some areas of these layers, the typical appearance of amorphous material gives way to the contrast of semi-organized and even crystallized material (Fig.6.39a). The white spots, spaced at about 3.9Å, show a hexagonal arrangement, which is consistent with the existence of BaO domains, similar to those observed in the $YBa_2 \cdot Cu_3O_{6.55}$ superconductor [6.13]. As in the previous sample, no confirmation can be given of this suggestion. Nevertheless, examination of the sample after exposure to air for several weeks produces results in agreement with this hypothesis: one observes the formation of barium carbonate. In the same way, amorphous areas are sometimes surrounded by disordered atomic layers (Fig.6.39b)

Very few superstructures have been observed in such samples, which is consistent with the rapid decrease of temperature preventing ordering of oxygen from occurring. However, variations in the contrast are clearly visible on the images. Keeping in mind the delicate problem of contrast interpretation related to variations of the oxygen content and the resulting ca-

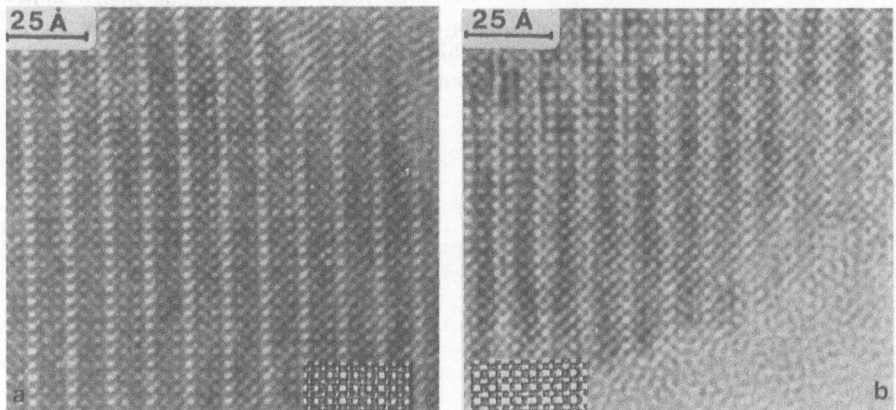

Fig.6.40a,b. [100] high-resolution image showing areas of different oxygen content: (a) oxygen-poor area, (b) oxygen-rich area. Insets: Images, calculated for $D_f = -150\,\text{Å}$, $t = 30\,\text{Å}$, corresponding to the formulae (a) $YBa_2Cu_3O_{6.2}$ and (b) $YBa_2Cu_3O_7$

tion displacements, as mentioned previously, a systematic investigation was performed. Careful examination of the images shows that, for numerous crystals, the variations of image contrast cannot be ascribed to variation of thickness, focus or orientation and are in favor of a significant change of the oxygen content from one crystal to another, or from one part of the crystal to another, as suggested by the ED study. This feature is clearly visible when the crystals are viewed along [001] or [100]. In the latter case, results of careful measurements of the ratio c/b are consistent with the values deduced from the X-ray data for different δ values. As an example of oxygen-poor and oxygen-rich areas, Fig.6.40 shows $-150\,\text{Å}$ focus images of two experimental series (Figs.6.40a and b, respectively). The images recorded along that orientation are systematically speckled with small domains of some tens of Å^2 (Fig.6.41), which correspond to local variations of the oxygen content, without any drastic disturbance of the cation framework. Such domains are also observed along [001], with some local ordering of the oxygen (Fig.6.42).

Fig.6.41. [100] high-resolution image of a crystal showing a speckled contrast, which corresponds to local variations of oxygen content

187

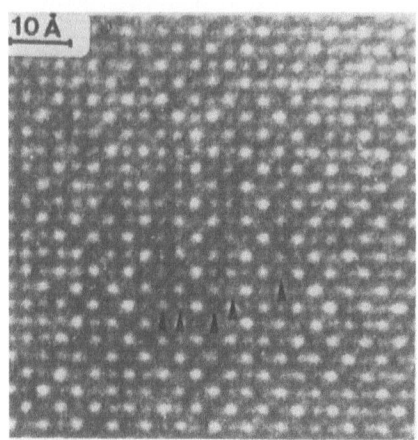

Fig.6.42. [001] high-resolution image showing dots spaced at 2.8Å. Local ordering, which appears in the form of much fainter spots (*arrowed*), is observed

e) Ordering of Oxygen Vacancies: Concluding Remarks

The large dispersion of the results dealing with the superconducting behavior and microstructural state of the nonstoichiometric $YBa_2Cu_3O_{7-\delta}$ oxides originates in the great complexity of the ordering of the oxygen vacancies. The results are only apparently conflicting, since they are related to "different" materials. The model of copper disproportionation $2Cu^{II} \rightarrow Cu^I + Cu^{III}$ [6.26] allows the more complex structural and microstructural features to be described, taking into account such essential parameters as copper coordination, through the coexistence of domains and/or chains. These domains are characterized by different oxidation states and coordination of copper: sticks, square groups, pyramids and octahedra [6.24, 26], which involve different oxygen contents: O_6, O_7, $O_{7.5}$ and O_8 respectively. The area or the length of these domains or chains depends on the thermal treatment and all the intermediate structural states can be generated:

- $YBa_2Cu_3O_7$ and $YBa_2Cu_3O_6$ are the simplest examples of perfectly ordered crystals.
- The alternating O_7 and O_6 rows can be roughly organized (Figs.6.35a and 6.32a) or established throughout a crystallite.
- Metastable superstructures can occur in the form of small domains some tens of angstroms wide [6.35].
- Frequent changes in direction of the chains due to the presence of octahedra and pyramids in the intermediate layer and absence of concerted ordering from one perovskite slice to the next, lead to "apparently" tetragonal phases, more or less disturbed.

The thermal treatment is the key to the microstructural state of the $YBa_2Cu_3O_{7-\delta}$ oxides and, as a consequence, to the physical behavior. No one compound is the archetype of these nonstoichiometric oxides; rather, each one is the product of a particular thermal history.

6.1.2 $La_{3-x}Ba_{3+x}Cu_6O_{14+y}$ Phases

The existence of a wide homogeneity range for the phases $La_{3-x}Ba_{3+x}\cdot Cu_6O_{14+y}$ ($0 \leq x \leq 1$) with a "123"-type structure is the first evidence of the different behaviors of the Y and La systems. The second indication is the easy achievement of an apparently tetragonal superconductor $LaBa_2Cu_3 \cdot O_{7-\delta}$ ($x = 1$) but the failure of attempts to prepare $La_3Ba_3Cu_6O_{14}$ with superconducting properties. Is the mechanism of superconductivity different from that in the yttrium system? The electron microscopy study of the two extreme compositions $LaBa_2Cu_3O_{7-\delta}$ ($x = 1$) and $La_3Ba_3Cu_6O_{14}$ ($x = 0$) gives an answer to this question.

a) $LaBa_2Cu_3O_{7-\delta}$

The main characteristic feature of the "tetragonal" superconducting samples concerns the existence of numerous domains, which, obviously, result from different structural phenomena.

While many crystals appear, from the ED investigation [6.47], to be truly tetragonal, others exhibit microdomains of some hundred angstroms width, undoubtedly related to the existence of microtwins: they exhibit the characteristic splitting of the reflections in the ED patterns (Fig.6.43) related to an orthorhombic distortion of the cell. It should be emphasized that, contrary to the yttrium system, the amplitude of these distortions varies from one crystal to another. An example is shown in Fig.6.43. The first crystal (Figs.6.43a and b) corresponds to a very slight orthorhombic distortion, where the splitting is only well observed on the enlargement of the 220 reflections, while the second (Figs.6.43c and d) exhibits a clearly visible splitting. In both cases, the twinning domains are easily dark-field imaged (Fig.6.43b and d). Thus, two sorts of crystals exist in this pseudo-tetragonal superconducting oxide $LaBa_2Cu_3O_{6.7}$: rigorous tetragonal and orthorhombic crystals. This deviation from the tetragonal character, which is variable, could not be observed by powder X-ray diffraction, except as a broadening of some reflections. The superconducting behavior of such a tetragonal sample is therefore explained by the existence, in the bulk, of orthorhombic and superconducting crystals that allow the percolation to be achieved but are not directly detectable in X-ray diffraction patterns.

The existence of such distorted crystals suggested the possibility of isolating the different phases by suitable thermal treatments. The "pure" phases were synthesized, either the tetragonal phase, whose behavior is semiconducting (untwinned crystals), or the orthorhombic one [6.48], which is superconducting with $T_c = 75K$.

The second type of domains are of a different nature: they are larger and arise in both the orthorhombic and tetragonal crystals. An image of two adjacent domains and the corresponding electron pattern is shown in Fig. 6.44. From the electron diffraction pattern (Fig.6.44b), it is clear that the c^* axes of the two domains (labeled I and II) are perpendicular. A schematic drawing of this ED pattern is depicted in Fig.6.44c. Circles correspond to the reflections of domain I and black points to the reflection of domain II;

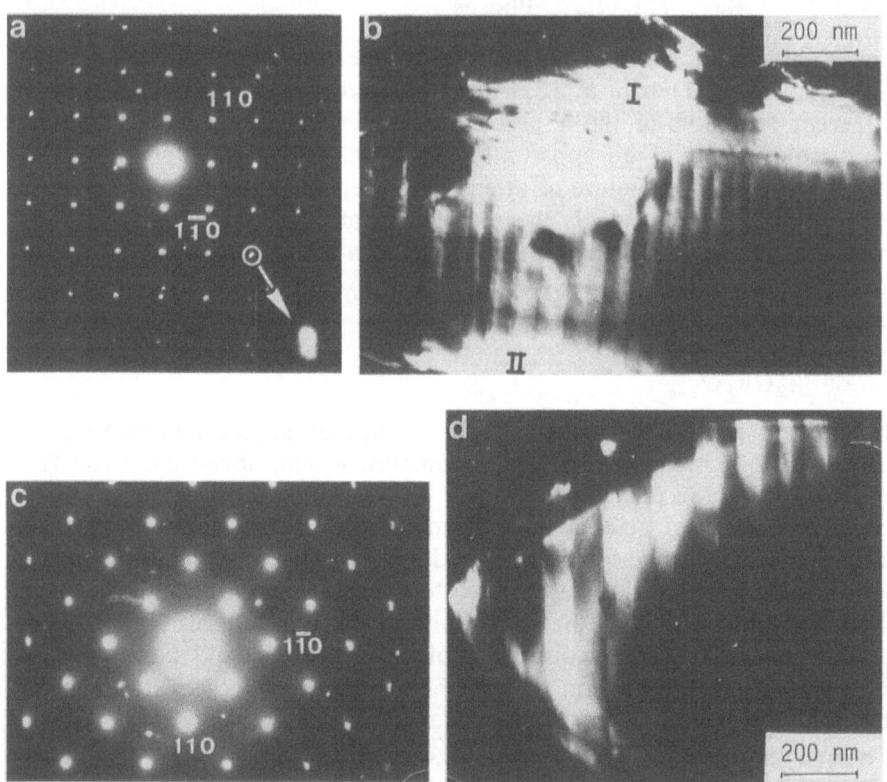

Fig.6.43a-d. Microtwins in $LaBa_2Cu_3O_7$. Crystal exhibiting a slight distortion: (a) [001] ED pattern and enlargement of the 220 reflection; (b) corresponding DF image. Crystal exhibiting a stronger distortion: (c) clearly visible splitting of the reflection on [001] pattern; (d) corresponding DF image

the stars illustrate the faint spots due to double diffraction phenomena. In this example, the [103] plane, deduced from the ED patterns, is common to both domains; the arrangement of the cells is displayed in Fig.6.44d. The low-resolution image (Fig.6.44a) shows that the boundary between the two domains does not correspond to a well-defined plane but wanders in the bulk. An idealized drawing of these oriented domains in the "123"-type structure of $LaBa_2Cu_3O_7$ is shown in Fig.6.45. The oxygen vacancies are not represented; they are located in the median copper plane (Chap.2). This scheme shows how the boundary can wander in the bulk. Two remarks may be made.

1) Such oriented domains can easily be formed, depending on the particular values of the parameter ratios a/b (equal or close to 1) and c/3a (very close to 1). They correspond to the setting up of the superstructure along the three equivalent perpendicular orientations with respect to the cubic perovskite subcell. This behavior is different from that observed for the superconductor $YBa_2Cu_3O_{7-\delta}$ for which a/b and c/3a are significantly different from 1, in both the orthorhombic and tetragonal phases. However,

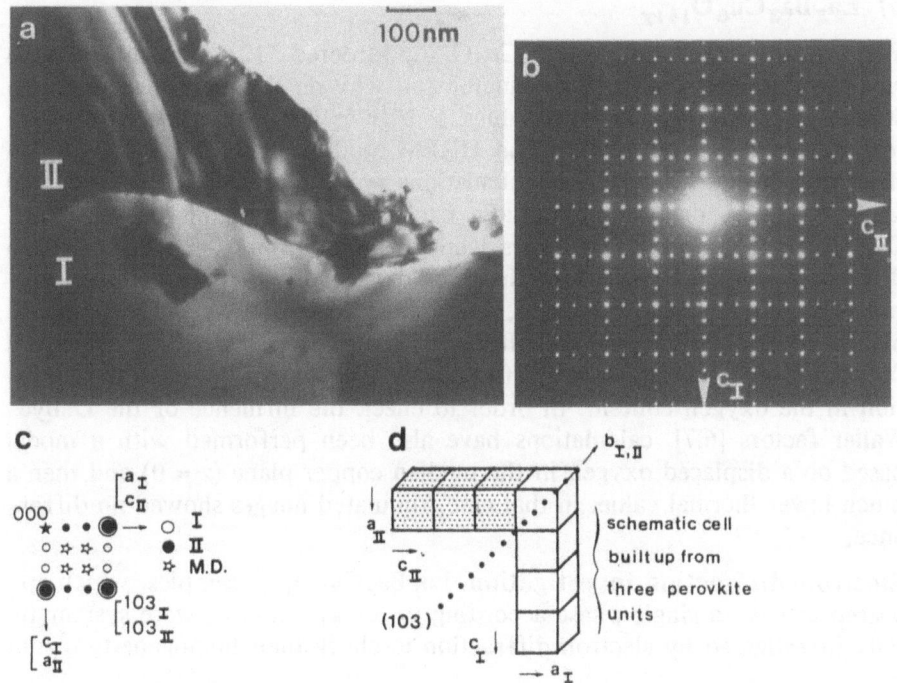

Fig.6.44a-d. Oriented domains in LaBa$_2$Cu$_3$O$_7$. (a) Low-resolution image; (b) ED pattern; (c) schematic drawing of the ED pattern; (d) relative cell orientations

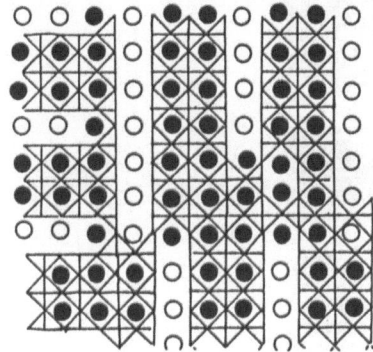

Fig.6.45. Example of an imaginary defect showing the way the oriented domains can occur and how the boundaries wander in the bulk. Oxygen vacancies are not represented. Circles are La atoms with eightfold coordination and dots are La or Ba

such phenomena were sometimes observed in the yttrium ceramics [6.15, 29], for low calcination temperatures.

2) The way the boundary wanders in the bulk can be compared to the shifting of the yttrium planes in YBa$_2$Cu$_3$O$_{7-\delta}$ (see the end of Sect. 6.1.1b and Figs.6.21, 22]. The implicated mechanism, which involves a reversal of the barium and lanthanum layers, is favored by the possible partial occupancy of barium ion sites by lanthanum.

b) $La_3Ba_3Cu_6O_{14+y}$

Why cannot the oxide $La_3Ba_3Cu_6O_{14+y}$ [ordered "123"-like and Cu(III) rich] be obtained as a superconductor and why do the neutron-diffraction studies [6.49, 50] show unusual values, as high as $11\text{\AA}^2$, for the thermal factors of some oxygen atoms? The HREM study of this oxide focuses on these two issues. The image calculations were performed for tetragonal $LaBa_2Cu_3O_{7-\delta}$, orthorhombic $LaBa_2Cu_3O_{7-\delta}$ and tetragonal $La_3Ba_3Cu_6 \cdot O_{14}$ (Fig.6.46), whose parameters were obtained from X-ray and neutron diffraction results. Several hypotheses of oxygen distribution were considered but, as previously reported, in such structures the contrast depends mainly on the cation positions. It results that, as in the yttrium system, a few defocus values allow the difference to be detected, for a strong variation in the oxygen content. In order to check the influence of the Debye-Waller factors [6.7], calculations have also been performed with a model based on a displaced oxygen in the median copper plane (z = 0) and then a much lower thermal value. In that case, simulated images showed no difference.

Electron-diffraction investigation. $La_3Ba_3Cu_6O_{14+y}$ samples, which appeared to have a single phase according to X-ray patterns, were systematically investigated by electron diffraction to check their homogeneity. From

$La_3Ba_3Cu_6O_{14}$

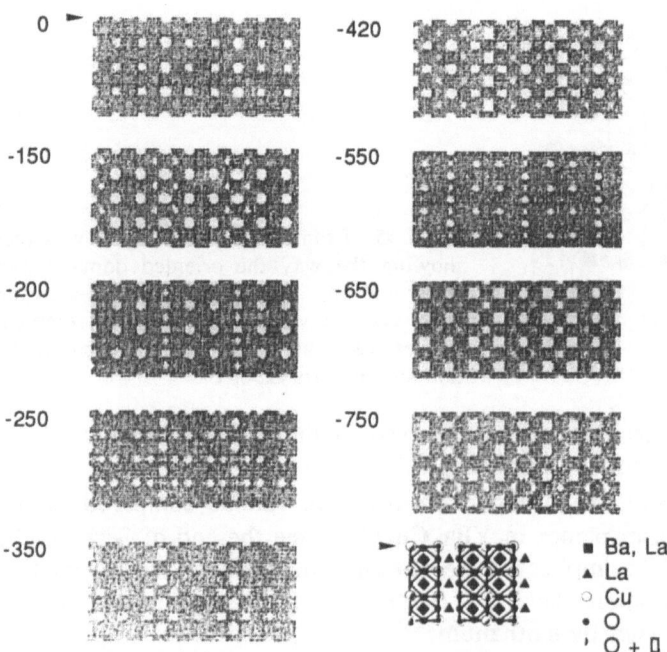

Fig.6.46. Example of calculated through-focus series for tetragonal $La_3Ba_3Cu_6O_{14}$. The focus values are in angstroms

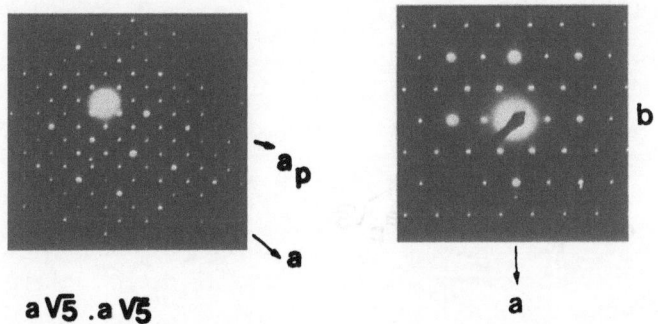

a√5 . a√5   a

Fig.6.47. ED patterns of crystals with a $\sqrt{5}a_p \times \sqrt{5}a_p$ supercell (~2% of the crystals)

Fig.6.48. ED pattern of crystals with a $\sqrt{2}\times2\sqrt{2}a_p$ supercell (~1.5% of the crystals [001] with h+k = 2n

about 300 crystals, the matrix can be characterized by the following distribution of the cell parameters.

i) 2% of the crystals are tetragonal with $\sqrt{5}a_p \times \sqrt{5}a_p \times a_p$ (Fig.6.47); they correspond to the tunnel structure of $BaLa_4Cu_5O_{13+\delta}$ previously described [6.51], which exhibits a metallic behavior. This oxygen-deficient perovskite will be labelled $\sqrt{5}\times\sqrt{5}$.

ii) 1.5% of the crystals are orthorhombic $\sqrt{2}a_p \times 2\sqrt{2}a_p \times a_p$ (Fig.6.48) symbolized $\sqrt{2}\times2\sqrt{2}$. The weak reflection corresponding to a 5.5Å periodicity which is often observed in the X-ray patterns may be generated by the presence of this phase. The structure of this compound is not yet known. Similar parameters were observed in manganese oxygen-deficient perovskites built up from MnO_5 pyramids [6.47, 48]; however, the conditions limiting the reflections (h+k = 2n) suggest that the superstructure is different in that system.

iii) 96.5% of the crystals are tetragonal with $a_p \times a_p \times 3a_p$, symbolized 1×3. Some diffraction patterns (three examples in Fig.6.49) show the existence of extra spots. Figure 6.49a corresponds to a coherent superposition of two superstructures: $\sqrt{5}\times\sqrt{5}$ and 1×3. Figure 6.49b is characterized by the superposition of $2\times2\sqrt{2}$ and 1×3 superstructure reflections with strong perovskite subcell spots forming arcs. Some spots of the second arc (P 002) coincide either with the 240 reflection of $\sqrt{2}\times2\sqrt{2}$ or with the 006 reflection of 1×3, in agreement with the directions of the superstructures with respect to the perovskite cell. Figure 6.49c is a typical pattern, corresponding to the superposition of two 1×3, oriented at 90°, similar to those observed in $LaBa_2Cu_3O_{7-\delta}$. They result because it is possible for the superstructure (c = $3a_p$) to be set up in the three equivalent orientations of the cubic perovskite subcell.

High-resolution study. The experimental images [6.50] reveal that a very large majority of the particles are not single crystals but consist of a mosaic structure. The size of the domains is highly inhomogeneous, ranging from some hundred $Å^2$ to almost 1 μm^2, the smallest domains being the most

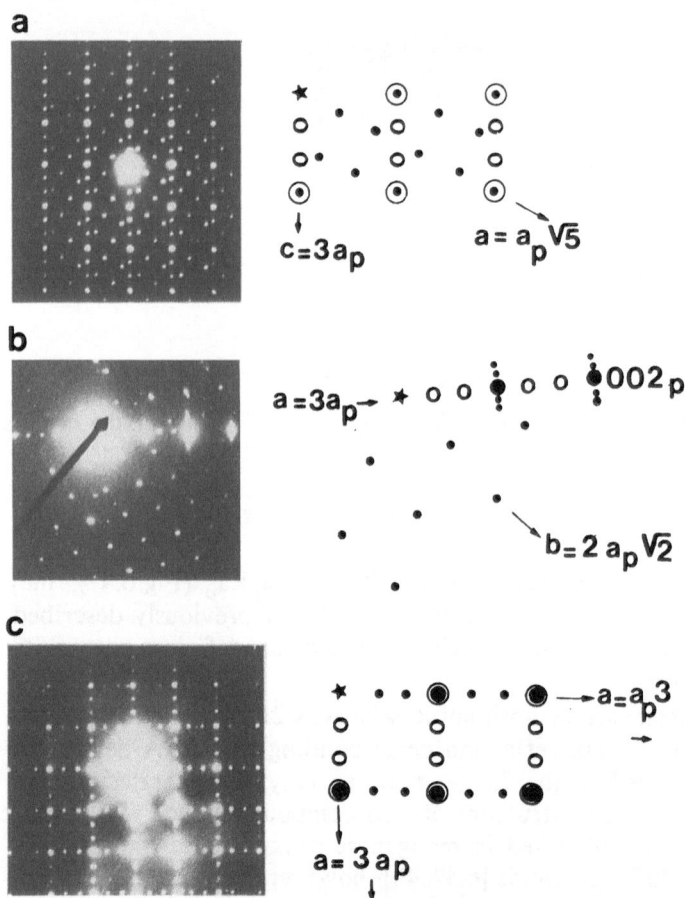

Fig.6.49a–c. Some ED patterns of the "123" $LaBa_2Cu_3O_7$ exhibit extra spots. (a) Superposition of $a_p \times 3a_p$ and $\sqrt{5}a_p \times \sqrt{5}a_p$. (b) Superposition of $a_p \times 3a_p$ and $\sqrt{2}a_p \times 2\sqrt{2}a_p$. (c) Superposition of two cells oriented at 90° to each other

numerous. Three selected typical images (Fig.6.50) clearly illustrate the nature of the samples. Figure 6.50a represents an image of an array of small domains, Fig.6.50b shows a domain area with a highly disturbed contrast and Fig.6.50c presents an image of a regular crystal. On the images, the periodicity with respect to the cubic perovskite subcell is specified: $\sqrt{5}a \times \sqrt{5}a$ refers to $BaLa_4Cu_5O_{13}$, and $a \times 3a$ to "123" type. The small arrow under "3a" indicates the orientation of the c axis of the "123"-type domains.

The nature of the domains is in complete agreement with the ED observations. Some areas characterized by a $\sqrt{5} \times \sqrt{5}$ structure are observed (Figs.6.50a and 51a). The corresponding diffraction pattern (Fig.6.51b) clearly shows that these domains take both possible orientations in the matrix, as previously shown (the superimposed $a \times 3a$ domains are indicated by black arrows).

The study of these domains shows that the interfaces are perfectly coherent but that the domain boundaries are not well defined and rarely

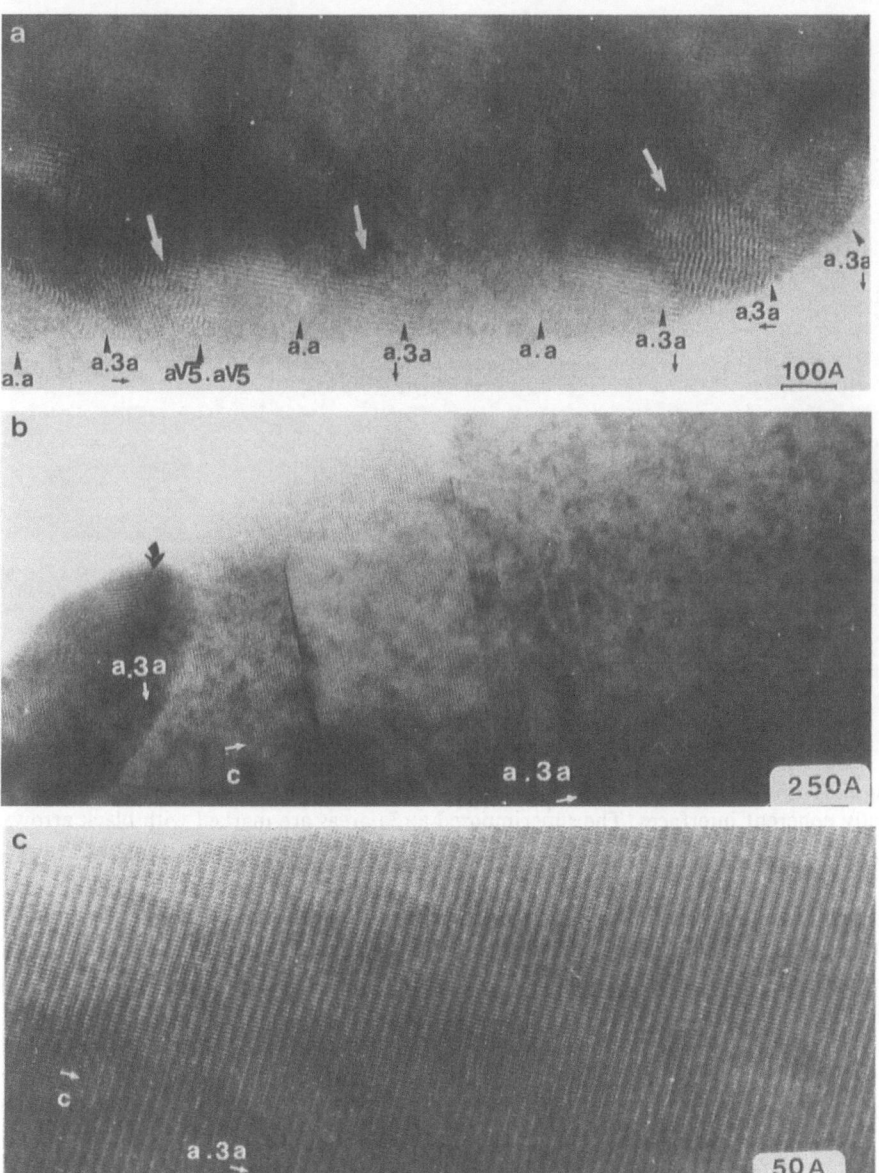

Fig.6.50a–c. Three typical HREM images of $LaBa_2Cu_3O_7$. (a) "Mosaic structure" built up of small domains corresponding to 90° oriented domains and $\sqrt{5}a_p \times \sqrt{5}a_p$ domains. They are marked with black arrows and the multiplicity of the a_p parameters is specified. The small arrow under the a×3a domains shows the orientation of the c axis. The white arrows indicate zones of multishifting. (b) Image of a disturbed crystal. Note the superposition of two oriented domains (*black arrow*), the speckled appearance of the crystal and stacking defects. (c) Perfect crystals are extremely rare; this image is an example of one

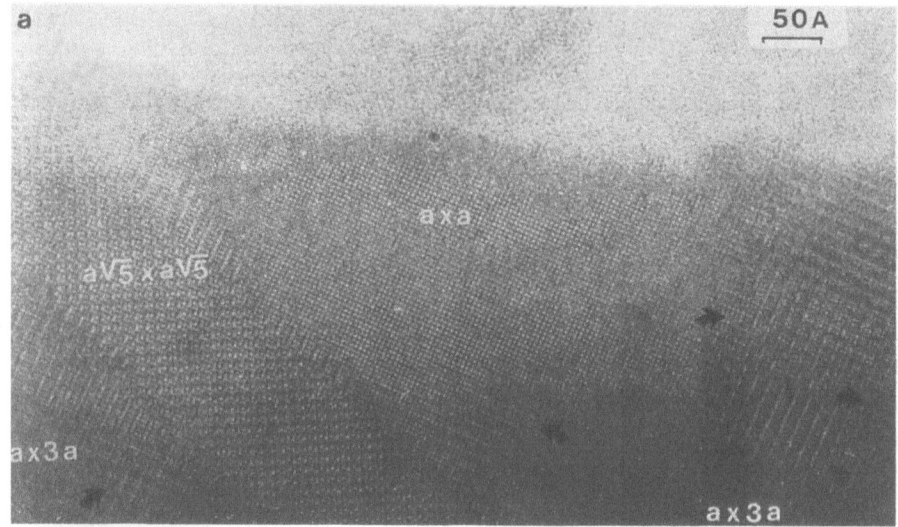

Fig.6.51a,b. Coexistence of $\sqrt{5}a_p \times \sqrt{5}a_p$ cells with the "123" structure. Note the perfectly coherent interfaces. The superimposed a×3a areas are marked with black arrows. (a) HREM image. (b) ED pattern

correspond to a particular plane. In the same way, the shifting of the layers, which was previously described (end of Sect.6.1.1b) becomes in this oxide a systematic feature in areas of small domains. Two examples are shown in Figs. 6.50a and 52.

These three phenomena - various superstructure types with coherent interfaces, oriented domains and easy multishiftings, which occur in the perovskite matrix without any drastic perturbation - are related to two structural characteristics: the easy migration of the oxygen vacancies and the size of lanthanum, which plays a double role. First it allows, besides the establishment of order (c $\simeq$ 3a_p), the partial replacement of barium in the perovskite layers. The idealized drawing in Fig.6.53 illustrates the way the different superstructures and interfaces occur in the matrix. But, second, this capacity of replacement is limited and the preservation of the La/Ba ratio will require areas richer in lanthanum. Lanthanum is assumed to tend either to be interleaved between the layers $[Cu_3O_7]_\infty$ or to occupy distorted hexagonal tunnels, like in $La_4BaCu_5O_{13+\delta}$ [6.51], owing to its size (smaller than barium's). Thus, the combination of the number of domains and multishifting, which allows a greater number of lanthanum ions to be interleaved between them, as well as the formation of $\sqrt{5} \times \sqrt{5}$ domains involving

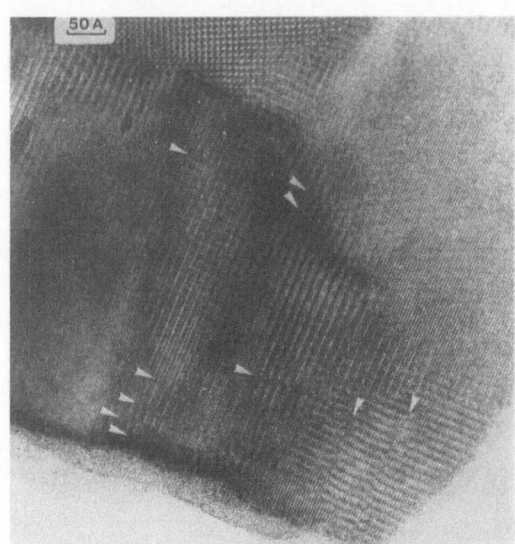

Fig.6.52. Multishifting of the layers is a common feature, indicated by white arrowheads in this HREM image

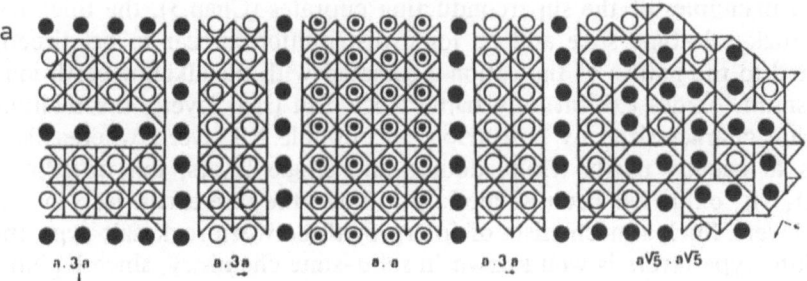

a.3a   a.3a   a.a   a.3a   a√5.a√5

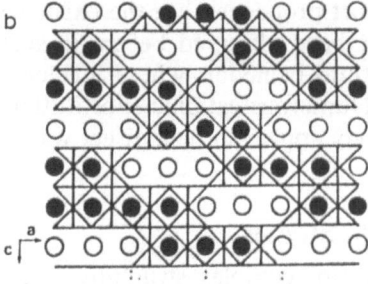

Fig.6.53. Idealized drawing showing the way (a) the different domains occur in the bulk (the multiplicity and the orientations of the c axis are specified) and (b) the layers are shifted

these tunnels, appears to be a way to accommodate an excess of lanthanum with respect to the idealized ordered structure $LaBa_2Cu_3O_{7-\delta}$ allowing the molar ratio La/Ba = 1 to be reached.

The oxide $La_3Ba_3Cu_6O_{14+y}$ is a typical example demonstrating the limit of the notion of a single phase. The average domain size observed by neutron diffraction is about 350Å (of the same order as that determined by *David* et al. [6.49] from the viewpoint of profile refinement) and coincides with the size of the domains observed by electron microscopy. Clearly, neutron diffraction "sees" only these domains, i.e. a biased average structure. The actual structure is, in fact, more complex and can be described as a juxtaposition of microphases, whose composition may vary and whose boundaries are coherent with the structure of the domains. It is also likely that the small size of these domains, induced by the presence of an excess of lanthanum, hinders the ordering of oxygen atoms. Consequently, $[CuO_2]_{\infty}$ chains formed of corner-sharing CuO_4 square planar groups similar to those observed in the superconductor $LaBa_2Cu_3O_{7-\delta}$ cannot be built for the composition $La_3Ba_3Cu_6O_{14+y}$. In the same way, the two-dimensional character of the structure is not achieved. Such a hypothesis would explain why the orthorhombic superconducting phase is never observed for the composition $La_3Ba_3Cu_6O_{14+y}$.

6.2 Nature and Ordering of the Stacked Layers: Intergrowth Mechanisms

If we consider the general formula $(AO)_n(A'CuO_{3-x})_m$ to express the building principles of the superconducting cuprates (Chap.5), the thickness of the rock-salt-type slice and its ionic distribution appear to be directly correlated to the nature of the cation associated with the alkaline earth ions. The bismuth oxides exhibit only triple Rock Salt (RS) layers; the thallium oxides are characterized by both triple and double RS layers, whereas lead appears to occupy partly triple, double and single layers, but no $(PbO)_n$ $(A'CuO_{3-x})_m$ oxide has been isolated up to now; the rare earths lead to single RS layers. Such a mechanism of intergrowth between rock-salt-type and perovskite-type layers is well known in solid-state chemistry, since K_2NiF_4 is the famous model with n = 1. However, the novel features of these families, besides their spectacular electron transport properties, are the existence of n values greater than 1, which has never been observed before in such structures and the complexity of their structural behavior, whose influence on physical properties could be significant. Numerous cationic substitutions have been performed in these oxides since they appear to be a direct way to vary the Cu^{III}/Cu_{tot} ratio.

6.2.1 Structural Considerations for HREM Studies

The role played by HREM in the recognition of complex structures is essential, as shown for example from the studies of intergrowth tungsten bronzes [6.54-57], phosphate tungsten bronzes [6.58-61] or "hexagonal" perovskites [6.62-64]. However, such an investigation needs a perfect knowledge of the basic structures as well as of the corresponding images.

Let us, as an example, consider the idealized structures built up from a double perovskite layer (m = 2) intergrown with one, two, three or more rock-salt-type layers, corresponding to n = 1, 2, 3,...,n, respectively. They are drawn schematically in Fig.6.54 without taking into account the nature of the A and A' cations. It appears clearly that when n is an even number

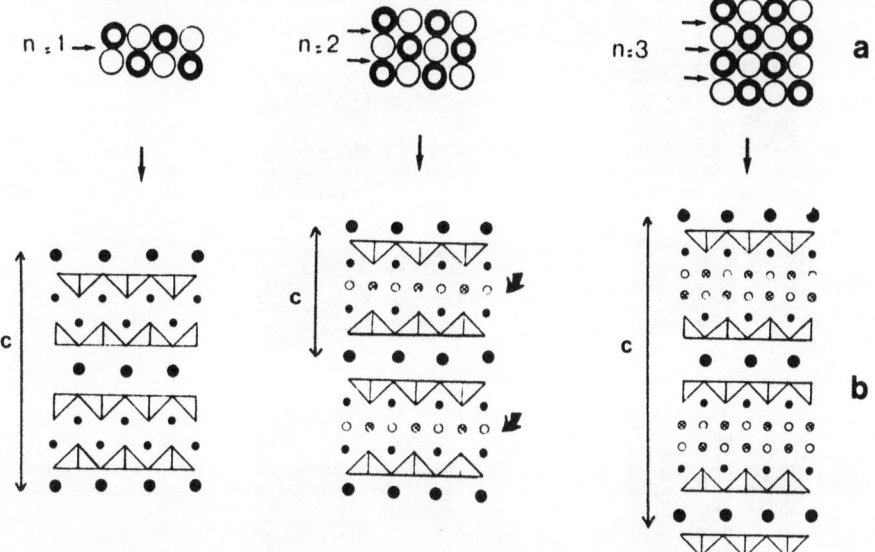

Fig.6.54. Idealized drawing showing (a) rock-salt-type layers built up from AO layers (n= 1,2,3), and (b) corresponding intergrowth phases built up from double oxygen-deficient perovskite layers (m = 2) and n rock-salt-type layers

the median AO layer (arrowed in Fig.6.54b) plays the part of a mirror and thus the c parameter can easily be calculated by considering every layer thickness. On the other hand, when n is an odd number, the mirror disappears and one observes a shifting, (a+b)/2, of alternate complete slices (rock-salt and perovskite-type layers): the c parameter is doubled and the atomic positions imply an I-type space group. If a_p is the thickness of a perovskite layer (~3.9Å) and a_R that if a rock-salt-type layer (~2.2Å), the parameters of the different members of the various families can be expressed as

even n values: $a \sim b \sim a_p$
                   $c = na_R + ma_p$
                   space group: P type

odd n values: $a \sim b \sim a_p$
                   $c = (na_R + ma_p) \times 2$
                   space group: I type

It results that the first indication of the n value in the HREM images will be obtained from the shifting, or not, of one slice with respect to the adjacent ones, when [100] images are considered.

It is more difficult to determine the m value in these structures, owing to the oxygen-deficient character of the perovskite layer and to the possible replacement of the calcium ions by either thallium or strontium. Such substitutions were observed from X-ray diffraction calculations (Chap.5).

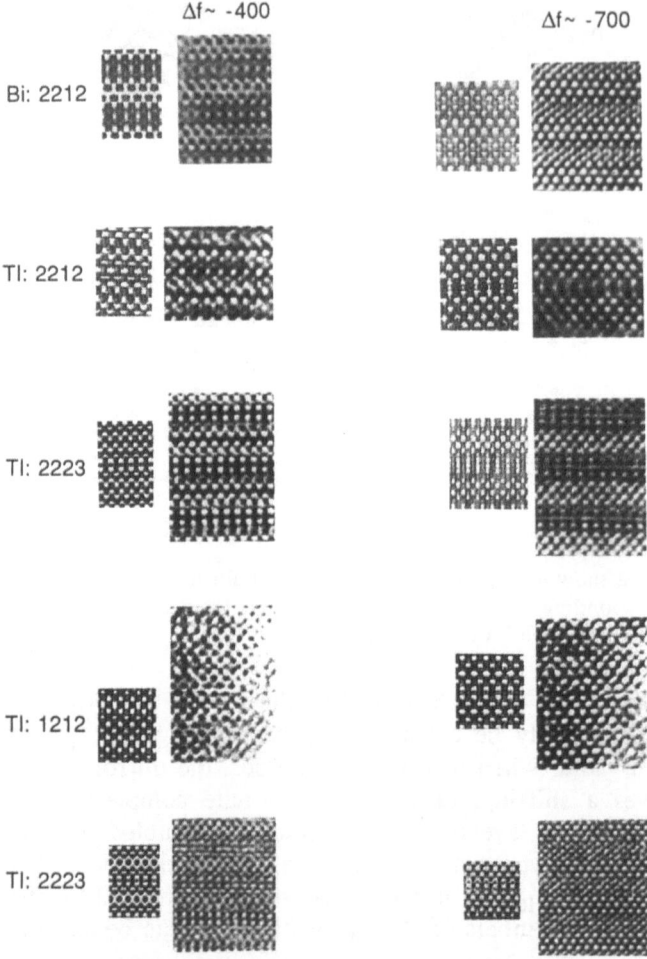

Fig.6.55. Significant images usually used during contrast interpretation. Right: focus value close to -780Å, where the white dots are correlated to the heavy atom positions. Left: close to the Scherzer value

However, two types of images are important and have to be recorded during the observations. They correspond to focus values close to the Scherzer value and reverse in both the bismuth and thallium families. These images are taken from the through-focus series shown in Chap.2 and some of them are collected together in Fig.6.55 for easy comparison. In the first column, for Df close to -400Å, the bismuth or thallium atoms appear as dark dots while the white dots correspond to the oxygen or vacancy positions. In the second column, for Df close to -780 Å, the cations are highlighted and heavier cations Bi, Tl and Ba appear brightest. In the n = 3 families the double rows of Bi or Tl with two adjacent Sr or Ba rows (respectively) are clearly visible, whereas in the n = 2 family of thallium just three staggered rows are observable. It should be noted that the calcium is not highlighted for this focus value. The lead and rare-earth cuprates exhibit, generally, a more complex contrast owing to the simultaneous occupation of the cation site in the rock-salt layer by two different atoms (Pb/Ca, Pb/Sr, Sr/Nd, etc.); the interpretation of the image contrast must then be considered for each family. Lastly, consideration of the [001] projection of these structures shows that no information about the nature of the layer stacking can be obtained, owing to the superposition of both cation and anion positions.

6.2.2 The Bismuth Family

The problems encountered in the bismuth oxides are numerous, Chap.2 deals with some of them. One of the important points is undoubtedly the characterization of the exact nature of the bismuth-oxygen double layer and the origin of the complex features observed in ED and single-crystal X-ray patterns. The existence of a double superstructure, commensurate with $a = \sqrt{2}a_p$ and $b = \sqrt{2}a_p$ with added incommensurate satellites (Fig. 6.56) along the a^* axis, was firstly observed in the $m = 1$ member, $Bi_2 \cdot Sr_2CuO_6$ [6.65]. These phenomena appear, afterwards, to be systematic [6.66]: they were observed also in the second member, $Bi_2Sr_2CaCu_2O_8$, by several researchers [6.67-74] and appear now to be a characteristic of the bismuth cuprates with double $[BiO]_2$ layers.

Are the HREM images able to give significant information about this point? Image calculations were carried out for different structural hypotheses: rock-salt-type layer (Tarascon model), Aurivillius-type layer (Sleight model) or perovskite-type layer (von Schnering model) and for different Ca/Sr ratios. Some examples of the calculated images are given in Fig.6.57. It is clear that the images are so similar that the correct hypothesis cannot be determined. The contrast obtained for the Aurivillius model, especially for defocus values close to -375Å, seems to exclude this model with an intercalated oxygen layer between the two bismuth layers. In the same way, some researchers [6.75] who observed the crystal edges (Fig.6.58a) conclude

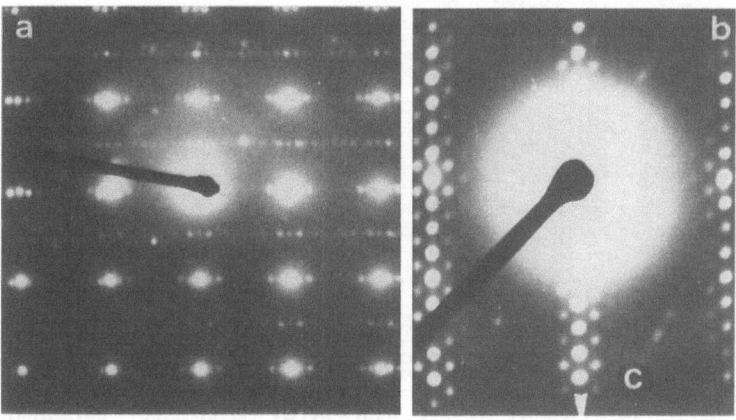

Fig.6.56. [001] (a) and [010] (b) ED patterns of the bismuth oxides. Satellites lie in incommensurate positions along a^*

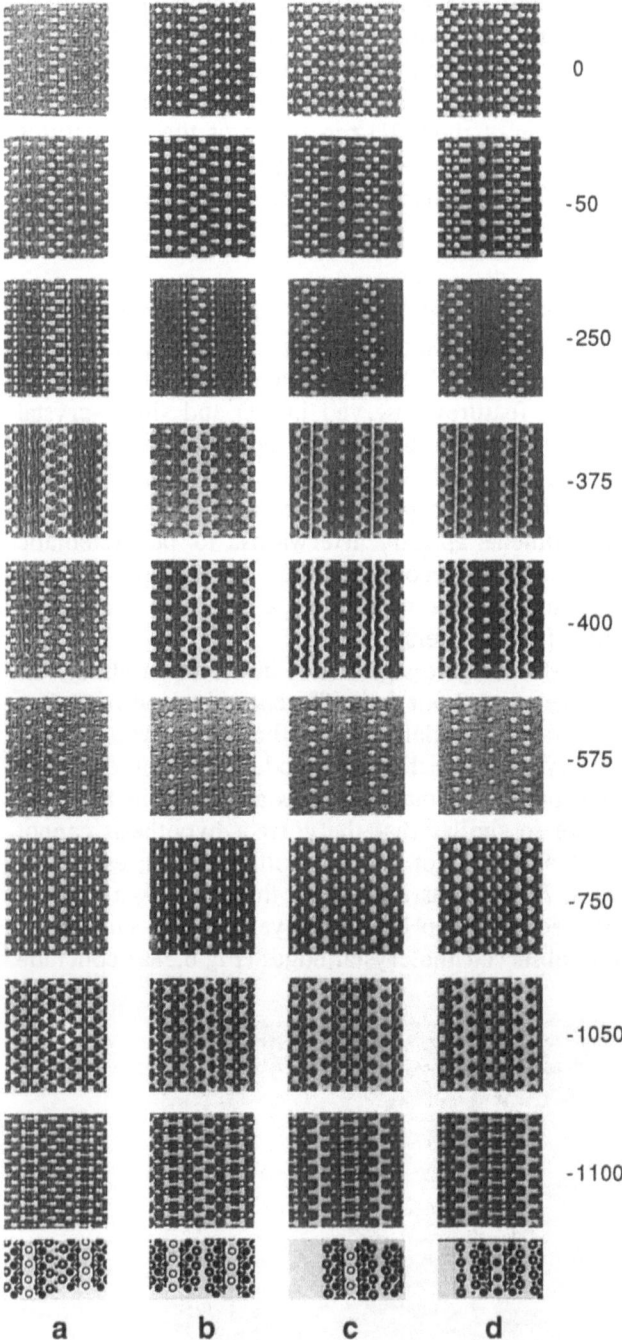

0

-50

-250

-375

-400

-575

-750

-1050

-1100

a   b   c   d

Fig.6.57a-d. $Bi_2Sr_2CaCu_2O_8$: examples of calculated through-focus series for four structural models and a thickness $t = 30.7 \text{Å} \sqrt{2}a_p \times \sqrt{2}a_p \times 30.7 \text{Å}$, zone axis [110], Space Group (SG) Amaa: **(a)** Sleight's model with Bi_2O_2 layers, **(b)** our model. $a_p \times a_p \times 30.7 \text{Å}$, zone axis [100], SG Immm: **(c)** Tarascon's model with BiO layers, **(d)** Tarascon's model with a mixed layer $Sr_{0.5}Ca_{0.5}$

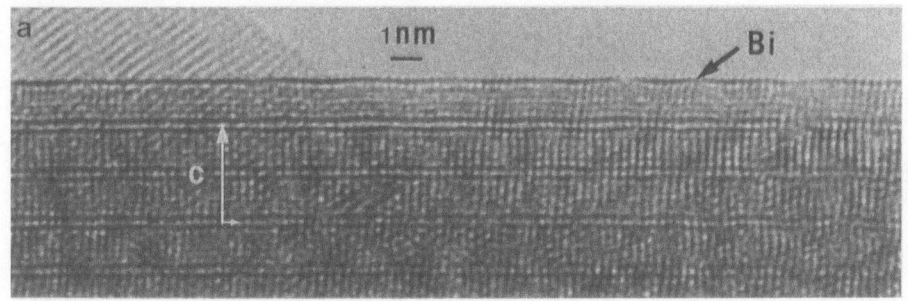

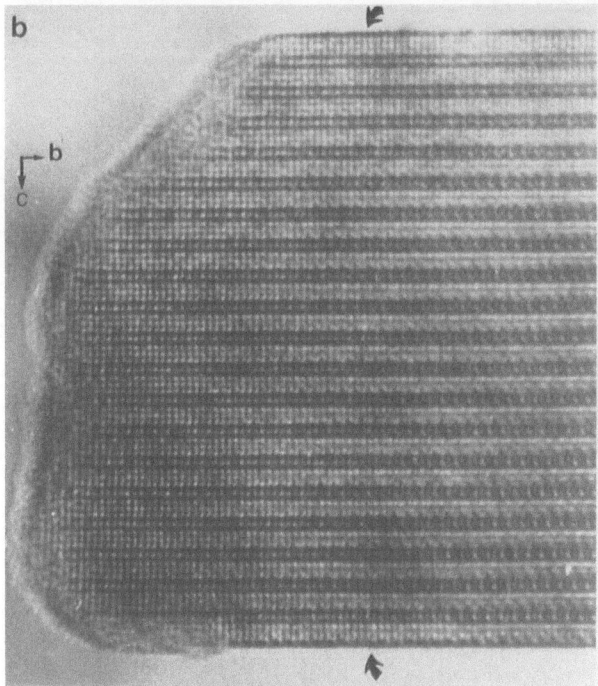

Fig.6.58. (a) Image of $Bi_2Sr_2CaCu_2O_8$ close to [010]. Note the undulations in the Bi layer and the single Bi plane at the surface [6.75]. (b) [100] HREM image of $Bi_2Sr_2CuO_6$. Note the perfect stacking of the bismuth bilayers but the single Bi plane at both crystal surfaces (*arrows*)

that an intermediate oxygen layer is not present. The systematic cleavage of the lamellae in the middle of the two bismuth layers is observed in all these bismuth oxides: two examples for $Bi_2Sr_2CaCu_2O_8$ and $Bi_2Sr_2CuO_6$, respectively, are shown in Fig.6.58a,b. This feature is probably related to the existence of the $6s^2$ lone pair of bismuth Bi(III) as discussed later.

The modulations of the contrast in the [001] images were observed by several groups [6.65-69, 72-77], see for example Fig.2.9 and Fig.6.59. In the [001] images (Fig.6.59a and b), the insertion of blocks of $4(\sqrt{2}/2)a_p$, $5(\sqrt{2}/2)a_p$ and $6(\sqrt{2}/2)a_p$ in a nonperiodic way is clearly visible. Other images along the [010] direction [6.72-77] show these phenomena in a striking

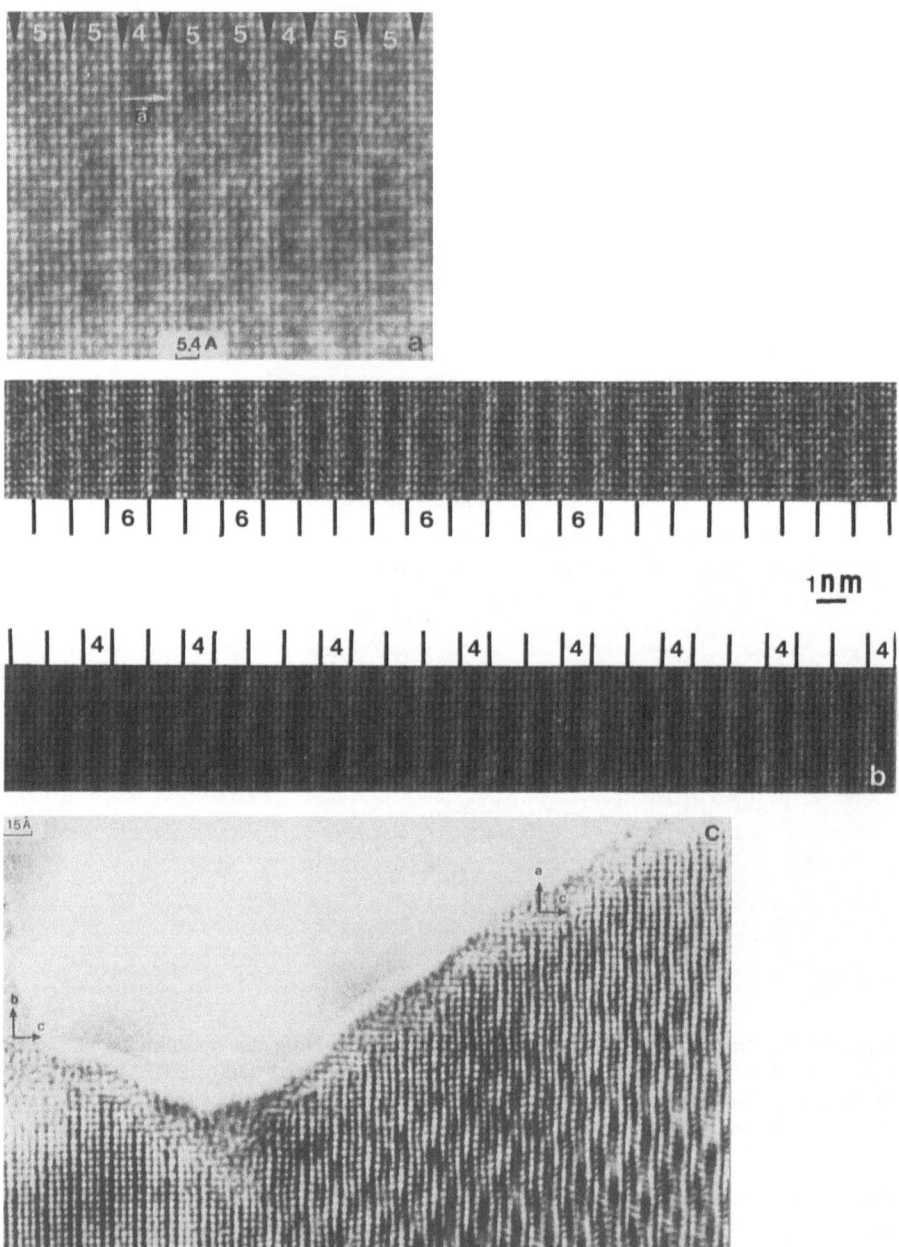

Fig.6.59. (a) Modulations in bismuth oxides. $Bi_2Sr_2CuO_6$: pseudoperiodicity of blocks corresponding to $4(\sqrt{2}/2)a_p$ and $5(\sqrt{2}/2)a_p$ in an [001] image. (b) [001] projections of $Bi_2Sr_2CaCuO_8$ crystals. The insertion of blocks of ×4 and ×6$(\sqrt{2}/2)a_p$ is not perfectly periodic. (c) HREM image of 90° oriented domains in $Bi_2Sr_2CuO_6$. The undulations of the bismuth layers are clearly visible at the right-hand side, in agreement with an [010] orientation of the crystals. The left part of the micrograph exhibits a contrast which agrees with a [100] orientation. Bismuth positions are correlated with the black spots of the image. Note the perfect interface

way. Such modulations along **a** are clearly visible on the right-hand side of the micrograph in Fig.6.59c; they are correlated with Bi ion displacements. On the left-hand side, the modulations are no longer observed, in agreement with a local [100] orientation of the lamellae, whereas the right-hand part is [010] oriented. A [001] observation of such a crystal built up from the superposition of two 90° oriented lamellae leads to ED patterns where the satellites appear to be set up along two perpendicular directions (Chap. 2). Note the perfect coherence of the interface between the two domains. As discussed in Chap.2, the real origin of these layer undulations and their possible influence on physical properties via additional oxygen or Bi or Sr vacancies are not yet understood. The influence of cationic sustitutions has allowed some points to be clarified. Substitution of iron for copper leads to the synthesis of new oxides $Bi_2Sr_{2+x}Ca_{1-x}Fe_2O_9$ [6,76] and $Bi_2Sr_4Fe_3O_{11}$ [6.78], isostructural with $Bi_2Sr_2CaCu_2O_8$ and $Bi_{2-x}Pb_xSr_2Ca_2Cu_3O_{10}$, respectively. These bismuth-iron oxides, the [n=3, m=2] and [n=3, m=3] members of the family, are characterized by the absence of oxygen vacancies in the perovskite layers, which are built up from two and three layers of FeO_6 octahedra, respectively. Despite the considerable modification in the perovskite slice, these oxides exhibit similar satellites in the ED patterns and similar modulations throughout the crystal. Examples of [001] ED patterns are given in Fig.6.60. In the same way, different substitutions were performed on Bi, Sr or Ca sites [6.79-81] and the resulting materials were studied by ED. The partial solid solutions corresponding to the couples $(Bi_{2-x}Pb_x)$ or $(Sr_{2-x}La_x)$ or $(Ca_{1-x}Y_x)$, in fact, involve an additional effect connected with a variation of the oxygen content; in these oxides a variation of the wavelength vector was observed, but no direct correlation could be established with the x value. On the other hand, the double substitution on Bi^{III} and Ca^{II} sites by Pb^{II} and Y^{III}, respectively, implies theoretically a constant oxygen content [6.82]. In the limit composition $BiPbSr_2YCu_2O_8$, the satellites in incommensurate positions do not exist any more.

These features will be discussed later (Sect.6.4.1), but they clearly show that the modulations have their origin in the double bismuth-oxygen layers.

Stacking faults are quite rare in the $Bi_2Sr_2CaCu_2O_8$ matrix [6.67,68]. One can however observe, in some crystals, defect members with a 24.4Å periodicity along the c axis (Fig.6.61a). This parameter and the observation of the contrast, correlated with the calculated images, suggest indeed that it corresponds to the $Bi_2Sr_2CuO_6$ intergrowth of triple rock-salt-type layers $[(BiO)_2 (SrO)]_\infty$ with a single perovskite layer $[SrCuO_3]_\infty$, whose structure is depicted in Fig.6.61b. The member m = 3 of the bismuth family with the formula $Bi_2Sr_2Ca_2Cu_3O_{10}$ appears to be extremely difficult to synthesize as a pure phase. Researchers [6.83,84] firstly reported a mixed matrix where there are m = 3 single crystals, and the HREM study provides evidence for the occurrence of numerous defects corresponding to different thicknesses of the perovskite layers. Several mechanisms have been proposed to explain the phase relations and the origin of the difficulty in stabilizing a regular triple perovskite layer in the Bi-(Sr,Ca)-CuO system

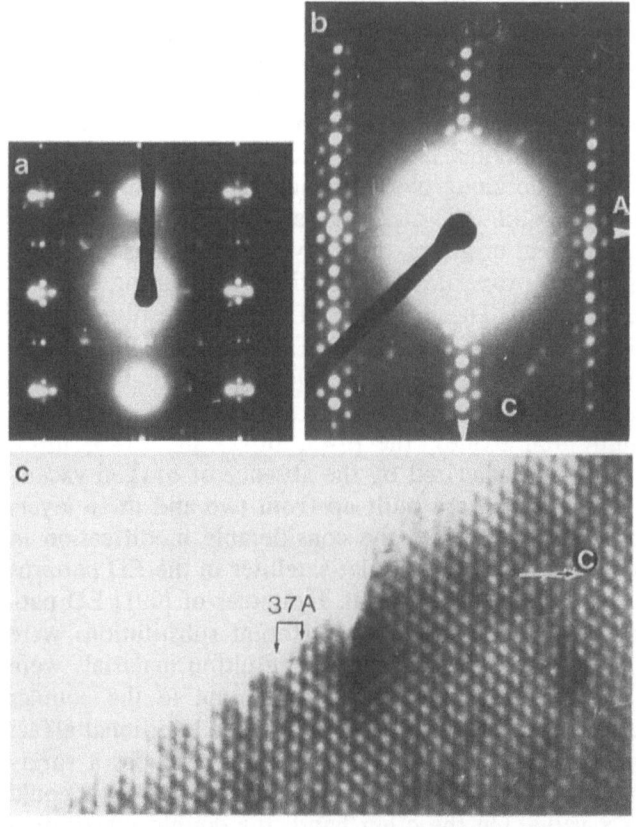

Fig.6.60a–c. $Bi_2Sr_4Fe_3O_{11}$. (a) Example of [001] ED pattern. (b) [010] ED pattern; the satellites are clearly visible. (c) Corresponding bright-field image

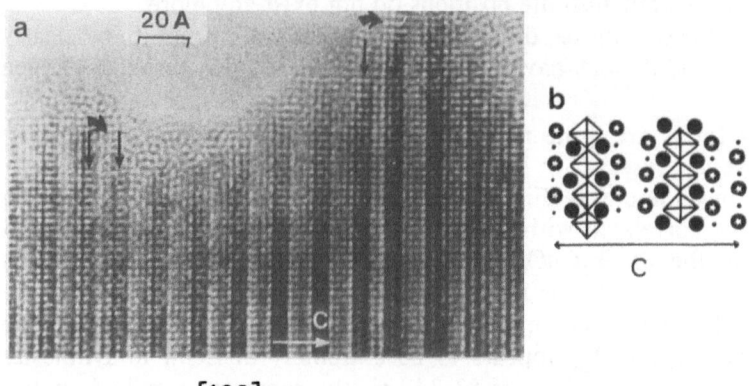

[100]

Fig.6.61. (a) [100] image of $Bi_2Sr_2CaCu_2O_8$ showing the formation of $Bi_2Sr_2CuO_6$ layers. (b) Idealized drawing of $Bi_2Sr_2CuO_6$ structure (n = 3, m = 1)

[6.85-89]. This member of the bismuth family is stabilized by a partial substitution of lead for bismuth [6.90].

Careful measurements performed on the images recorded from m = 2 samples, $Bi_2Sr_2CaCu_2O_8$, show a small monoclinic distortion in some slices. Such an effect could be related to an artefact resulting from slight bending of the lamellae but the occurence of the effects in opposite directions ($90+\epsilon$ or $90-\epsilon$) in adjacent layers is rather suggestive of an intrinsic structural phenomenon, related to the bismuth displacements.

One striking feature is the entire absence of stacking defects in the BiO layers (Fig.6.58) in contrast to the thallium compounds, as seen in the next section. This feature is probably related to the $6s^2$ lone pairs of the Bi(III) ions, which would impose a pairing of the bismuth layers.

6.2.3 Thallium Families: The Classical Defects

The two thallium families, corresponding to n = 3 and n = 2, respectively, can be considered as both simpler, owing to their structure (localization of the oxygen in the TlO layer and absence of systematic modulations), and more complex, owing to their wealth of nonstoichiometry mechanisms, than the bismuth family.

The origins of these complex structural features are multiple. Some of them can be considered as classical; they are directly related to the particular nature of these phases, i.e. intergrowth phases and oxygen-deficient perovskites, and were previously observed in other similar systems. The others are related to the existence or the absence of satellites in the electron diffraction patterns: bismuth, thallium and lead oxides can exhibit these phenomena, which will be further examined.

The "classical" mechanisms affect both the perovskite-type layers and the rock-salt-type layers, whatever the nominal n and m values, in the thallium oxides.

a) Perovskite Layers

In the perovskite layers two types of features can be considered. The first is related to the thickness of a perovskite slice in relation to the nominal composition. The second is related to the nonstoichiometry mechanisms and deals with the oxygen content and the ion ordering.

Intergrowth defects. In the structures built up from the intergrowth of two different structural types, the occurrence of layers whose thickness is different from that expected from the nominal composition is a common feature; the thallium oxides are governed by this mechanism [6.91-96]. The nature and the frequency of such a defect are generally functions of the thickness of the considered layer, i.e., in this case, they are functions of the m value. They can be summarized by the following rules:

- For low m values, the faulty layer corresponds generally to m′ values close the nominal composition, i.e. $m' = m \pm 1$.
- The m′ value can differ by a large amount when the nominal m value increases.
- The defects are more numerous as m increases.

Examples of such variations are shown in Fig.6.62, where faulty layers appear as isolated defects in a nearly perfect matrix:

1) They can be m′=2, double perovskite layers, i.e. $[BaCa(CuO_{2.5})_2]_\infty$, in a nominally m=3 matrix, $TlBa_2Ca_2Cu_3O_9$ and $Tl_2Ba_2Ca_2Cu_3O_{10}$, respectively (Fig.6.62a, b).

2) m′=2 and m′=4 in a $TlBa_2Ca_2Cu_3O_9$ matrix ($m' = m-1$ and $m' = m+1$) (Fig.6.62c).

3) In a nominally m=4 matrix, $Tl_2Ba_2Ca_3Cu_4O_{12}$, where the perovskite slice is built up from four perovskite layers, the faulty layers can be much larger. One observes besides m′=5 layers, $[BaCa_4(CuO_{2.5})_2(CuO_2)_3]_\infty$, intergrowth defects built up from seven perovskite layers, $m' = 7$, corresponding to the composition $[BaCa_6(CuO_{2.5})_2(CuO_2)_5]_\infty$, i.e. $m' = m+3$, as shown in Fig.6.62d. Note the variation of the thickness of a rock salt slice (white arrows), which will be discussed further.

Oxygen nonstoichiometry. Although the role of the oxygen pressure in the synthesis of these oxides appears to be a very important factor, the exact oxygen content and the influence of its localization in the perovskite layers on the physical behavior has yet to be explained. From the HREM images, small variations of the c parameter and of the contrast in the middle of some perovskite layers are sometimes observed at the thin edges of the crystals. Variations of the contrast were clearly observed in this way in the "1223" matrix in Fig.6.63. This variation is assumed to be due to a local change of the copper coordination and thus of the oxygen content. However, such a variation is difficult to interpret, owing to the small contribution of oxygen to the potential. In such a case, the presence of either an oxygen atom or an anionic vacancy implies a strong variation of the copper position, which would then be responsible for the contrast variation. It should be noticed that this observation is consistent with the partial occupancy of the oxygen sites in the calcium planes suggested by the X-ray diffraction data.

Ion ordering. Other variations in the contrast of the perovskite layers are observed. An example of a double defect is shown in the image of a "2212" $Tl_2Ba_2CaCu_2O_8$ matrix (Fig.6.64), where the focus value is assumed to be close to -680Å, i.e. where the white dots are correlated to the positions of the cations, except for the calcium ions. The normal sequence of the regular matrix is thus six rows of white spots corresponding to the cation sequence Cu-Ba-Tl-Tl-Ba-Cu. The defect appears as eleven adjacent rows of shifted white spots. A careful examination of the distances between the rows and of the images recorded in the through-focus series shows that this feature corresponds to a double defect, related to both rock-salt-type and

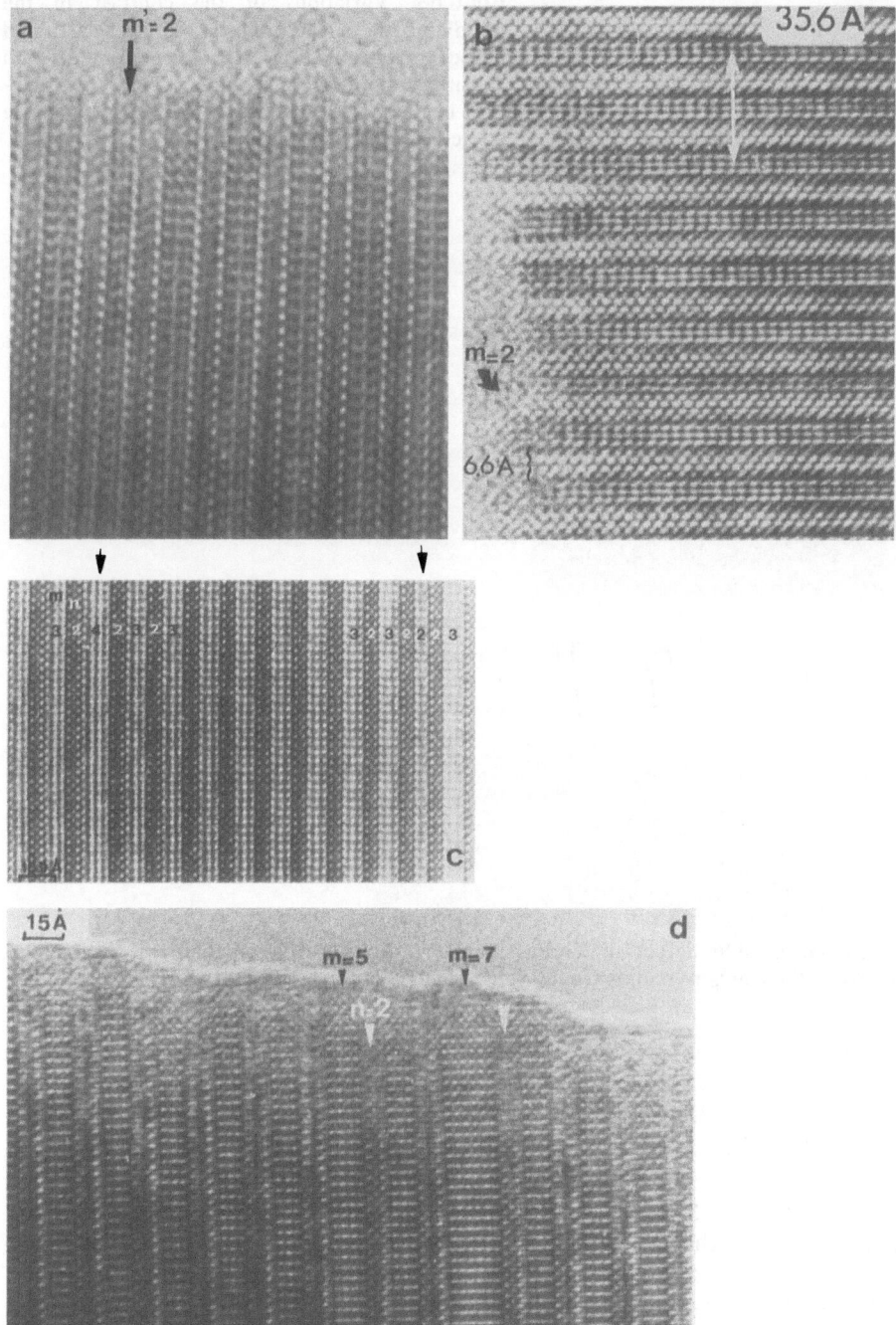

Fig.6.62a-d. Perovskite layers with defects in the thallium oxides. The defects can appear in isolation: $m' = 2$ in a nominally $m = 3$ matrix: (a) $TlBa_2Ca_2Cu_3O_9$ and (b) $Tl_2Ba_2Ca_2Cu_3O_{10}$. (c) For the first members: $m' = m \pm 1$; $m' = 2$ and $m' = 4$ are arrowed. (d) The number of defects increases with m and the m' value can then deviate greatly from the nominal composition: example of $m' = 7$ in a nominally $m = 4$ matrix

Fig.6.63. Variations of the contrast in the middle of the perovskite layer. They are assumed to be correlated with cation displacements and variations of the oxygen content. Note the variation between the two arrows with respect to the adjacent triple perovskite layer

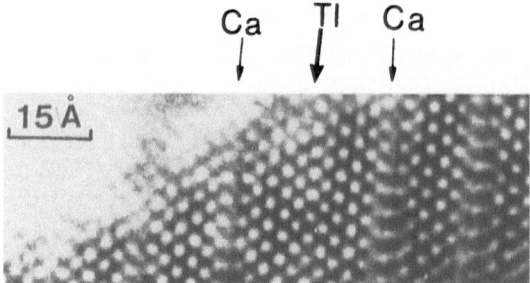

Fig.6.64. Double defect in $Tl_2Ba_2CaCu_2O_8$: Substitution of thallium for calcium between the CuO_5 pyramids (focus value $-680\,\text{Å}$), and variation of the n value

perovskite-type layers. The bright contrast of the middle row suggests that a calcium layer is replaced by a thallium layer, leading to a double oxygen-deficient perovskite layer $[BaTl(CuO_{2.5})_2]_\infty$. The latter is sandwiched between 2 thallium-deficient rock-salt-type layers $[(BaO)(TlO)]_\infty$, i.e. n′ = 2, instead of n = 3 for the nominal composition. Calculated images of both perovskite-type layers, $[BaCa(CuO_{2.5})_2]_\infty$ and $[BaTl(CuO_{2.5})_2]_\infty$, are superimposed on the experimental image: the difference in the contrast of the Ca row (very weak dots) and the Tl row (white dots) is clearly seen in both images. The eleven-row sequence can thus be represented as shown in Fig.6.65. It should be pointed out that this observation is in agreement with the double nonstoichiometry of thallium and calcium observed from the X-ray powder and single-crystal data (Chap.2).

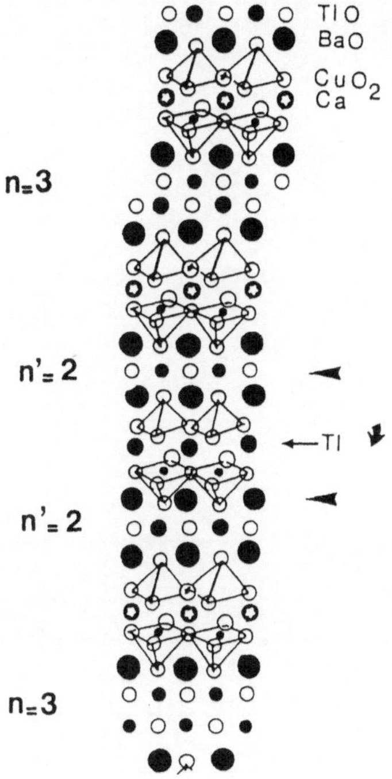

Fig.6.65. Idealized drawing of the double defect imaged in Fig.6.64

n=3

n'=2

n'=2

n=3

TlO
BaO
CuO_2
Ca

b) Rock-Salt-Type Layers

From the previous examples (Figs.6.62, 64) a very original behavior of the thallium oxides can be inferred. In both cases, we indeed observe the arising of defects which affect simultaneously the perovskite-type and rock-salt-type layers. The nonstoichiometry mechanisms that affect the rock-salt-type layers are rather more numerous than those observed in the perovskite layers [6.91,94-97]. They are of two types:

1) **Intergrowth defects**. The first type corresponds to a well-characterized variation of the n value. A typical example is exhibited in Fig.6.66a, where white dots are correlated to the positions of the cations. The nominal matrix is n = 3, i.e. $[(BaO)(TlO)_2]_\infty$ layers, and so the normal contrast corresponds to four rows of bright dots for the sequence Ba-Tl-Tl-Ba. On some images, faulty groups of only three rows of bright dots are observed. They were interpreted as the intergrowth defect n' = 2, i.e. $[(TlO)(BaO)]_\infty$ layers involving thallium monolayers corresponding to the cation sequence Ba-Tl-Ba (Fig.6.66b). This type of defect arose sometimes in such great numbers in the n = 3 matrix of the oxide $Tl_2Ba_2CaCu_2O_8$ that it suggested that a pure phase corresponding to the formulation $TlBa_2CaCu_2O_7$, n = 2, could be synthesized; this was the origin of the second family of thallium oxides. Other examples of such defects are shown in Figs. 6.67 and 62d, where thallium monolayers corresponding to $[(TlO)(BaO)]_\infty$ appear in a "2234" matrix (n = 3), i.e. in $[(TlO)_2(BaO)]_\infty$ rock-salt-type layers.

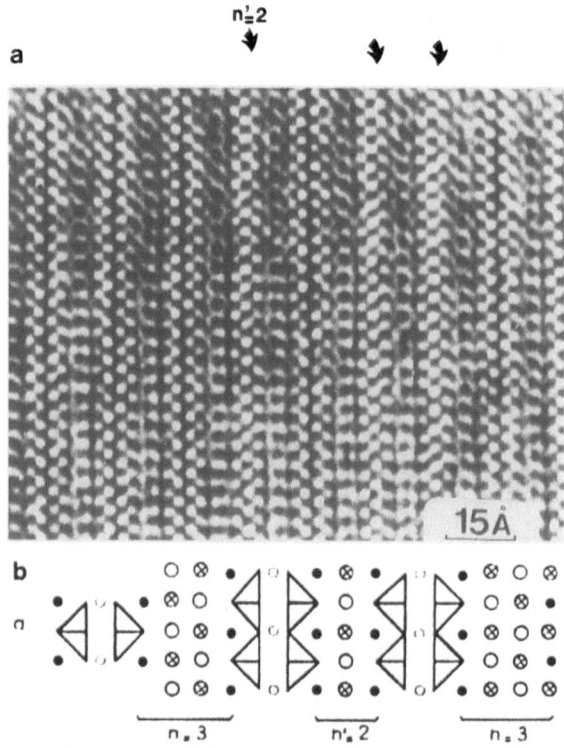

Fig.6.66. (a) Defective rock-salt-type layer stacking. $n' = 2$ in an $n = 3$ nominal matrix are arrowed. (b) Idealized drawing of the defect

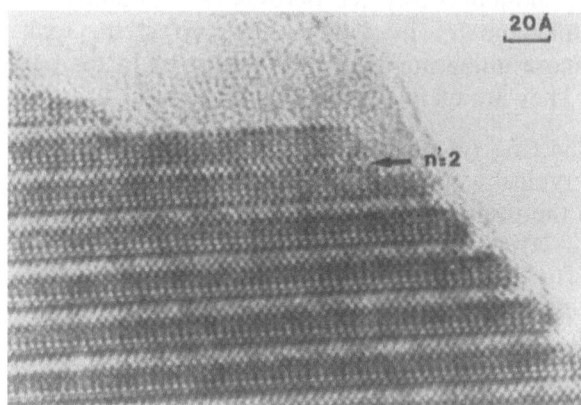

Fig.6.67. Thallium monolayer ($n' = 2$) in a nominal matrix $Tl_2Ba_2Ca_3Cu_4O_{12}$

This mechanism is also observed in a matrix built up from double rock-salt-type $[(TlO)(BaO)]_\infty$ layers ($n = 2$), where we observe $n' = 1$ and $n' = 3$ layers. An example of $n' = 1$ is shown in Fig.6.68a, where the faulty layer appears as a double row of white dots. In that case, it is indeed impossible to know whether the double row of cations which forms the single

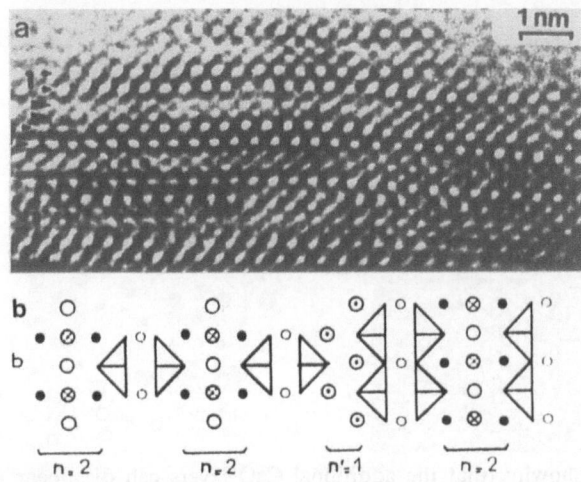

Fig.6.68. (a) Single rock-salt-type layers appear as defects in $TlBa_2CaCu_2O_7$. **(b)** Idealized drawing of the defect

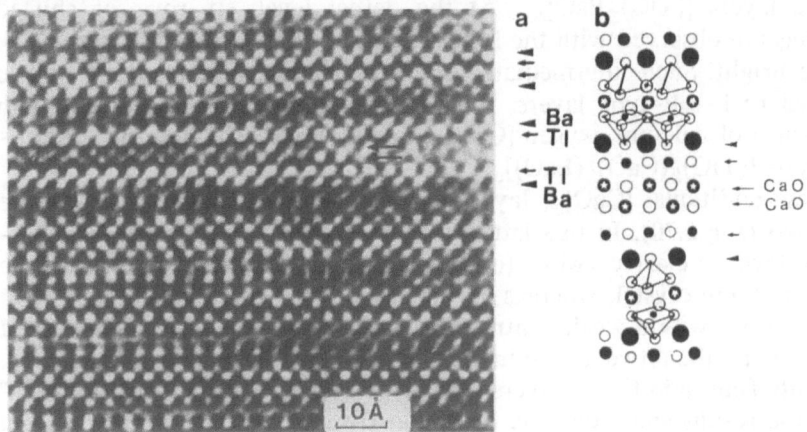

Fig.6.69. (a) Additional AO layers between the two TlO layers. **(b)** Idealized model of the defect

$[AO]_\infty$ rock-salt-type layer ($n' = 1$) is made of two BaO rows, two TlO rows or one TlO row and one BaO row. This type of defect, $n' = 1$, is quite rare because the $n = 1$ family is not a stable phase of the system Tl-Ba-Cu-O. The frequency of the defects which affect the rock-salt-type layers depends on the existence of the corresponding phase.

2) Intercalated layers. The second type of defect which affects the rock-salt-type layers corresponds to the intercalation of additional $[AO]_\infty$ layers in their middle. An example is depicted in Fig.6.69. Considering that the images where the cations appear as white dots are the best for such an interpretation, we observe in the regular matrix "2212" four rows of bright

213

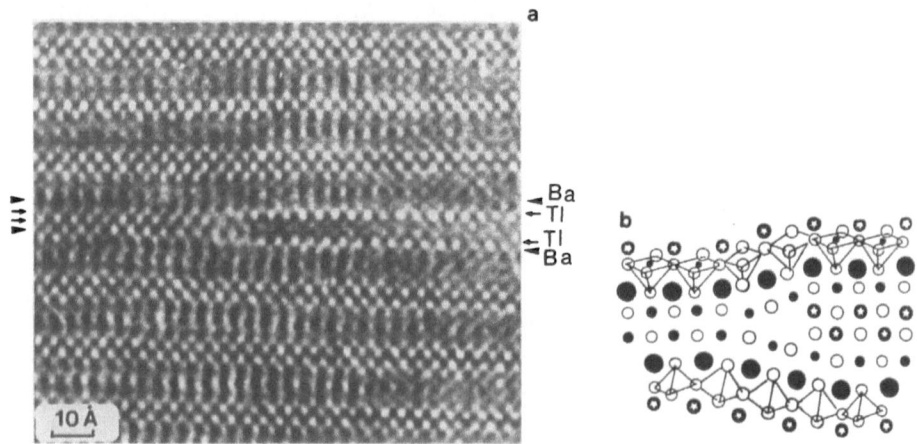

Fig.6.70. (a) HREM image showing that the additional CaO layers can disappear in the bulk in a "superdislocation"-like mechanism. (b) Idealized model of the defect

dots corresponding to the cation sequence Ba-Tl-Tl-Ba, i.e. to triple rock-salt-type layers $[(TlO)_2 BaO]_\infty$. At the defect level, six rows of shifted white dots are observed with the following intensities: two bright, two weak and two bright. Such intermediate rows are assumed to correspond to two additional rock-salt-type layers: their weak contrast is in agreement with the presence of calcium oxygen $[CaO]_\infty$ layers (Fig.6.69b), leading to quin-tuple layers $[(TlO)_2(CaO)_2(BaO)]_\infty$.

These additional $[CaO]_\infty$ layers can be infinite (Fig.6.69a) or can be interrupted (Fig.6.70). In this latter case, the normal contrast of the rock-salt-type layer is clearly visible (on the left side of the image), even in the thick part of the crystal, whereas on the edge of the crystal (right side) two additional rows with a weak contrast are observed. It should also be pointed out that more than two additional rows are sometimes intercalated (Fig. 6.71) with four additional layers quickly stopped in a "superdislocation" way. These results show that the deviation from stoichiometry for thallium

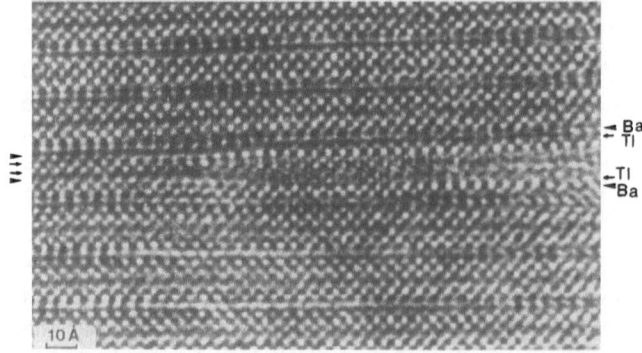

Fig.6.71. More than two additional CaO layers are observed between the TlO layers and "disappear" in the matrix

214

and calcium that can be observed by X-ray diffraction (Chap.5) results from intergrowth phenomena which affect the rock-salt-type layers.

Consequently, this deviation from stoichiometry may play an important role in the superconducting properties of these thallium oxides. It can indeed increase the Cu(III) content by changing the charge balance in the bulk but it can also affect the thallium layers and thus influence the role of hole reservoir that such layers play.

From the solid-state chemistry viewpoint, the structural mechanisms that govern the thallium oxides are important, since a random stacking of the two structural types that form the intergrowth is observed. The stacking defects lead to the formation of either isolated defects or polytypes, depending on the generating mechanism. In a nominal matrix [n, m], characterized by n rock-salt-type layers and m copper polyhedra layers, polytypes [n, m | n', m], [n, m | n, m'] or more complex [n, m | n'', m''] are observed. The simultaneous variation of both units in an intergrowth phase had previously been observed only in the intergrowth tungsten bronzes (ITB), whose structure was characterized by *Hussain* and *Kihlborg* [6.54]. In that case, HREM studies [6.54-57] showed that 1, 2 or 3 rows of hexagonal tunnel rows can, in the same way, interleave $(WO_3)_m$ layers whose width is variable. In these structures, the role of a mirror played by the tunnel rows is similar to that of the TlO layers in the thallium oxides. However, up to now, no similar features have been observed in the structures built up from the intergrowth of perovskite-type layers and rock-salt-type layers. Indeed, until the discovery of the thallium copper superconductors, n = 1 was the only characterized family, whatever the nature of the A cations located in the NaCl layer: La, rare earth or alkali ions. Are these phases stabilized by the highly oxygen-deficient character of the perovskite? Such an oxygen-deficient perovskite has never been observed for such thicknesses (m>1). The limitation of these series probably results from the stability of the perovskite layer, high m members (m > s) being observed only as faulty layers in a matrix of lower m nominal composition.

6.2.4 The Nonsuperconducting TlBa$_2$NdCu$_2$O$_7$: A New Mechanism

TlBa$_2$ACu$_2$O$_7$, A = Y,Nd, was found to be isostructural [6.98] with the 60 K superconductor TlBa$_2$CaCu$_2$O$_7$ [6.48, 78, 90, 92-101]. Although no superconductivity was detected in these oxides, they crystallize in a tetragonal cell corresponding to the member of the series with m = n = 2. Nevertheless, a remarkable structural difference was observed between calcium and neodymium (or yttrium) cuprates from the X-ray powder calculations: it concerns the distance between two $[CuO_2]_\infty$ layers, which is much shorter (2.71 Å) than that observed in the calcium cuprate (3.17 Å) (Fig.6.72). Consequently, the NdO$_8$ cages are more similar to a fluorite cage, tending to a cubic symmetry, than to the CaO$_8$ cage, which is strongly elongated along c. Such a feature would indeed imply important variations in the nonstoichiometry mechanisms.

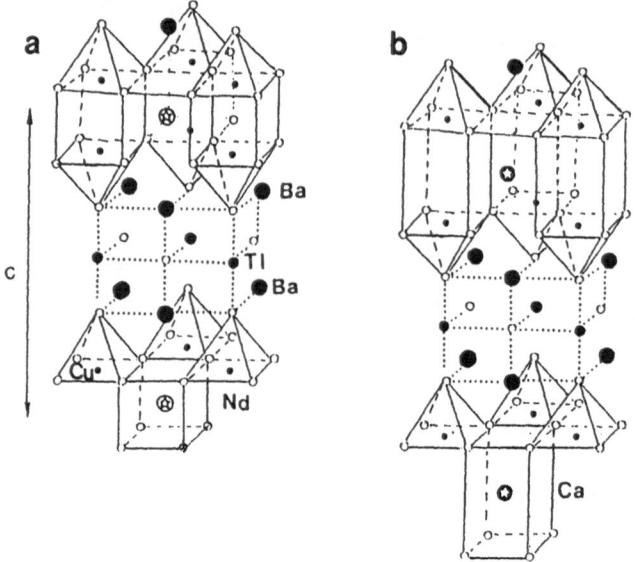

Fig.6.72. Perspective drawings of (a) $TlBa_2NdCu_2O_7$ and (b) $TlBa_2CaCu_2O_7$. The two structures differ only in the fluorite-type cage symmetry

From the high resolution image point of view, two essential features must be considered. First, the neodymium substitution for calcium involves an important variation of the scattering factor of the atoms localized in the fluorite cage. This feature, coupled with the second feature, the strong displacement of the $[CuO_2]_\infty$ layers, should imply an important variation in the contrast with respect to the calcium phase. Extensive image calculations have been carried out for both oxides and compared with the calcium ones [6.100]. Some [010] images are shown in Fig.6.73a, obtained with the atomic parameters previously determined [6.99].

It appears clearly that images obtained for focus values close to -400 and -700 Å are very similar; the first ones correspond to the Scherzer value where oxygen and vacancy positions are highlighted, whereas for the second the cation positions appear as white dots. The most significant images are obtained for intermediate focus values, for example ranging from 0 to -200 Å These results show that the through-focus series are needed to interpret the experimental contrast.

Some experimental images are reported in Fig.6.73b, for Df = 50, -75 and -350 Å. The experimental and calculated images of Fig.6.73 show that the triple rows of white dots are correlated with the three successive layers BaO-TlO-BaO (i.e., a double rock-salt-type layer, n = 2). [001] images exhibit the classical contrast of these phases, consisting of an array of white dots spaced at 3.8 Å (Fig.6.74); no contrast modulation was observed along this direction. Most of the crystals exhibited large regions of perfect layer stacking (Fig.6.75), in agreement with the X-ray diffraction results. However, the systematic HREM investigation shows the formation of extended defects in a great number of these crystals.

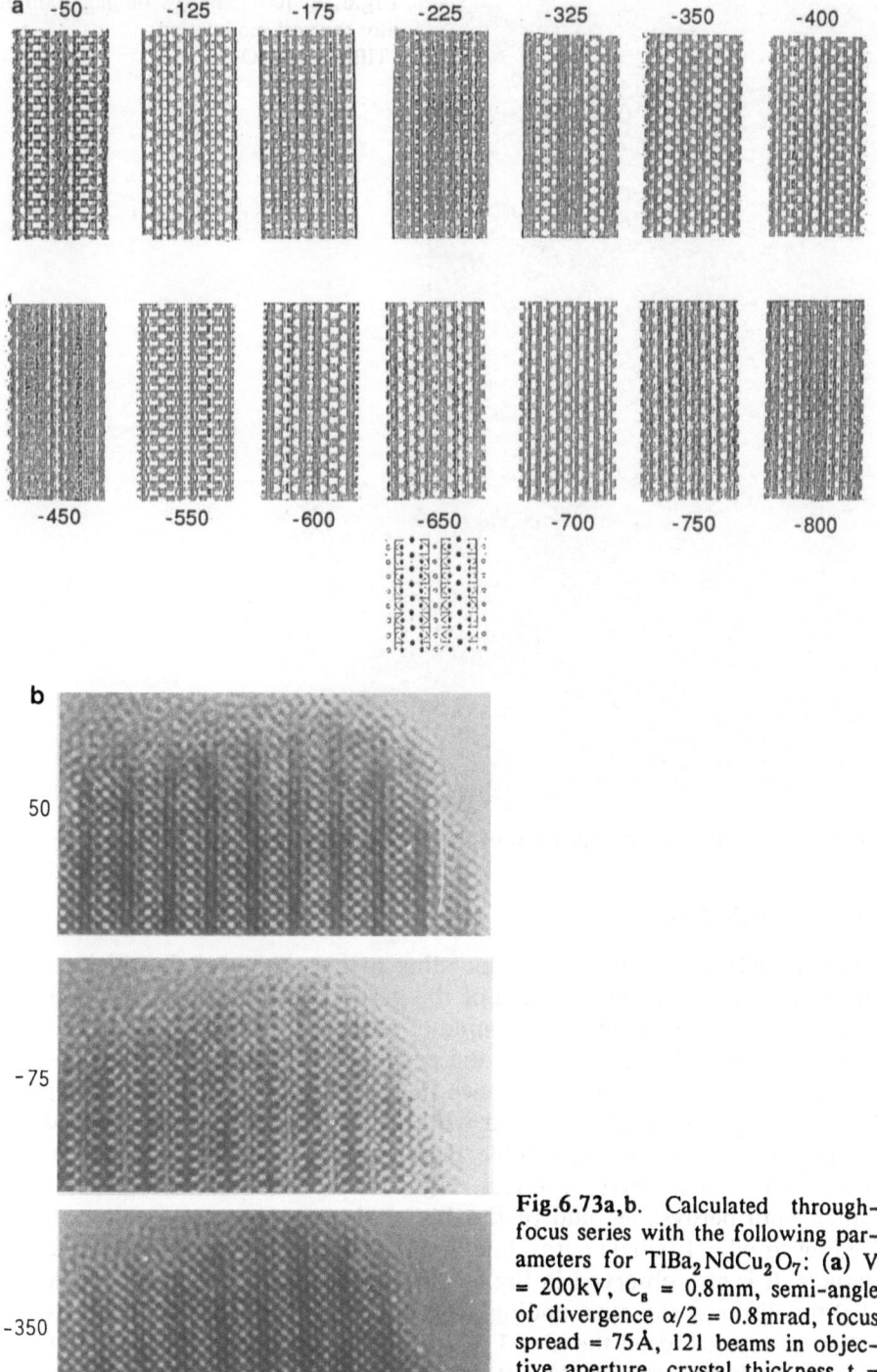

Fig.6.73a,b. Calculated through-focus series with the following parameters for $TlBa_2NdCu_2O_7$: **(a)** V = 200kV, C_s = 0.8mm, semi-angle of divergence $\alpha/2$ = 0.8mrad, focus spread = 75Å, 121 beams in objective aperture, crystal thickness t = 31.2Å, defocus values are given in Å **(b)** Experimental [010] images for D_f = 50Å, -75Å and -350Å

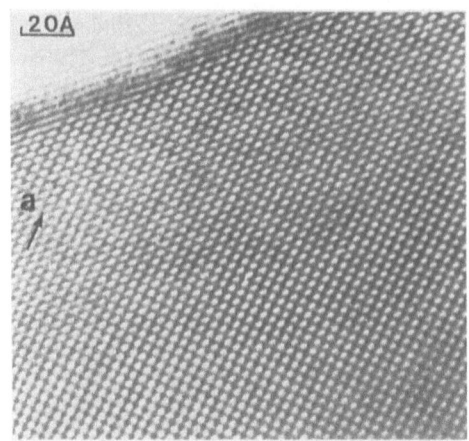

Fig.6.74. [001] images do not exhibit any contrast modulation ($TlBa_2NdCu_2O_7$)

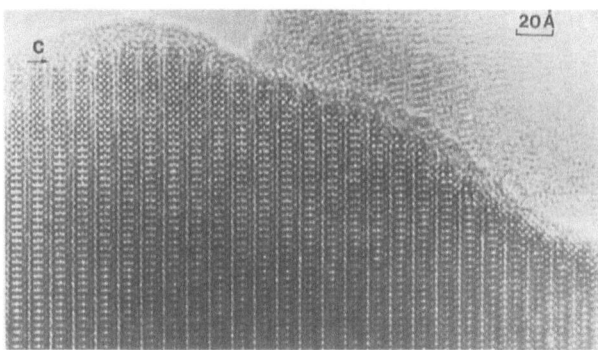

Fig.6.75. Crystals exhibit large areas of perfect layer stacking ([010] image)

a) Classical Defects

Classical defects, i.e. those corresponding to a variation of the thickness of the rock-salt-type layers (n′) and of the perovskite-type layers (m′) are observed. Figure 6.76 shows an example of an n′ = 3 defect, i.e. a triple rock-salt-type layer characterized by the sequence BaO-TlO-TlO-BaO which appears in the n=2 matrix (sequence BaO-TlO-BaO): The contrast of the defective layer is clearly observed with the four rows of white dots and the shifting of the perovskite double layer, owing to the loss of the mirror plane in the rock-salt-type layers. Figure 6.77 shows another example, in which both perovskite- and rock-salt-type layers can vary simultaneously, leading to m′ = 1 (single perovskite-layer) and n′ = 3 (triple rock-salt-layer) defects. It is also observed in Fig.6.77 that the presence of an m′ = 1 defect slightly disturbs the surrounding matrix, contrary to the defective rock-salt-type layer. Indeed the n′ = 3 defect does not involve contrast modification in the bulk, whereas the five perovskite double layers surrounded by a single perovskite layer (on the right) correspond to a disturbed area of the crystal.

218

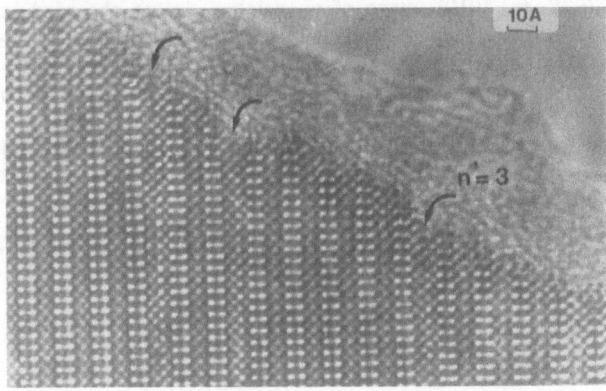

Fig.6.76. n′ = 3 defective rock-salt-type layers are sometimes observed in the (n=2, m=2) matrix ([010] image)

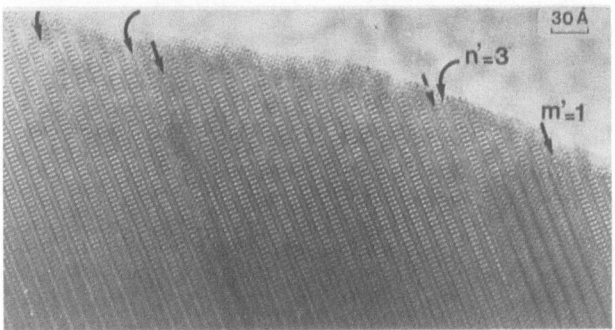

Fig.6.77. Both rock-salt-type and perovskite-type layer widths can vary simultaneously: example of n′ = 3 and m′ = 1 ([010] image)

b) Variations in the Fluorite-Type Layers

Original defects are observed which occur in the neodymium layers. A first example is shown in Fig.6.78. Two rows of double perovskite layers are clearly observed; they are arrowed (Fig.6.78a). Adjacent to these perovskite layers, two series of three rows of white dots (three small black arrows) are correlated to the three BaO - TlO - BaO layers (Fig.6.78a). Between these two triple series of bright dots, four rows of less-bright dots (labelled F) are observed. The contrast observed in this area (for different focus values), the relative positions of the rows of dots and their distances do not allow a hypothesis of defective n′ or m′ layers to be considered. A model is proposed in Fig. 6.78c,d. It corresponds to the intercalation of an additional neodymium layer between two successive $[CuO_2]_\infty$ layers. The structure of such a defect can be described as consisting of double fluorite-type layers, leading to the formulation $TlBa_2 Nd_2 Cu_2 O_9$, whereas only single fluorite-type layers are observed in $TlBa_2 NdCu_2 O_7$.

Another example is shown in Fig.6.79a. In this micrograph, the three rows of white dots correlated to the oxygen atoms of the sequence

Fig.6.78. (a) [010] image of an original defect observed in TlBa$_2$NdCu$_2$O$_7$. Symbols are DP for the Double Perovskite layers, three small arrows for the BaO-TlO-BaO layers and F in a square bracket for the structural unit (Nd$_2$Cu$_2$O$_6$). (b) Schematic representation of the image contrast. The sizes of the white spots are not strictly proportional to those observed in the experimental image; they only indicate the nature of the different rows. (c) Perspective drawing of the model of the defect. (d) Schematic projection along [010] of the structural model

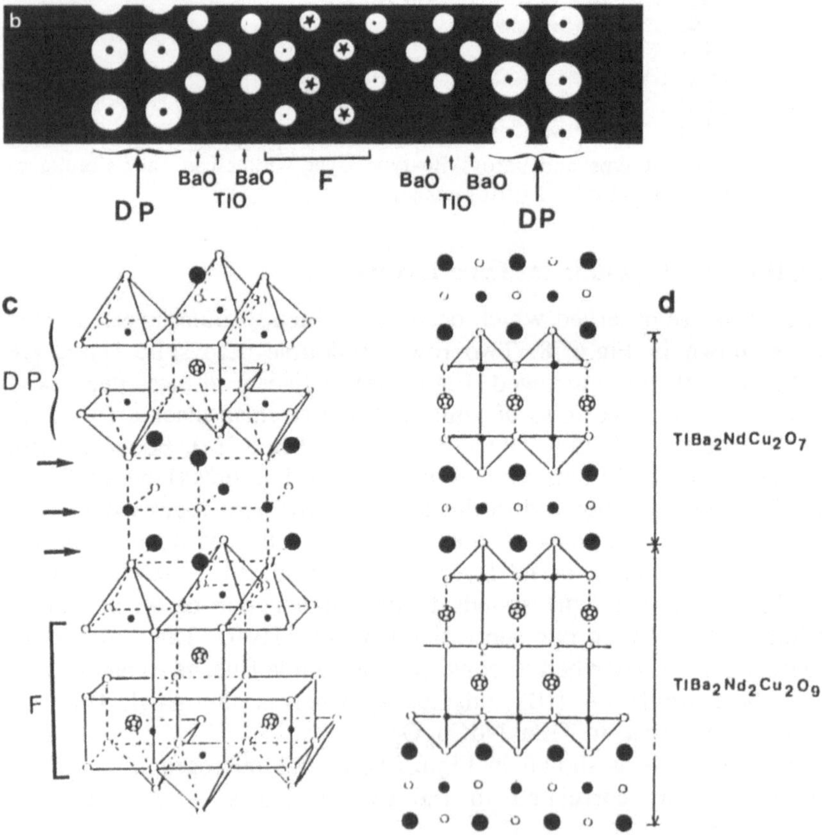

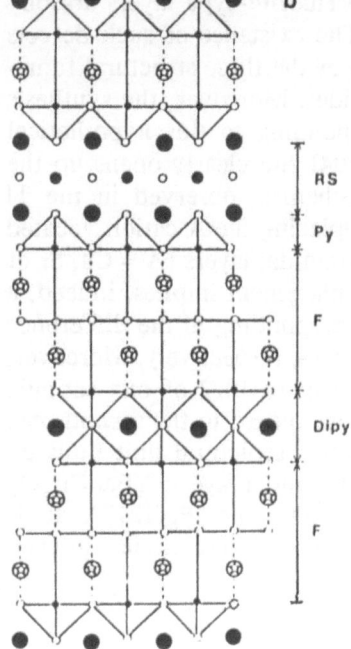

Fig.6.79. (a) [010] image showing another defect observed in $TlBa_2NdCu_2O_7$. The three small arrows correspond to the Rock-Salt-type layers (RS), i.e. to the sequence BaO-TlO-BaO. The white arrows, labelled F, correspond to the Fluorite-type layers. (b) Idealized drawing of the proposed model

BaO-TlO-BaO are separated by dots corresponding to the oxygen atoms and vacancies of the two copper layers. Between two slices of Rock-Salt-type layers (RS), a sequence of nine rows is observed. The first row on either side (large black arrows) correspond to $[CuO_2]_\infty$ layers, and the adjacent double layers (white arrows) to the same type of double fluorite-type layers as previously observed (Fig.6.78c).

Thus a structural model of this defect could be proposed (Fig.6.79b) according to the following sequence of the heavy atoms: $[Ba-Tl-Ba]_{RS}$ $[Cu]_{py}$ - $[Nd-Nd]_F$ - $[Cu-A'-Cu]_{dipy}$ + $[Nd-Nd]_F$ - $[Cu]_{py}$ - $[Ba-Tl-Ba]_{RS}$. Such a scheme corresponds to a complex intergrowth of Rock-Salt-type

Fig.6.80. The new mechanism observed in the Tl"1212" oxides consists, in a simple way, in the replacement of the $[Ca]_\infty$ layer sandwiched between two rows of copper pyramids by a double fluorite layer $[Ln_2O_2]$

layers (RS), double Fluorite-type layers (F), double copper layers (dipy) similar to those observed in $YBaFeCuO_5$ [6.88] and single pyramidal copper layers (py).

It should be pointed out that such double fluorite-type layers are observed in the Nd_2CuO_4-type oxides [6.103]. The existence of such defects clearly shows the possibility of the intergrowth of the three structural families: perovskite, fluorite and rock-salt-type oxides. Moreover, the synthesis of oxides with a nominal composition corresponding to this hypothetical structure (Fig.6.78c) leads to a pure phase [6.104] and clearly opens up the series for new structures, considering the mechanism observed in the Tl "1212" oxides. It consists, in a simple way, in replacing the A cation, located in the AO_8 cages built up from two rows of pyramidal layers (A = Ca, Sr or RE), by a double fluorite (Fig.6.80). Such a replacement implies, indeed, a variation in the periodicity along the c axis corresponding to the difference of thickness between the two types of layers, ~2.5Å, respectively. Moreover, it implies change in the symmetry since a shift of (a+b)/2 of one cationic layer of the double fluorite layer is observed with regard to the second one. In the cuprate families, several oxides exhibit this structural unit built up from an $[A]_\infty$ layer sandwiched between two pyramidal layers: $YBa_2Cu_3O_7$ Tl"2212", Bi"2212", $Pb_2Sr_2ACu_3O_8$, $NdBaFeCuO_5$ and Pb"1212". It is remarkable that this new mechanism, when applied to these oxides, in each case led to new materials, as seen in Chap.5

6.2.5 Lead Oxides

The issue of lead oxides is different from that of bismuth or thallium oxides since they do not correspond to real families. In fact, as discussed in different sections, lead cuprates always exhibit mixed (AO) layers as shown in Table 6.1, where only those oxides are reported in which lead atoms occupy at least half of the sites of a rock salt layer. One of the difficulties encountered in these oxides is that it is possible for lead to adopt a mixed valence, with the size and environment of the cations being strongly modified. These features could indeed result in the local ordering and distortions which are discussed in Chaps.5,8.

The partial replacement of lead by bismuth in the [1,2; 1,1] oxide, $Pb_2Sr_2Ca_{0.5}Y_{0.5}Cu_3O_8$ [6.110], leads to a great improvement of the super-

Table 6.1. Composition and charateristics of the mixed lead cuprates with Pb/A = 1

Compositions	n	m	Sequence of the adjacent $(AO)_\infty$ layers	References
$BiPbSr_2YCu_2O_8$	3	2	SrO-(Bi, Pb)O-(Bi, Pb)O-SrO	6.82
$Pb_{0.5}Sr_{2.5}Ca_{0.5}Y_{0.5}Cu_2O_{7-\delta}$	2	2	SrO-(Pb, Sr)O-SrO	6.105,106
$Pb_{0.5}Sr_2CaY_{0.5}Cu_2O_{7-\delta}$	2	2	SrO-(Pb, Ca)O-SrO	6.107
$Tl_{0.5}Pb_{0.5}Sr_2CaCu_2O_7$	2	2	SrO-(Pb, Tl)O-SrO	6.108,109
$Pb_2Sr_2Y_{0.5}Ca_{0.5}Cu_3O_8$	1	1.5	SrO-PbO	6.110
$Pb_{1.4}Bi_{0.6}Sr_2Y_{0.5}Ca_{0.5}Cu_3O_8$	1	1.5	SrO-(Pb, Bi)O	6.111
$PbBaSrYCu_3O_{8-\delta}$	1	3	(Sr, Ba)O-PbO	6.112

conducting properties [6.113]. Detailed HREM studies of the lead oxide [6.114-117] and Bi-Pb oxide [6.111] have allowed their complex crystal chemistry to be understood. The electron diffraction study provides evidence for the existence of a C-type subcell (hkl, h+k = 2n) where, however, strong variations are observed in the relative intensities of the reflections, along with frequent multisplitting of the spots (Fig.6.81). In the first pattern (Fig.6.82a) the reflections corresponding to h and k values are weak,

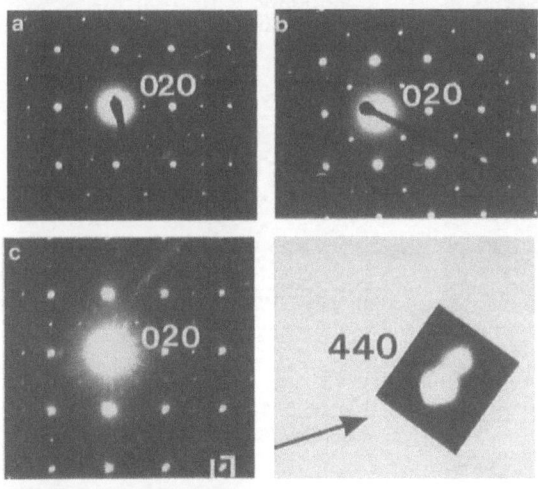

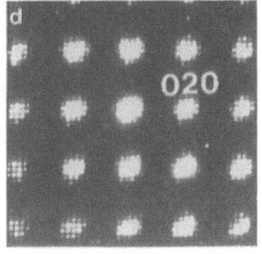

Fig.6.81a-d. Examples of [001] ED patterns; hk0 reflections with odd h and k values show strong variations in relative intensity, from weak (**a**) to strong (**b**). Characteristic splitting of the twinned crystals is observed (**c**); the 440 reflection is enlarged; (110) is the plane common to both components. Some patterns are characterized by numerous satellites (**d**), correlated with modulation features and misoriented areas

whereas they are intense in the second one (Fig.6.81b). Some patterns exhibit extra spots violating the C centering conditions (Fig.6.81a,c). The ED patterns are often characterized by a multisplitting of their reflections resulting from twinning phenomena (Fig.6.81c) and tilted domains with respect to each other. Lastly, some patterns are similar to those of bismuth cuprates substituted with lead; an example is shown in Fig.6.81d, corresponding to the existence of a modulation of the structure.

The experimental through-focus series were registered in regular zones of the crystals and compared with the calculated ones for different crystal thicknesses. An example of a calculated [110] series is shown in Fig.6.82a for a thickness of 20 Å, and part of the corresponding experimental series is shown in Fig.6.83a for D_1 = -480,-800 and -1050 Å, respectively. The most

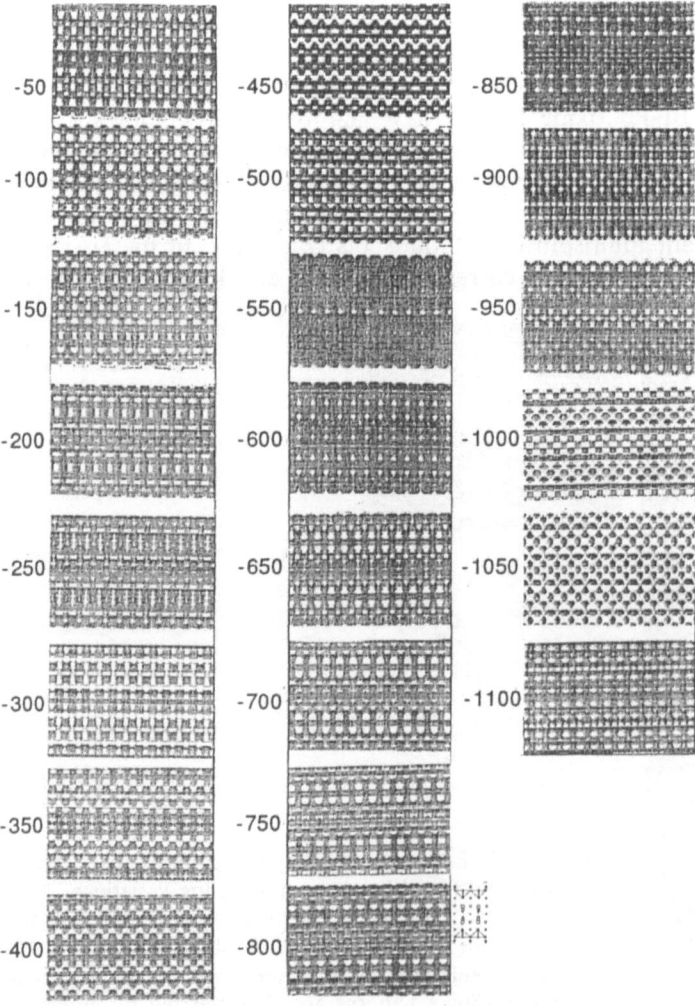

Fig.6.82. $Pb_{1.4}Bi_{0.6}Sr_2Ca_{0.5}Y_{0.5}Cu_3O_8$: Calculated [110] through-focus series. The focus values are in angstroms

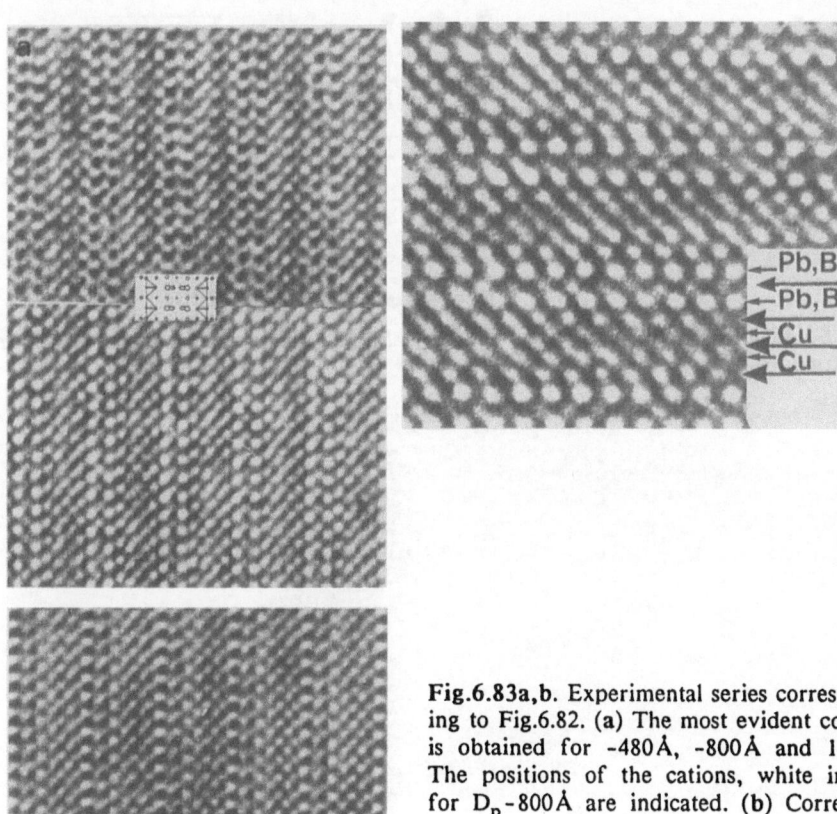

Fig.6.83a,b. Experimental series corresponding to Fig.6.82. (a) The most evident contrast is obtained for -480Å, -800Å and 1050Å. The positions of the cations, white imaged for D_p-800Å are indicated. (b) Correlation between the white dots and the cation positions

evident contrast is indeed for defocus values close to -430 and -800Å where the two rows of black spots and white spots are correlated with the positions of the columns of lead and bismuth atoms respectively; the positions of the different cations, which appear as white dots for D_f = 800 Å, are indicated in Fig.6.83b. It should be noted that the images registered and calculated for the bismuth-substituted oxide are very similar to those obtained for the undoped compound $Pb_2Sr_2Ca_{0.5}Y_{0.5}Cu_3O_{8+\delta}$ (our work and [6. 117]). That was expected considering the scattering powers of Pb(II) and Bi(III) and the very similar cells and positional parameters.

Investigation of the undoped lead cuprate $Pb_2Sr_yY_{1-x}Ca_xCu_3O_8$ [6. 114,115,117] showed that the formation of intergrowth defects is a common feature, slices of 11.7 or 22.4Å being reported. Such a phenomenon was not detected for the bismuth-doped phase; a perfect regularity of the layer stacking along the c axis is observed.

In the PbBaSrYCu$_3$O$_{8-\delta}$ oxide, "0223" (Fig.6.84b), and in $Pb_{0.5}Sr_{2.5}\cdot Ca_{0.5}Y_{0.5}Cu_2O_{7-\delta}$, "1212" (Fig.6.84c) the [AO] layers of the rock salt slice can be easily imaged. Two examples are shown in Fig.6.85. The double [PbO][(Ba,S$_2$)O] layers of the "0223" structure appear as two rows of white

225

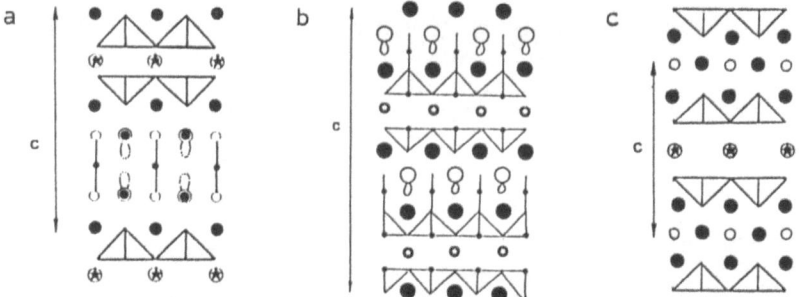

Fig.6.84. Idealized drawing of (a) [n = 1, m = 1.5], (b) [n = 1, m = 3] and (c) [n = 2, m = 2] structures

Fig.6.85a,b. Typical [100] images of the lead cuprates: (a) $PbBaYSrCu_3O_{8-\delta}$ [6.112] with a "0223"-type structure. (b) $Pb_{0.5}Sr_{2.5}Ca_{0.5}Y_{0.5}Cu_2O_{6.75}$ [6.105] with a "1212"-type structure. The rows of brightest dots are correlated with the [AO] layers. The mixed lead layers, [$(Pb_{0.5}Sr_{0.5})O$] and [PbO] layers, are labelled P, and the [SrO] layers, S

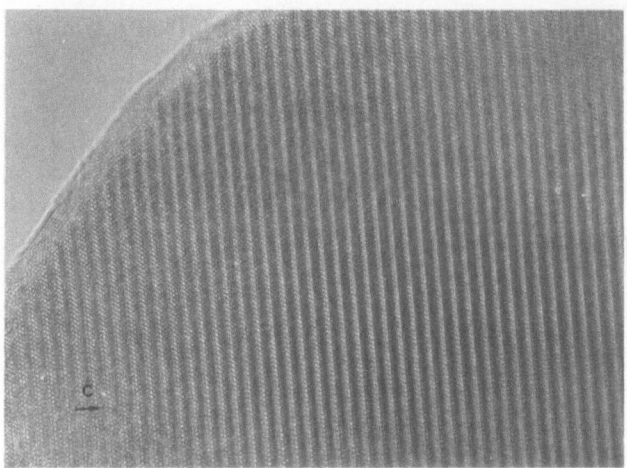

Fig.6.86. Overall view of a "1212" crystal showing the perfect stacking of the rock salt and perovskite layers

spots (Fig.6.85a) and, in the same way, the triple layers [SrO] [$Sr_{0.5}Pb_{0.5}O$] [SrO] of "1212" structure appear as three rows of bright staggered spots (Fig.6.85b). In both samples, we observe a highly regular stacking of the layers, as shown for example in Fig.6.86 for the "1212" oxide.

6.2.6 The Rare Earth Oxides

Consider the cuprates whose structure is built up from the intergrowth of rock salt and perovskite units. The rare earth oxides belong to the $n = 1$ family; the physical behavior of the $m = 2$ members of this family is rather striking since, although they exhibit both mixed valence of copper and two-dimensional structure, they do not superconduct when prepared under oxygen flow: they need oxygen pressure to be obtained as superconductors (Chap.3). One interesting oxide of this group is $Sr_2NdCu_2O_{5.75}$ [6.118], owing to the complex distribution of oxygen and vacancies. The structure of this oxide is described in Chap.3, so that here we only quickly recall the essential features. Four CuO_5 pyramids exhibit basal planes parallel to the c axis and two oxygen sites are partially filled, O_4 and O_8 (Fig.6.87). The occupancy of these anionic sites implies that the copper polyhedron located between the two pyramids is either a distorted CuO_4 group or a CuO_5 pyramid whose basal plane is perpendicular to the c axis. along the b axis a superstructure $b = 3a_p$ is observed, resulting from the ordering of the oxygen vacancies. The parameters of the orthorhombic cell are thus $a = a_p$, $b = 3a_p$ and $c \sim 20$ Å and the space group, Immm. The [100], [010] and [001] ED patterns are shown in Fig.6.88. The high resolution images [6.119] along [100] exhibit a striking contrast variation, the most typical contrast corresponding to a honeycomb-like image as shown in Fig.6.89. The calculated (Fig.6.90) and experimental images for a crystal thickness close to 30Å show that this honeycomb-like contrast results from both cationic and anionic sublattices and that a direct interpretation is not straightforward. On the other hand,

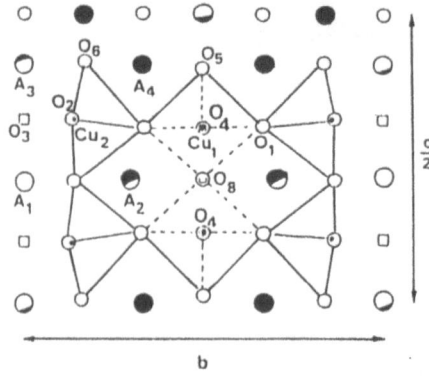

Fig.6.87. Structure of $Sr_2NdCu_2O_{5.75}$ projected along [100]

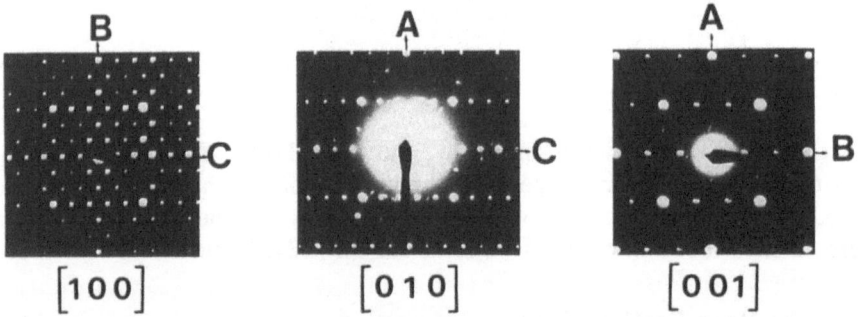

Fig.6.88. $Sr_2NdCu_2O_{5.75}$ – (a) [100], (b) [010] and (c) [001] ED patterns

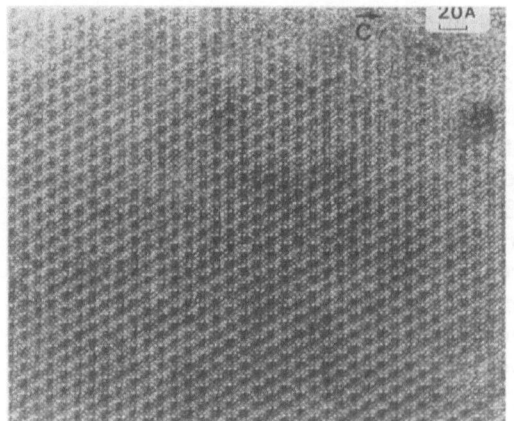

Fig.6.89. $Sr_2NdCu_2O_{5.75}$: [100] HREM image. A honeycomb-like contrast is observed, resulting from the ion ordering

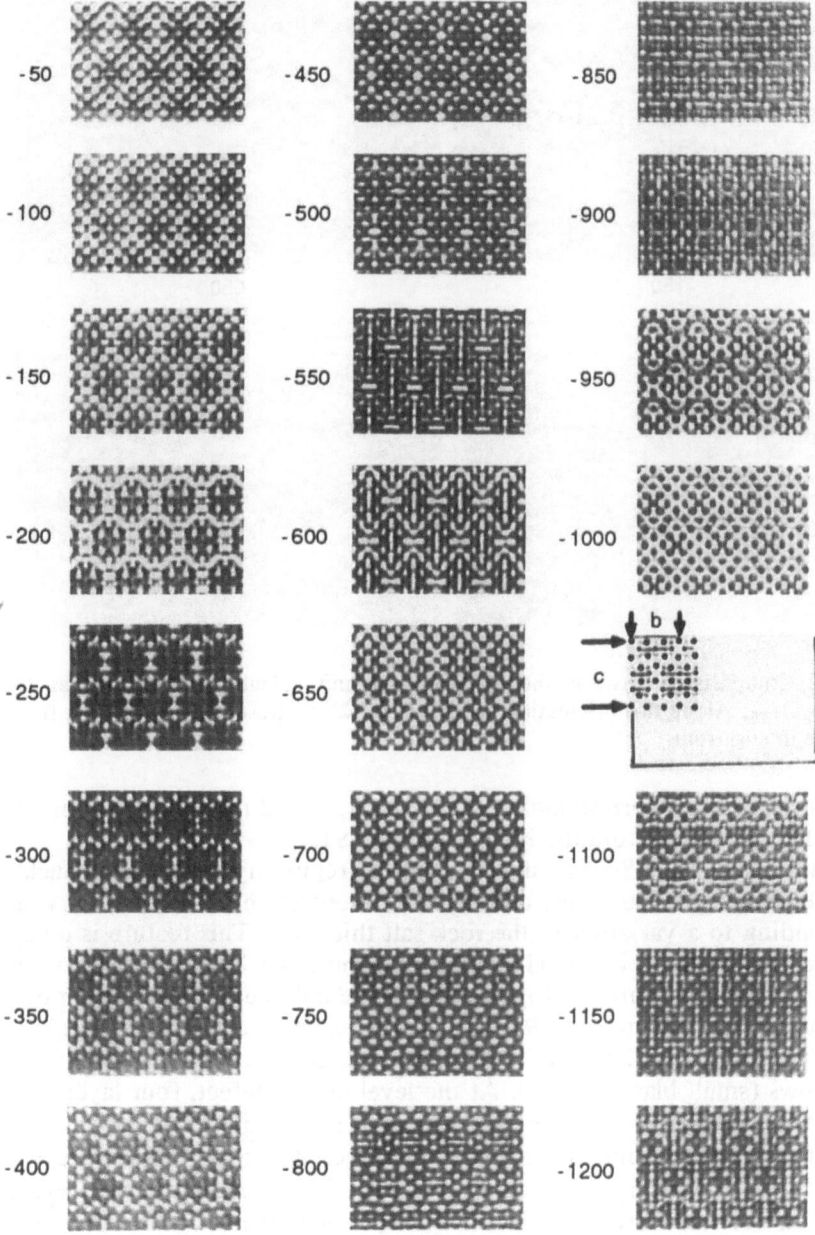

Fig.6.90. $Sr_2NdCu_2O_{5.75}$ [100] calculated through-focus series for a crystal thickness of 30Å, V = 200kV and C_s = 0.8mm. The focus values are in Ångstroms

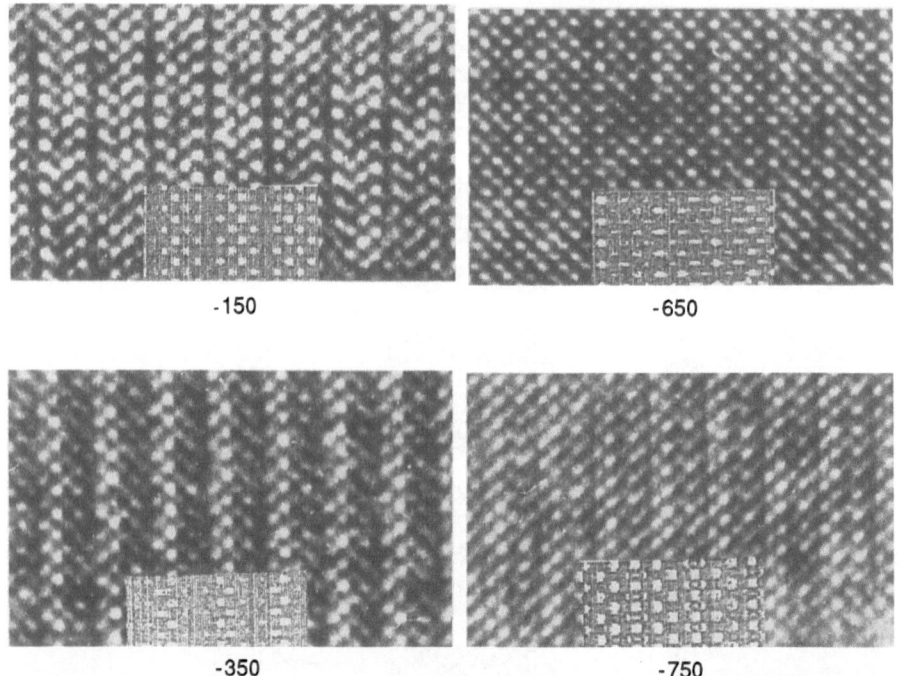

-150

-650

-350

-750

Fig.6.91. Comparison between the experimental and calculated [010] images for $Sr_2NdCu_2O_{5.75}$. Along this orientation, a typical "0212" contrast is observed. The focus values are in angstroms

the classical contrast corresponding to the n = 1, m = 2 members is observed along [010], as shown from the images in Fig.6.91.

In the case of the Sr-Nd cuprates, a great regularity in the layer stacking is observed too. However, one type of intergrowth defect is observed, corresponding to a variation in the rock salt thickness. This feature is especially clearly observed in the [100] images, where a white strip appears in the honeycomb-like contrast (Fig.6.92a). On the enlarged image of such defects (curved arrows in Fig.6.92b), the phenomenon can be easily interpreted. The two [AO] layers of the single rock salt layer(n = 1) appear as whiter rows (small black arrows). At the level of the defect, four layers are observed as the translation of the two adjacent perovskite slices (labeled P); this is in agreement with the existence of larger defective rock salt layers (n' = 3) as shown in Fig.6.92c. along [010], the interpretation of the defect is more difficult to carry out. In the example shown in Fig.6.93 the perovskite slices "P" are moved away from each other and translated by a/2 in the same way. Only n' = 3 intergrowth defects have been observed in that sample. It should be noted that this type of intergrowth is unique since in the phases related to Sr_2TiO_4 or $Sr_3Ti_2O_7$ only variations of the perovskite slice thickness have been reported.

To conclude, it appears that the cuprates whose structure is built up from the intergrowth of rock-salt-type and perovskite-type layers do inde-

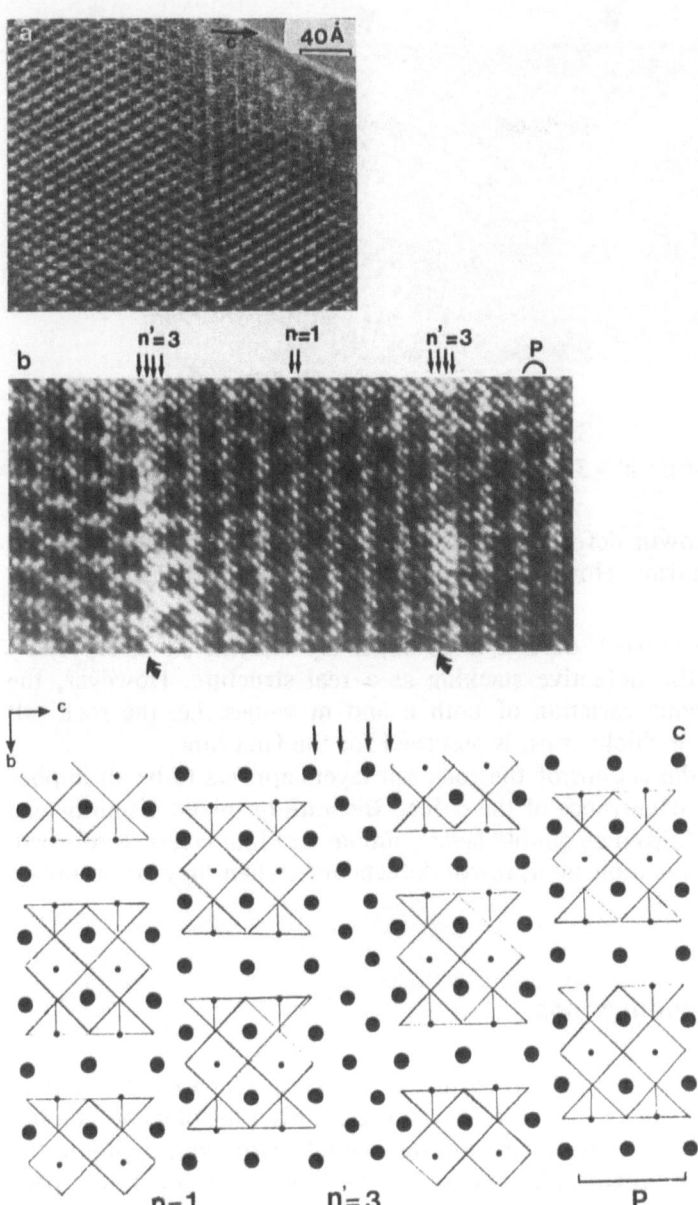

Fig.6.92a-c. Intergrowth defect observed along [100] in $Sr_2NdCu_2O_{7.5}$ corresponding to the setting up of a triple rock salt layer. (a) The defect appears as a white strip in the honeycomb-like contrast. (b) Enlarged image of two $n' = 3$ defective layers. (c) Idealized model of the defect

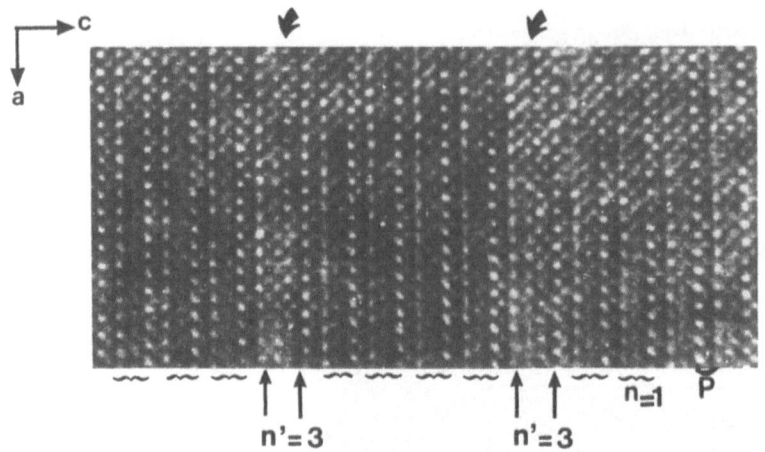

Fig.6.93. Contrast of the $n' = 3$ defect observed along [010] in $Sr_2NdCu_2O_{6.75}$

ed exhibit intergrowth defects as observed in all the compounds generated by such a mechanism. However, two important characteristics should be mentioned:

1) The formation of such defects is governed by classical laws, especially the existence of the defective stacking as a real structure. However, the possible simultaneous variation of both n and m values, i.e. the rock salt and perovskite layer thicknesses, is observed for the first time.

2) The nature of the cations of the rock salt layers appears to be an important factor for the occurrence of the defect. Bismuth or mixed bismuth-lead layers are only stabilized as double layers, unlike thallium layers. Rare earth or strontium layers exhibit intergrowth defects only when they are in mixed layers.

6.3 Layer Interconnections

A distinctive mechanism is observed in the cuprates, which corresponds to the connection of layers of different natures. This phenomenon implies generally the existence of boundaries in the crystals. However, owing to the complex atomic rearrangements it requires, the boundaries are rarely perfect.

The first example [6.111] was observed in the $Pb_{1.4}Bi_{0.6}Sr_2Ca_{0.5}Y_{0.5} \cdot Cu_3O_8$ superconductor whose structure is shown in Fig.6.81a. For [100] and [010] orientations of the crystals "antiphase boundaries" are observed whose direction and shifting mode can change. A first image is shown in Fig. 6.94a. In this image, the contrast mainly consists of two rows of intense white spots, correlated to the positions of lead and bismuth atoms and labeled P on the image. A shift of about c/2 is observed across the boundary (marked on the micrograph by a row of small arrows). In this shift, it ap-

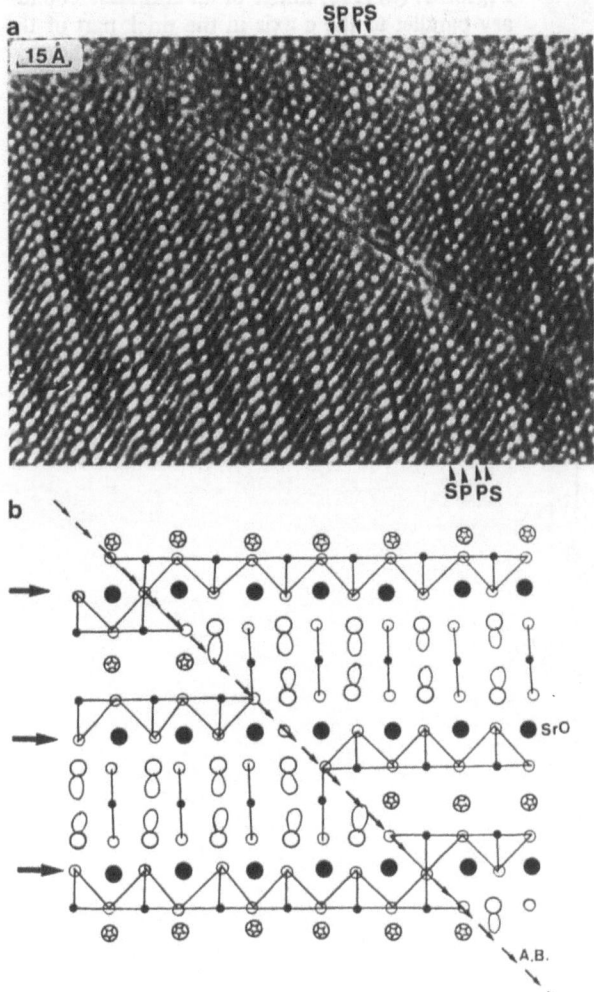

Fig.6.94a,b. $Pb_{1.4}Bi_{0.6}Sr_2Ca_{0.5}Y_{0.5}Cu_3O_8$: **(a)** [1̄10] HREM image of a 45° antiphase boundary (AB). The nature of the cationic strontium and lead layers is indicated as S and P respectively; the boundary is marked by a row of small black arrows. **(b)** Idealized drawing of the connection of the different layers through the 45° antiphase boundary. The SrO layers, which remain unchanged across the boundary, are shown with large black arrows. The 45° boundary is parallel to an oxygen plane (*plane of small arrows*)

pears that one of the two SrO layers (labeled S) coincides in the two domains. The Antiphase Boundary (AB) is inclined, in that case at 45° with respect to the c axis. A structural model of this antiphase boundary can be proposed (Fig.6.94b), taking into consideration the fact that this structure can be described as built up from pure oxygen planes alternating with metallic planes parallel to (441), i.e. $(101)_p$, inclined at 45° with respect to c. Thus, the idealized antiphase boundary consists of an oxygen plane

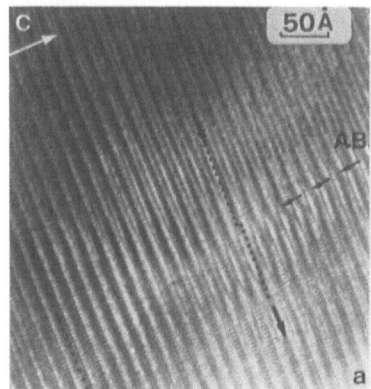

Fig.6.95. (a) [110] image of an antiphase boundary parallel to the c axis in the thick part of the crystal. The shift of the layers across the boundary is close to c/4 (indicated by rows of small dots parallel to the layers). (b) Idealized drawing of the connection of the layers through the antiphase boundary parallel to (110). The unchanged planes; $[PbO]_\infty$ and $[CuO_2]_\infty$, are indicated by large and medium arrows. The boundary appears in an [AO] plane (*row of small arrows*)

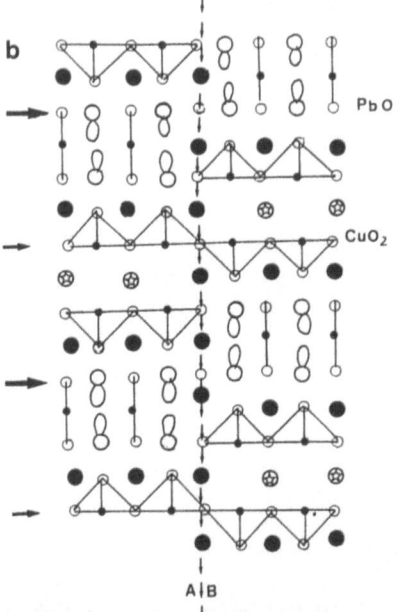

parallel to (441). across the boundary, the $[SrO]_\infty$ layers remain unchanged, whereas the $[CuO_2]_\infty$ layers are connected to the $[PbO]_\infty$ layers; moreover, it can be seen from the model that the planes where the oxygen vacancies are localized, i.e. $[Ca,Y]_\infty$ and $[Cu^I]_\infty$ layers are "connected" and thus the oxygen-deficient plane is unbroken. It should be noted on the HREM image that this boundary is not perfectly straight; this is because the layers connected through this boundary exhibit slightly different thicknesses.

A second example is shown in Fig.6.95a, with an antiphase boundary almost parallel to (110). The image, taken in the thick part of the crystals, shows a shift of about 4 Å across the boundary. In the same way as before, a model can be proposed (Fig.6.95b). In this idealized model the antiphase boundary, parallel to (110), appears in an (AO) plane (A = Sr, Ca or Y). It can be seen that two types of layers are unchanged, one $[CuO_2]_\infty$ and one

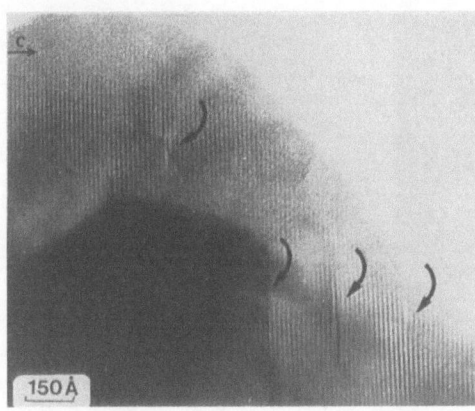

Fig.6.96. Typical [110] overall image showing the absence of intergrowth defects and the existence of defects parallel to (001) which affect only one slice

150 Å

$[PbO]_\infty$ layer out of two, whereas the second $[CuO_2]_\infty$ and $[PbO]_\infty$ are connected through the boundary. This type of boundary was observed in the pure lead phase $Pb_2Sr_2Ca_{0.5}Y_{0.5}Cu_3O_8$.

As mentioned in Sect.6.25, no intergrowth defect was observed in the $Pb_{1.4}Bi_{0.6}Sr_2Ca_{0.5}Y_{0.5}Cu_3O_8$ oxide but overall images of the microcrystals (Fig.6.96) show the existence of defects which affect only one slice parallel to (001) in the crystals. Some of them can be easily explained by taking into account the structural mechanisms involved in the formation of antiphase boundaries, i.e. the interconnection of different layers. Two examples are shown in Figs.6.97 and 6.98. In Fig.6.97a, identification of the contrast at the edge of the crystal shows that the defect affects a triple PbO-Cu-PbO slice, which appears as a double row of white dots in the thick part of the crystal (white arrows); the double row disappears and is replaced by a triple row of small dots. This is easily explained by the fact that the 45° antiphase boundary can be interrupted at the level of the "SrO" planes. Consequently, such an "elementary" extended defect would consist of the replacement of a triple "PbO-Cu-PbO" layer by a "CuO_2-Y-CuO_2" triple layer (Fig.6.97b). The second example (Fig.6.98a) shows a strong variation in the contrast in the middle of the slice, between the double row of $[PbO]_\infty$ layers (white rows on the image). In the same way, this elementary extended defect can be explained by a local connection of a "PbO-Cu-PbO" layer to a "CuO_2-Y-CuO_2" triple layer (Fig.6.98b).

Although the thickness of the crystals and the disturbance of the contrast do not allow a reliable interpretation to be given for the defects observed in the bulk, it appears that most of them arise from such mechanisms. It should be pointed out that these defects, which generally affect only one slice, never exceed 100 Å in length.

Sometimes the variations in the contrast are very localized at the level of the $[SrO_2]_\infty$ layers. This suggests that Sr-Pb substitution is involved in these defects; such mixed (Sr-Pb)O layers are easily formed and were recently observed in the superconductor $Pb_{0.5}Sr_{2.5}Ca_{1-x}Y_xCu_2O_{7-\delta}$.

Similar phenomena are observed in the Sr-Nd cuprates. Boundaries are observed in some crystals: a translation of the network is observed through

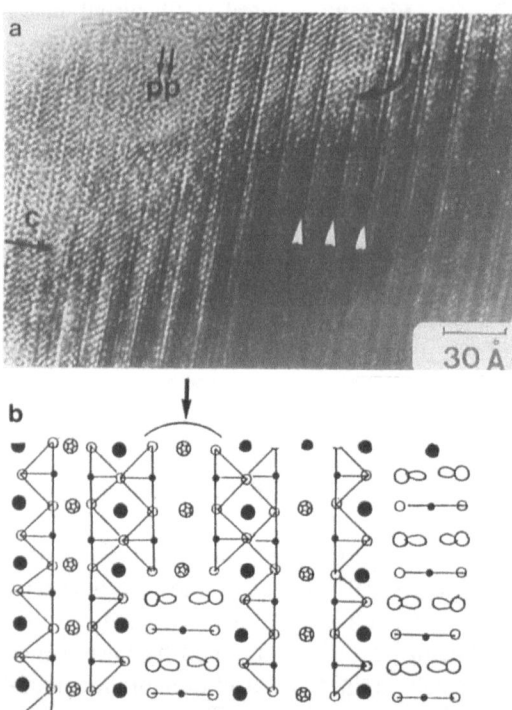

Fig.6.97. (a) HREM image of a defect where the two rows of lead atoms, which appear as double white rows in the thick part of the crystal (*white arrows*), disappeared; they are replaced by three rows of staggered dots. (b) Drawing of the idealized structural model proposed: the defect would consist of the local replacement of a "PbO-Cu-PbO" triple layer by a "CuO_2-Y-CuO_2" triple layer. The defective layer is marked by a black arrow

the boundaries which are not straight but only roughly parallel to [100]. The striking feature is that the translation vector is not identical from one boundary to the other. For instance, in Fig.6.99a we observe a translation of about 3.8Å This value and the identification of the layers in the sequences allow a model to be proposed (Fig.6.99b). along the [010] direction the oxygen atoms and vacancies are superimposed, so that the copper polyhedra are drawn as squares, without taking into account their real nature (CuO_5 pyramids or CuO_4 groups). We observe that one row of copper $[CuO_{2-x}]_\infty$ (black arrows) out of two remains unchanged across the boundary whereas the other is connected to a $[AO]_\infty$ layer. As a consequence, the two other layers of Sr-Nd atoms cross over the boundary without a change in the nature of the ions but with a change in their coordination (from the rock-salt-type to the perovskite-type environment). In the second example (Fig.6.100a) the translation ($t\simeq2Å$) is shorter so that the interconnection of the layers is different; an idealized model of the feature, drawn with the same symbols for copper polyhedra as in Fig.6.99 is shown in Fig.6.100b. In that model, it can be seen that every copper layer is connected to a Sr-Nd

236

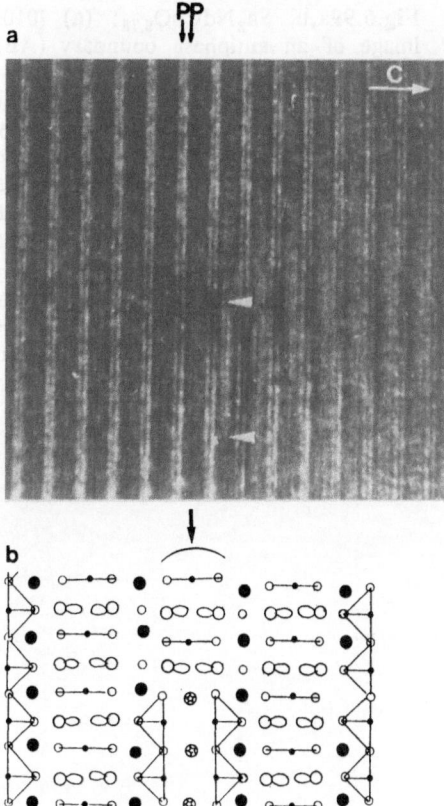

Fig.6.98. (a) Example of the strong variation of the contrast in the middle of the slice, at the level of the pyramidal layers. **(b)** Drawing of the idealized structural model proposed to explain this defect: a "PbO-Cu-PbO" triple layer locally substitues the "CuO_2-Y-CuO_2" triple layer. The defective part of the layer marked by a black arrow

layer, that one $[AO]_\infty$ rock salt layer (out of two) remains unchanged across the boundary, and that the boundary consists of an undulating rock salt layer.

In both families, a slight misorientation of the adjacent domains (close to 3° in this example) is observed: it was correlated to the difference in thickness of the rock salt and perovskite layers, which does not allow a perfect junction to be ensured.

It should be noted that similar features were observed in the bismuth cuprates [6.120]. In the case of the bismuth compounds, interconnection of layers leads either to antiphase boundaries or to dislocation mechanisms, as shown by *Eibl*, Fig.6.101.

These examples of stacking faults involving the connection of layers of different types are observed in lead and bismuth as well as in rare-earth oxides. It thus appears that the interconnection of layers is a general structural feature of oxides built up from the intergrowth of rock salt and perovskite layers.

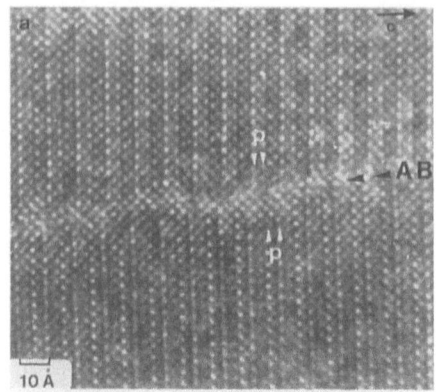

Fig.6.99a,b. $Sn_2NdCuO_{5.75}$: (a) [010] image of an antiphase boundary (AB) roughly parallel to (100). In this example, the networks are translated by about $3.8\,\text{Å}\sim a_p$, as may be clearly observed by considering the perovskite slices P (*white arrows*). Note that the boundary area is not disturbed. (b) Idealized model of both domains and the boundary. In this model, one copper polyhedra layer remains unchanged across the boundary (*black arrows*) whereas the other is connected to a $[AO]_\infty$ layer. The translation t of the network is close to 3.8Å

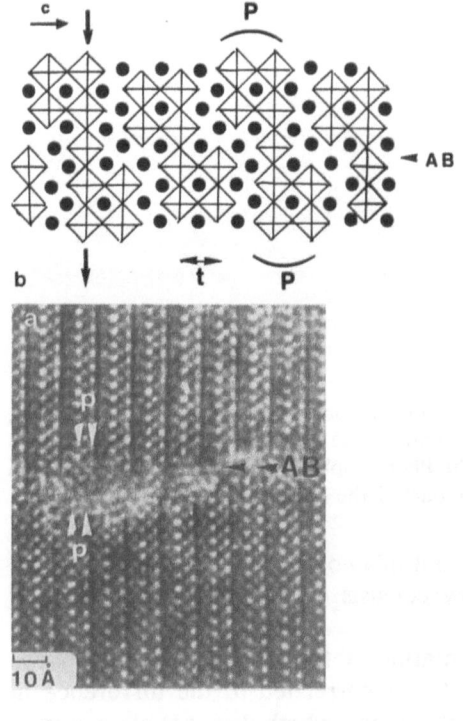

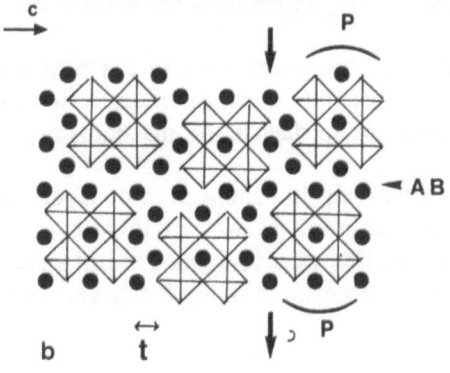

Fig.6.100. (a) [010] image of an antiphase boundary where a shorter translation, $t \sim 1.9$ Å is observed. (b) Idealized model of the defect: only one $[AO]_\infty$ layers remains unchanged across the boundary (*black arrows*) whereas the two $[CuO_2]$ layers are connected to Sr Nd layers

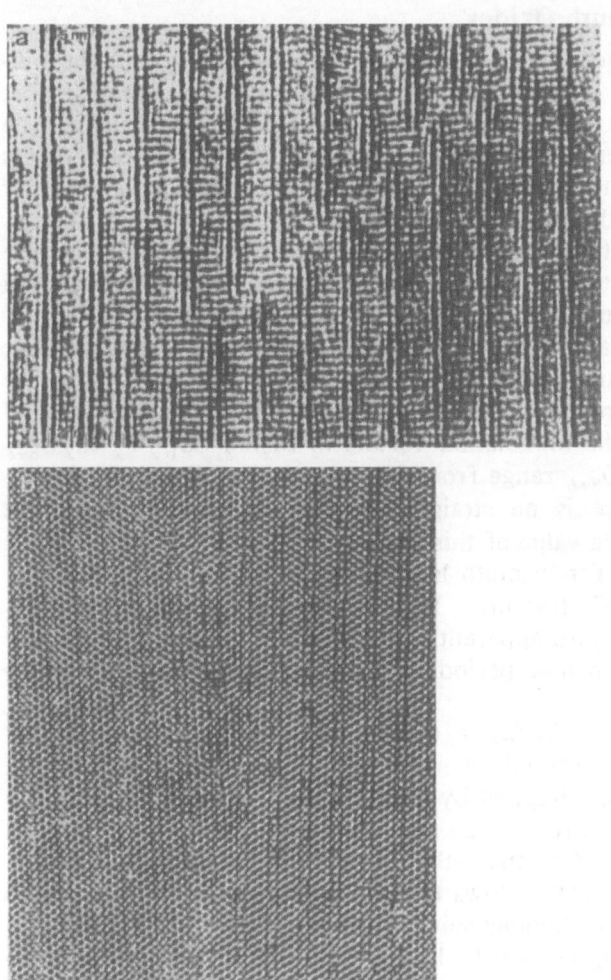

Fig.6.101. (a) Stacking fault in $Bi_2Sr_2CaCu_2O_8$ and (b) a small angle tilt boundary in $Bi_2Sr_2CaCu_2O_8$ [6.120].

6.4 Extra Spots in ED Patterns: An Amazing Variety

Extra spots of ED patterns are, in a general way, indications of long-range ordering and complex modulations arising in the bulk. An amazing variety of satellites are observed in several superconducting cuprates and isostructural phases, which can sometimes hinder the interpretation of the structural and physical properties. The discrimination between all the factors that are responsible for these phenomena is indeed an essential task for an understanding of the behavior of these materials. These features have been reported in different sections and chapters; they are regouped and discussed here.

6.4.1 Substituted Bismuth Oxides

The most spectacular phenomena are obviously those observed in the bismuth cuprates and the bismuth iron oxides: $Bi_2Sr_2CuO_6$, $Bi_2Sr_{3-x}Ca_x$ Cu_2O_8, $Bi_2Sr_3Fe_2O_9$ and $Bi_2Sr_4Fe_3O_{12+\delta}$ [6.65-67, 121-124]. They lie along the a axis in generally incommensurate positions. All the researchers who observed these phenomena agree with an average value of the q factor close to 4.7 in the bismuth copper oxides.

An interesting point in these phases is related to the evolution of the modulation vector with the cationic substitution. Two types of modulation vectors were observed in $Bi_2Sr_4Fe_3O_{12+\delta}$; the first type is slightly larger than that of the Bi cuprates, close to 5, whereas the second one is smaller, close to 2.6. [6.78] Similarly, the rare earth substitutions for Ca^{2+} that were performed by several research groups [6.121-124] lead to isostructural phases. It appears that the modulation vectors in Bi_2Sr_2 $Ca_{1-x}Y_xCu_2O_{8+\delta}$ and $Bi_2Sr_2Y_{1-x}Ce_xCu_2O_{8+\delta}$ range from 3.8a to 4.4a [6.124] but the investigations show that there is no straightforward correlation between the actual composition and the value of this vector.

Substitution of lead for bismuth leads to limited solid solutions in the "2201", "2212" and "2223" structures. ED study of $Bi_{2-x}Pb_xSr_2$ $CaCu_2O_8$ shows that the superstructure apparently remains constant but that the degree of order varies, with new periodicities corresponding to 4.2a and 7a [6.125].

In all these phases - $Bi_2Sr_4Fe_3O_{12+\delta}$, $Bi_2Sr_2Ca_{1-x}$ $YCu_2O_{8+\delta}$ and $Bi_{2-x}Pb_xSr_2CaCu_2O_8$ - it should be noted that a variation of the oxygen content is, in a formal way, implied by the difference in the substituted cation charges so that two effects, size and charge (correlated with oxygen content), are combined. On the other hand, the double substitution Pb(II)/Bi(III) and Y(III)/Ca(II) allows the oxygen content to be maintained [6.82] and leads to a larger homogeneity range for the phase $Bi_{2-x}Pb_xSr_2 \cdot$ $Ca_{1-x}Y_xCu_2O_8$, x varying from 0 to 1. The evolution of the c parameters and T_c vs x is similar to that observed in the lead-free oxide, a drastic decrease being observed for x close to 0.5. The limit composition $BiPbSr_2 \cdot$ YCu_2O_8 is not superconductive and the X-ray powder calculations [6.82] showed that this phase is isostructural with $Bi_2Sr_2CaCu_2O_8$. However, the electron diffraction study provided evidence of various and complex features, only the most frequent and striking of which are reported here.

1) The reconstruction of the reciprocal space attests that the space group is no longer Amaa as for $Bi_2Sr_2CaCu_2O_8$ but Pnam. Three ED patterns, [100], [010] and [110], are shown in Fig.6.101a-c. The h0l reflections with l = 2n+1 are clearly visible but 00l reflections with l = 2n+1 arise from double diffraction phenomena.

2) The intensities of the reflections hk0 with respect to the basic spots of the F subcell vary from one crystal to another. Very faint spots are observed in Fig.6.102a, whereas quite strong spots are found in Fig.6.102b; but, whatever their relative intensities, they are always observed. Some of the ED patterns show a splitting of the reflections (Fig.6.103) characteristic

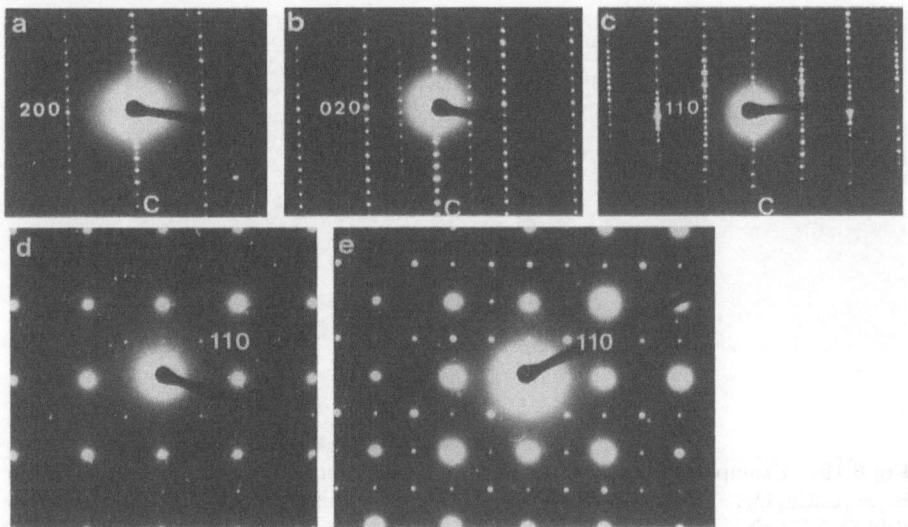

Fig.6.102a-e. BiPbSr$_2$YCu$_2$O$_8$ ED patterns. (a) [100], (b) [010] and (c) [110]. (d, e) Variation from one crystal to another of the intensity of the extra spots of the P-type space group with respect to the F space group

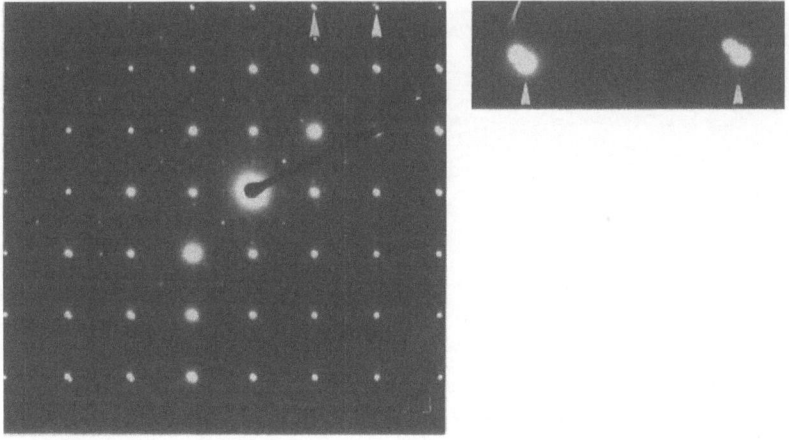

Fig.6.103. Splitting of the reflections characteristic of twinned crystals ([100] ED patterns) and enlargement of the arrowed reflections

of the existence of twins. The twin plane is [010], i.e. [110]$_p$ of the perovskite subcell.

3) Contrary to ED patterns of Bi$_2$Sr$_2$CaCu$_2$O$_8$ crystals, for lead-substituted bismuth oxides, satellites along the a* axis are rare. Two examples are depicted in Fig.6.104. In both examples, they are in incommensurate positions. This feature is similar to that observed in Bi$_2$Sr$_2$CaCu$_2$O$_8$ [6.7, 15-18], but in this case, the satellites are always weak and we observed generally only the satellites of first order. The q value, close to 5.25, is different from that of pure Bi phases. However, it should be noted that these

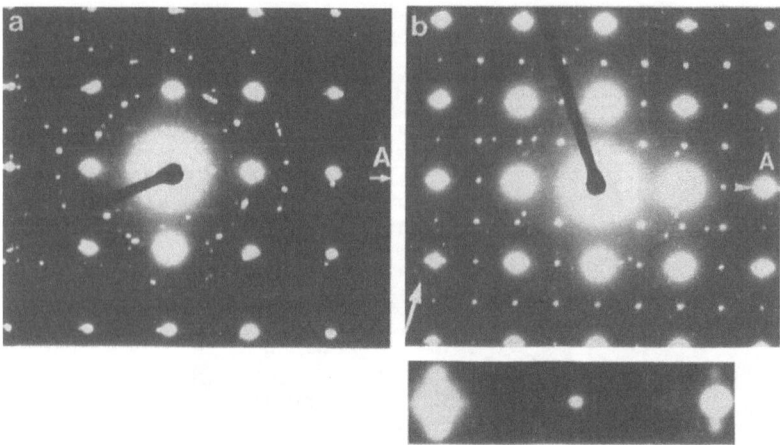

Fig.6.104. Examples of satellites along the a^* direction similar to those observed in $Bi_2Sr_2CaCu_2O_8$. The q factor is close to 5.25. This type of crystal is rare in $BiPbSr_2YCu_2O_8$

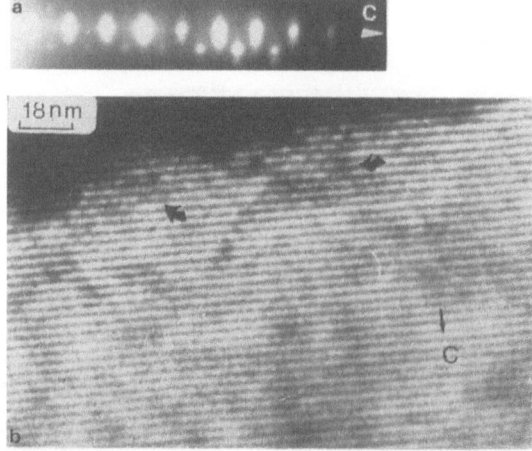

Fig.6.105a,b. $BiPbSr_2YCu_2O_8$: (a) Weak satellites can be observed in slightly tilted (001) planes. (b) The corresponding image reveals the existence of local modulation (*indicated by arrows*)

satellites are not observed in positions close to the reflections characteristic of the P space group but only close to the strong reflection of the A subcell. This can be interpreted as the superposition of both types of areas. It should indeed be noted that, in the same way, weak satellites can be seen in slightly tilted h0l planes (Fig.6.105). The corresponding bright field image shows the existence of modulations similar to those observed in the pure Bi oxides but they appear only in small areas of the crystals. This suggests that the rare crystals (or crystal areas) which exhibit this phenomenon correspond to a larger Bi/Pb ratio than that of the nominal composition.

4) Another type of satellite is shown in Fig.6.106a; similar systems of satellites were previously reported for $Bi_{1.8}Pb_{0.2}Sr_2CaCu_2O_8$ [6.126] and $Tl_{0.5}$

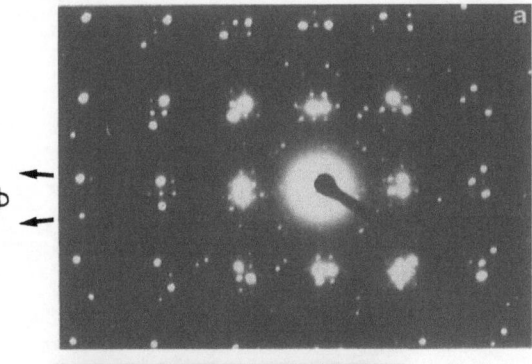

ϕ

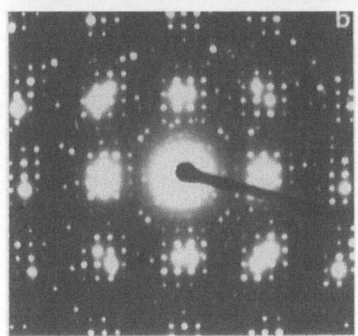

Fig.6.106a,b. Examples of satellites arising from the superposition of slightly misoriented lamellae. (a) The basic spots of the two components are clearly visible in the first pattern, contrary to the second one (b) where the satellites are intense

$Pb_{0.5}$ $Sr_2CaCu_2O_7$ [6.108]. These systems are bi-dimensional and incommensurate. They can be described by the single disorientation and superimposition of two lamellae in crystals. The two components are clearly visible in Fig.6.106a, where the spots of the basic cell are especially bright. The weaker spots, whose intensity varies from one crystal to another, arise from double diffraction phenomena. The angle between the two cells, labelled ϕ, remains quite constant from one crystal to another and close to 9°9'. This ϕ value can thus be considered as significant and suggests a structural feature as the origin of the misorientation.

5) The inhomogeneity of the arrangement of the structural features responsible for the satellites is illustrated in Fig.6.107, where five types of ED pattern were observed from different areas of the same crystal:

i) Satellites in incommensurate positions (q = 6.25) are set up along a direction roughly parallel to $[210]^*$, i.e. $[310]_p^*$ of the perovskite subcell. A schematic drawing is shown beside the micrograph (Fig.6.107a).

ii) Misoriented areas with $\phi = 9°9'$ (Fig.6.107b).

iii) Satellites which arise along $[110]^*$ (i.e., $[100]_p^*$ of the perovskite subcell) with q = 3.65 (Fig.6.107c).

iv) The previous satellites disappear, but streaks remain along the $[110]^*$ direction (Fig.6.87d).

v) Multisplitting of the spots and modulations with first-order satellites, as shown in the enlargement; they signal small variations of parameters, distortions and microtwins (Fig.6.107e).

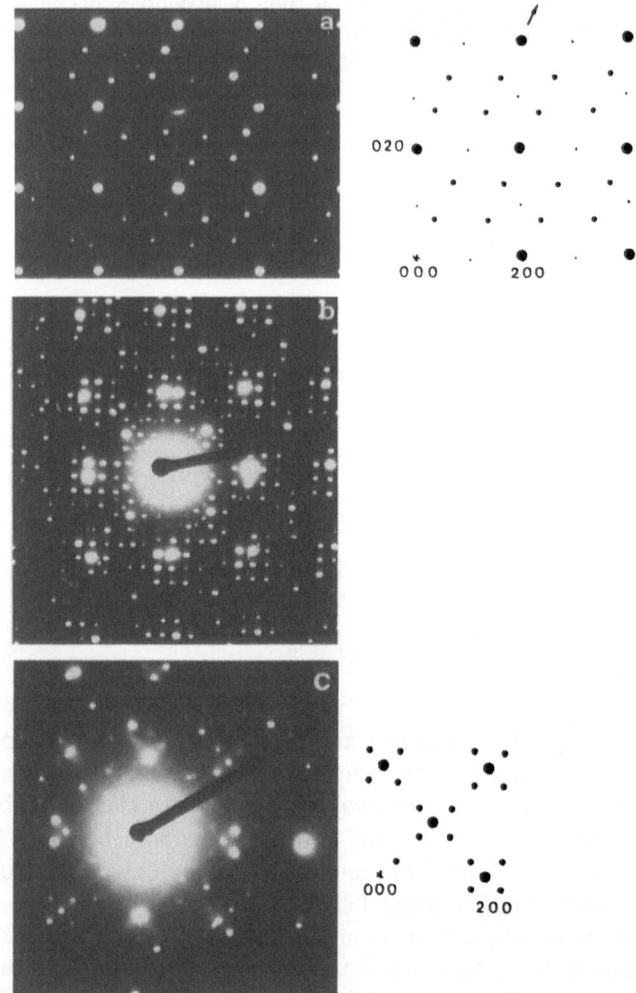

Fig.6.107a–c. The caption is given on the opposite page

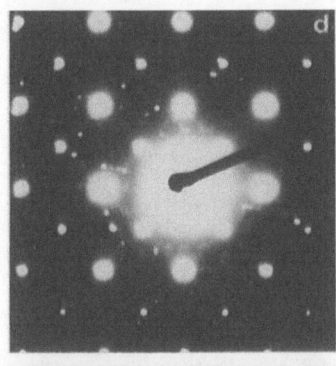

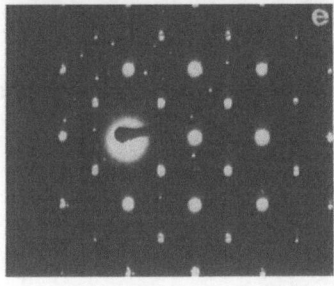

Fig.6.107a–e. [001] E.D. patterns registered in different areas of one $BiPbSr_2YCu_3O_8$ crystal. (a) Satellites set up along a direction roughly parallel to $[210]^*$. (b) Satellites arising from the misorientation of two superposed lamellae. (c) Satellites along $[110]^*$. (d) Streaks are observed along $[100]^*$. (e) Multisplitting of the reflections hkO, arising from small variations of the parameters and microtwinning

The disappearance of the system of satellites (along a^*) characteristic of the bismuth oxides indicates that bismuth can really be replaced by lead in a wide homogeneity range. This feature can be considered as additional proof of the role of the bismuth bilayers in the modulations. These results show, as well, that the introduction of lead involves original complex phenomena.

Between the two extreme compositions, a solid solution is observed, where the essential issue is to find out if the microstructural evolution is as continuous as suggested by the X-ray diffraction pattern. The investigation by ED and HR microscopy [6.108] of $Bi_{1.5}Pb_{0.5}Sr_2 \cdot Y_{0.5}Ca_{0.5}Cu_2O_8$, i.e. x = 0.5, confirms that Bi(III) and Pb(II) cations are simultaneously localized in the double layers. But these studies show that they both impose their own structural characteristics, which results in a plurality of the complex phenomena. The high-resolution images (Fig.6.108 and 109) allow the real nature of these oxides to be characterized:

a) [100] images show that the stacking defects are rare, especially at the level of the rock-salt-type layers, although no pure lead oxide is stabilized with n=3. Thus "2212"-type stacking of the cationic layers is not implied in these features (Fig.6.108). Note that the cleavage arises, as in the Bi oxides, in the middle of the double layer (black arrows).

b) [001] images show that modulations and superstructures are not set up over the whole crystal and that they arise in the form of small domains some tens of Ångstroms wide (Fig.6.109).

It should be noted that this influence is confirmed by the results obtained for $BiPbSr_2FeO_{6+\delta}$ [6.127, 128]. The regulation of the oxygen content is expected to be the same as in the above example: $Bi^{III}Pb^{II}$ and $Cu^{II}Fe^{III}$.

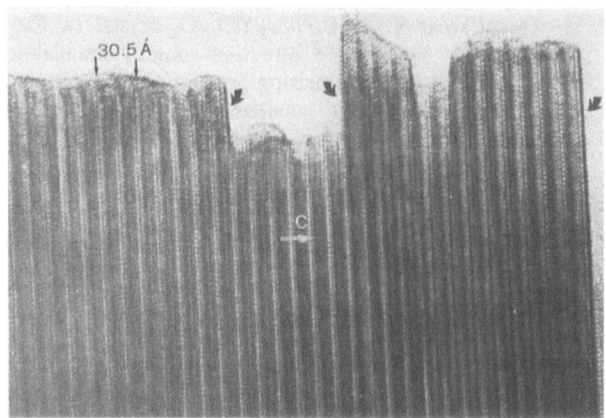

Fig.6.108. $Bi_{1.5}Pb_{0.5}Sr_2Y_{0.5}Ca_{0.5}Cu_2O_8$ [100] image. Stacking defects are rare and a preferential cleavage arises, as in the pure bismuth oxides, in the middle of the double $[Bi,PbO_y]_2$ layers

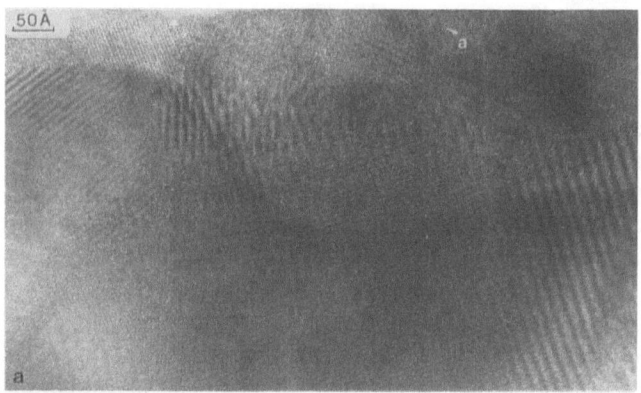

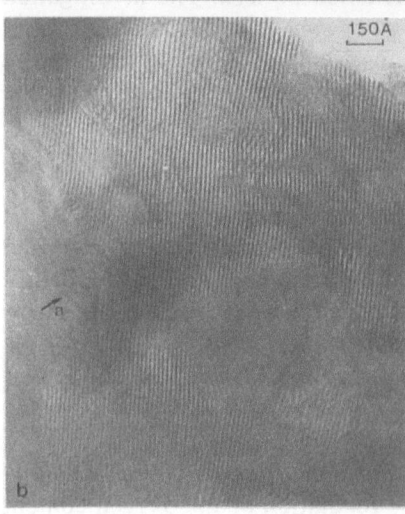

Fig.6.109. Examples of [001] HREM images in $Bi_{1.5}Pb_{0.5}Sr_2Y_{0.5}Ca_{0.5}Cu_2O_8$. Numerous modulations and superstructures appear in the form of small domains; moiré patterns are observed, resulting from the superposition and misorientation of lamellae

In the same way, the satellites characteristic of the double bismuth layers have disappeared and are replaced by others. Some examples of [001] ED patterns are depicted in Fig.6.110, they show that at least two networks are systematically superposed. The networks are labelled from a to d; they differ in the reflection conditions and the periodicity. Considering the subcell with a and b ~ $a_p\sqrt{2}$, their specific actions are:

a: $a_p\sqrt{2} \perp a_p\sqrt{2}$; hk0 : k = 2n ;
b: $a_p\sqrt{2} \perp a_p\sqrt{2}$; hk0 : h+k = 2n ;
c: $2a_p\sqrt{2} \perp a_p\sqrt{2}$; hk0 : h+k = 2n ;
d: $2a_p\sqrt{2} \perp a_p\sqrt{2}$; hk0 : no conditions.

The ability of iron to exhibit different coordinations, such as octahedra, pyramids or tetrahedra, could be at the origin of such a complexity.

6.4.2 Thallium Oxides

Contrary to the reports on the bismuth oxides, which all confirm a systematic existence of satellites, publications dealing with the thallium oxides ($Tl_2Ba_2CaCu_2O_8$ and $TlBa_2CaCu_2O_7$) report different results regarding the satellites. Some researchers [6.129, 130] indeed showed that systems of satellites appear in a systematic way in the thallium oxides, whereas others [6.93-95, 108, 109, 131-134] did not observe these phenomena, except in rare crystals. The role of the synthesis and thermal treatments was evoked to explain these differences [6.134].

The important role of oxygen pressure, holding temperature and cooling time is especially illustrated in the study of the "2201" oxide $Tl_2Ba_2CuO_6$. Much work has been devoted to this oxide owing to its complex superconducting behavior [6.135-139]. It appears that the cell exhibits either an orthorhombic or a tetragonal symmetry according to the thermal treatment. The two ED patterns shown in Fig.6.111 illustrate this variation. The first one (Fig.6.111a) corresponds to the tetragonal phase prepared in silica tubes under oxygen pressure and furnace cooling, whereas the second one (Fig.6.111b) corresponds to the orthorhombic phase, $a_p\sqrt{2} \times a_p\sqrt{2} \times c$.

In the same way, a variation in the cation distribution can produce satellites. This is the case of mixed valence Tl(III)/Tl(I) cuprates, $Tl^{III}_{2-x/3}$ ($Tl^I_{1-x}Ba_{1+x}$)$LnCu_2O_8$ [6.140]. The X-ray powder calculations and high resolution electron microscopy study confirmed that these phases are of "2212" type. The calculated and experimental images are shown in Fig. 6.112. One can also point out that these images are different from those obtained for $Tl_2Ba_2CaCu_2O_8$ owing to the variation in the cation distribution. The greatest difference is observed, as expected, for focus values close to -700Å and -1000Å, where the cation positions are highlighted.

The electron diffraction study showed that the patterns of all the crystals systematically exhibit satellites.

From the cation distribution, it is clear that an important fraction of the barium sites are occupied by thallium in this oxide. The charge balance

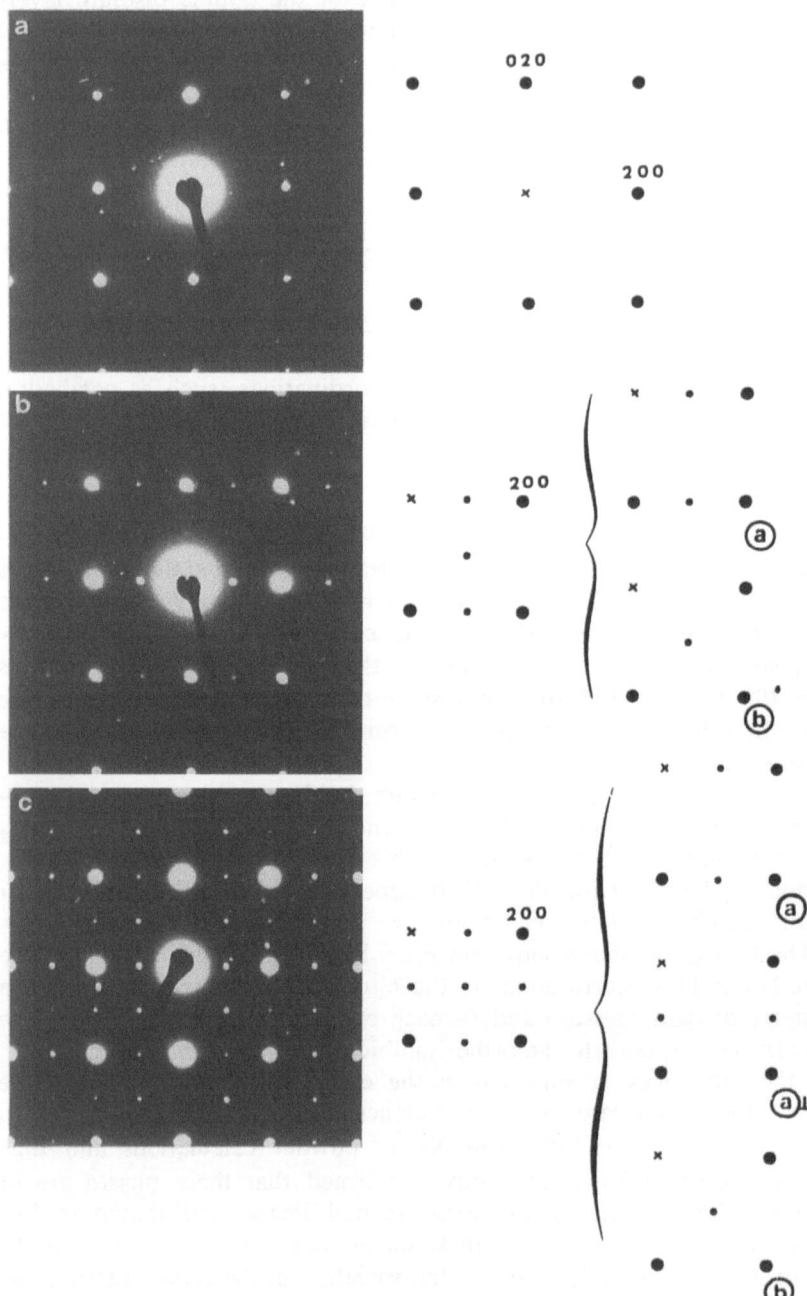

Fig.6.110a-c. The caption is given on the opposite page

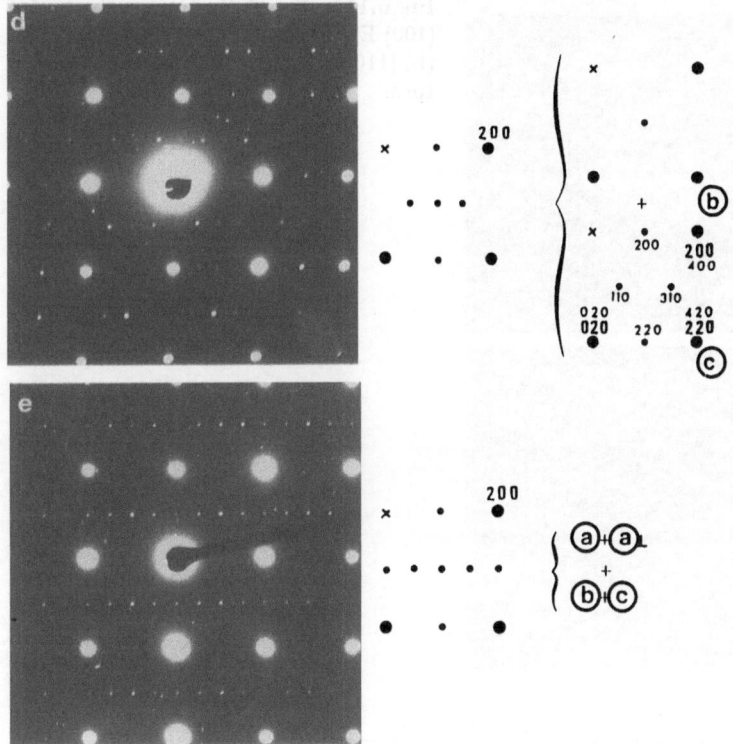

Fig.6.110a–e. $BiPbSr_2FeO_6$. The ED patterns show the systematic existence of at least two superposed networks; two domains oriented at 90° are labelled with the same letter and the symbol $\perp$ (a) Reflections of the subcell. (b) a and b (weak) are superposed. (c) a, the 90°-oriented $a_\perp$ domain and b are superposed. (d) b and c are superposed. (e) a (weak), $a_\perp$, b and c are superposed

and geometrical considerations show that Tl(I) is likely to be present in the barium layers [6.141]. In the same way, an examination of the interatomic distances suggests that the lone pair of Tl(I) is stereoactive. This hypothesis is moreover in agreement with the observations of satellites in the ED patterns and the HREM images which give evidence of a tendancy to cleave parallel to the layers (Fig.6.113). Moreover, it appears that the cleavage arises at the level of the $[Ba, Tl^IO]_\infty$ layers.

6.4.3 Lead Oxides

Numerous extra spots are observed too in the ED patterns of lead oxides. Owing to the absence of a real lead family and to the ability of lead rather to substitute Bi, Tl or alkaline-earth ions, the nature of these additional reflections is varied. From various publications [6.114–117] it appears that the $Pb_2Sr_2Y_{0.5}Ca_{0.5}Cu_3O_8$ phase, [1,2;1,1], is the rare one which does not exhibit satellites.

On the other hand, the "1212" phase, $Pb_{0.5}Sr_{2.5}Ca_{0.5}Y_{0.5}Cu_2O_{6.75}$, exhibits satellites whose intensity varies with the thermal treatment

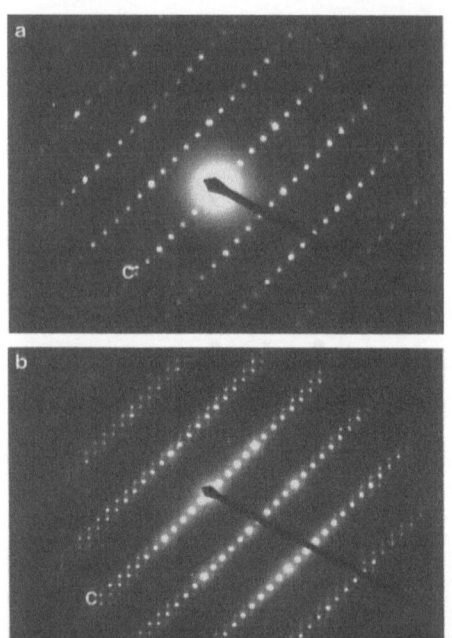

Fig.6.111a,b. $Tl_2Ba_2CuO_6$ phase. (a) [100] ED pattern of the tetragonal phase. (b) [110] ED pattern of the orthorhombic form

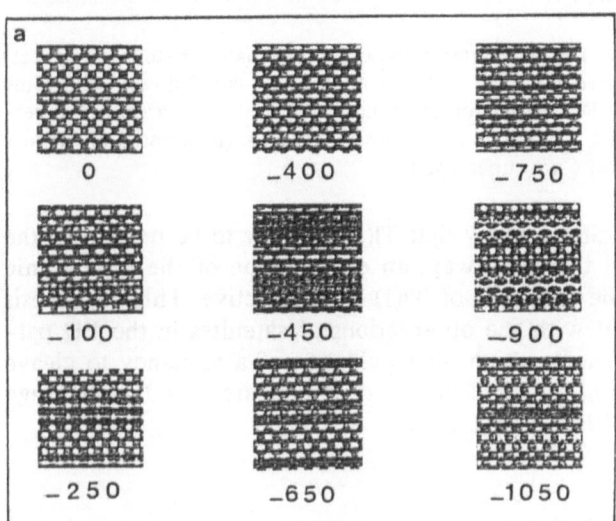

Fig.6.112a,b. $(Tl_{2-x/3})(Tl_{1-x}Ba_{1+x})LnCu_2O_8$: (a) calculated and (b, *see opposite page*) experimental through a focus series

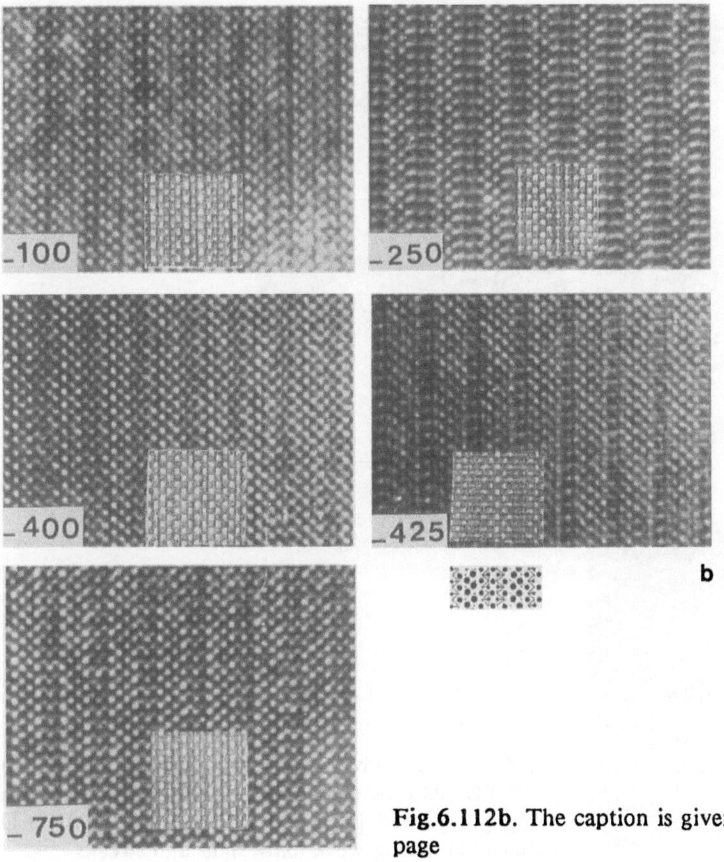

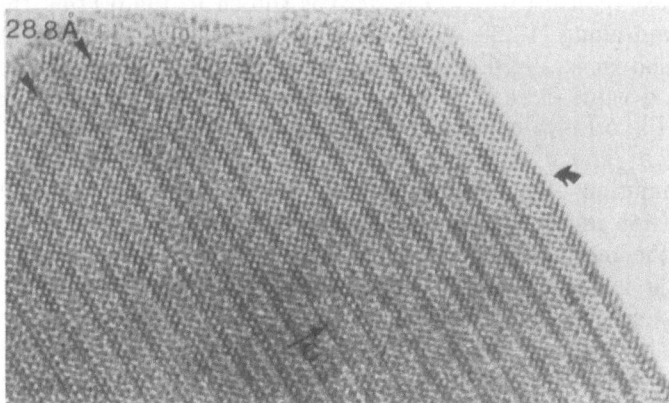

Fig.6.112b. The caption is given on the opposite page

Fig.6.113. The crystals exhibit a systematical cleavage parallel to the layers

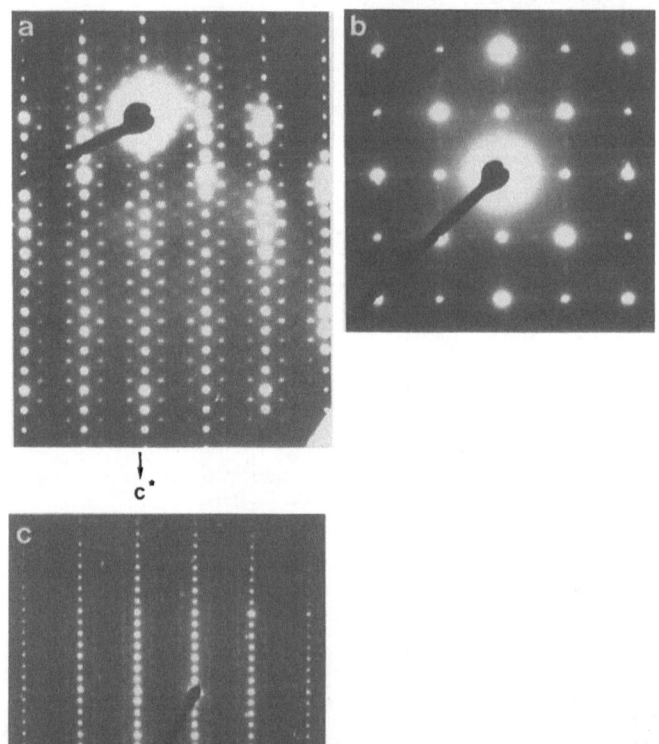

c^*

Fig.6.114a–c. $Pb_{0.5}Sr_{2.5}Y_{0.5}Ca_{0.5}Cu_2O_{6.75}$.   (a, b)
[010] ED patterns showing the existence of satellites
along $[102]^*$ and $[203]^*$, respectively. (c) [001] ED
pattern exhibiting weak extra spots and streaks

[6.105, 106]. The most typical [010] ED pattern is shown in Fig.6.114a: the
satellites are observed along $[102]^*$ with a modulation factor equal to 4. A
second example is shown in Fig.6.114b, where weak satellites are observed
along $[203]^*$ with a q value close to 6. In agreement with these observations
[001] ED patterns (Fig.6.114c) exhibit weak spots in 1/2 00 or 0 1/2 0 and
weak streaks along a^* and c^*. The corresponding high resolution images
show contrast modulations which generally appear in extended domains
(Fig.6.115a). From this image and the enlargement (Fig.6.115b) it is clear
that the origin of the modulations is different from that observed in the
bismuth oxides. If it is assumed that the row of bright dots is correlated
with the $[Pb_{0.5}Sr_{0.5}O]$ rows (black arrows in Fig.6.115b), it can be shown
that the variations in contrast arise at the level of $[Sr_2ACu_2O_{6-x}]$ slices
rather than at the level of the lead row. The origin of these modulations is
probably related to ordering of oxygen vacancies, taking into account the
oxygen-deficient character of the phase with respect to the theoretical O_7
oxygen content. This hypothesis is in agreement with the variations of intensity of the satellites with the thermal treatment. Thus, the modulations
observed in the "1212" lead phase should be compared to those observed in
the $Sr_2LnCu_2O_{6-x}$ phase rather than to those of the bismuth oxides.

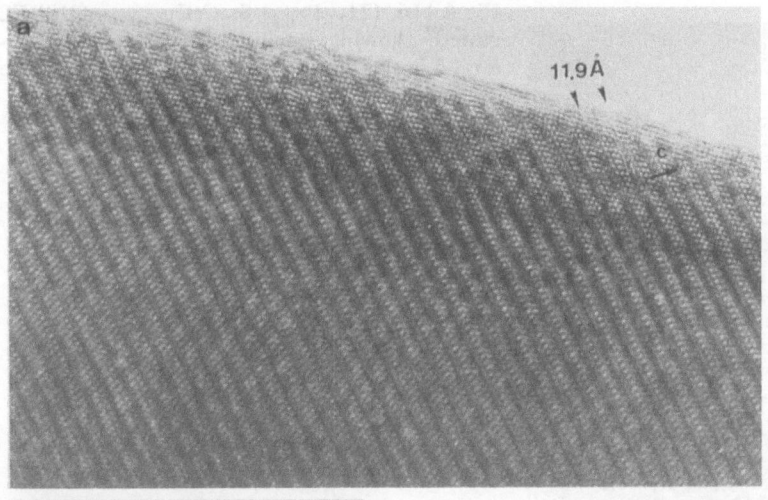

Fig.6.115. (a) [010] image showing an extended domain where the contrast modulations appear. (b) Enlargement of a modulated area (the [$Pb_{0.5}$ $Sr_{0.5}$O] (*layers are indicated by black arrows*)

Another interesting comparison can be made with another "1212" phase: the mixed $Tl_{0.5}Pb_{0.5}Sr_2CaCu_2O_7$ oxide. Diffuse satellites are observed in some [010] ED patterns. An example is shown in Fig.6.116, with satellites lying along [203]* and a q value close to six. This type of ED pattern is very similar to those encountered in $Pb_{0.5}Sr_{2.5}Ca_{0.5}$ $Y_{0.5}Cu_2O_{6.75}$ (Fig.6.114b). The corresponding images are, in the same way, similar (Fig.6.117), with the [$Tl_{0.5}Pb_{0.5}$O] layer correlated with the row of brightest spots; however, in the thallium-lead images, the modulations are so weak that they cannot be observed significantly. The scarcity of this type of modulations can be related to the oxygen content, which is assumed higher than in the strontium-lead phase.

Another type of additional spots is observed in the thallium-lead phase, leading to a $4 \times a_p$ superstructure along the b axis (Fig.6.118a); these superstructures often occur along the two equivalent directions of the perovskite subcell. The corresponding bright field images (Fig.6.118b) show

253

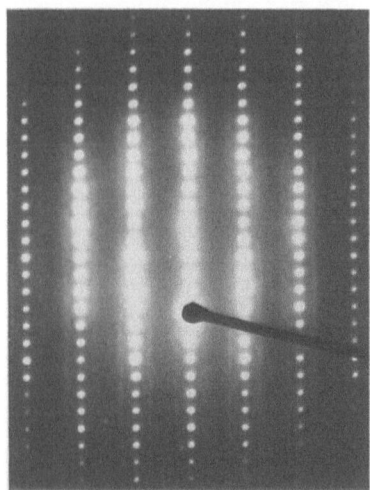

Fig.6.116. $(Tl_{0.5}Pb_{0.5})\,Sr_2CaCu_2O_{7-\epsilon}$. [010] ED pattern showing satellites similar to those observed in the $Pb_{0.5}Sr_{2.5}Ca_{0.5}\,Y_{0.5}Cu_2O_{6.75}$ oxide

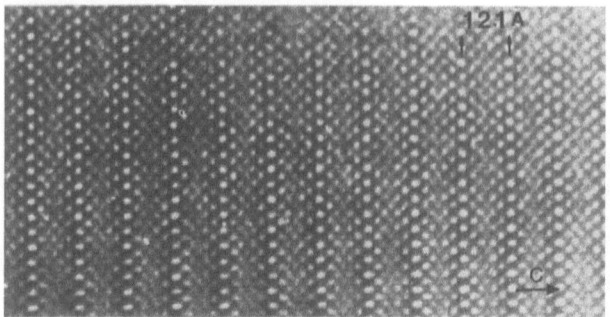

Fig.6.117. [010] HR image corresponding to Fig.6.116. The rows of brightest dots correlate with the $[Tl_{0.5}Pb_{0.5}O]$ rows. The contrast modulations are very weak

perpendicular domains. The domain boundaries are mainly parallel to ⟨110⟩ and ⟨1̄10⟩ of the perovskite cell and the 90°-orientation of the 4a superstructures is clearly visible (Fig.6.118c) through the 15 Å fringes. Another system of fringes, poorly resolved, is observed in these images; their direction is roughly parallel to the b axis of the perovskite subcell all over the crystal, whatever the orientation of the superstructure may be. It can be related to a fairly well established long-period structure; unfortunately, the contrast of the corresponding high resolution images is too disturbed to allow an interpretation to be proposed. Similar additional spots, which imply a periodicity along the a axis, were observed in another cuprate characterized by mixed rock salt layers: $PbBaSrYCu_3O_{8-\delta}$ [6.112]; an ion ordering in the mixed [AO] layers is thought to be responsible for this type of superstructure.

Ordering of oxygen and vacancies in the $[Sr_2A'Cu_2O_{6-x}]$ slices and ordering of lead with a second type of atom in the mixed $[Pb_{0.5}M_{0.5}O]$ rock salt layers appear then to be the two main properties responsible for the superstructure phenomena in the lead cuprates.

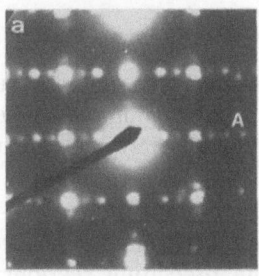

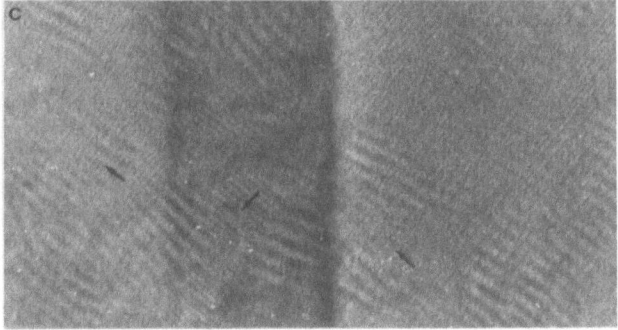

Fig.6.118a–c. $Tl_{0.5}Pb_{0.5}Sr_2CaCu_2O_{7-\epsilon}$. (a) [001] ED pattern showing a $4a_p$ super-structure along two perpendicular directions ($[100]_p$). (b) Bright field image of the 90°-oriented domains. (c) The enlargement shows clearly that the $4 \times a_p$ superstructure changes direction across the domain boundaries (*black arrows*). Note the second system of fringes, ill-resolved

6.5 Domains and Boundaries

The various structural features presented in this chapter which occur in the high T_c superconductors can often give rise to large or small domains. The variety of their characteristics (orientation, size, frequency, etc.) is as great as the variety of the phenomena they come from.

Ordering of oxygens/vacancies and cations is undoubtedly the phenomenon most often implicated in the formation of the domains. Most mechanisms of formation have been described in the preceding sections; they are summarized in Table 6.2. They are essentially observed in the n = 0 phase since the cationic network (A cations as well as copper ions) is exactly the same as in the perovskite.

Ordering of oxygen/vacancies in the "123"-type oxides leading to twinning domains with (110) boundaries is indeed the most spectacular phenomenon (Sect.6.1). Although the actual coordination of copper at the level of the boundary is not known, the HREM images show that the twin

Table 6.2. Domains and boundaries in the cuprates

Origin of the domain	Boundary orientation	Structural type	Example	Symmetry	Frequency or domain size
90°-oriented domain	(110)	n=0,m=∞	$LnBa_2Cu_3O_{7-x}$	O	Systematic some
O ordering	(001)	n=0,m=∞	$LnBa_2Cu_3O_{7-x}$	O	thousand Å to some tens Å vs x
90°-oriented domain; cation ordering	(001)or(010)	n=0,m=0	$YBa_2Cu_3O_{6+x}$ $LaBa_2Cu_3O_{7-\delta}$	T T	From rare to systematic
	(100)or(010) (110)	n=1,m=2 n=2,m=2	$Sr_2NdCu_2O_{5.7}$ $Tl_{0.5}Pb_{0.5}Sr_2CaCu_2O_7$	O T	Very rare Very rare
	(110)or(1$\bar{1}$0)	n=3,m=2	$Bi_2Sr_2CaCu_2O_8$ $Bi_2Sr_3Fe_2O_9$	O O	Systematic Systematic
		n=3,m=3	$Bi_2Sr_2Ca_2Cu_3O_{10}$	O	Systematic
Low angle "dislocation"	Tilt boundary	n=1,m=1.5	$Pb_{1.4}Bi_{0.6}Sr_2Y_{0.5}Ca_{0.5}Cu_3O_8$	O	Rare
		n=3,m=2	$Bi_2Sr_2CaCu_2O_8$	O	Rare
Low angle	Twist boundary	n=1,m=1.5	$Pb_{1.4}Bi_{0.6}Sr_2Y_{0.5}Ca_{0.5}Cu_3O_8$	O	Frequent
		n=3,m=1 n=3,m=2 n=3,m=3 n=2,m=2	$Bi_2Sr_2CuO_6$ $Bi_2Sr_2CaCu_2O_8$ $Bi_2Sr_2Ca_2Cu_3O_{10}$ $Tl_{0.5}Pb_{0.5}Sr_2CaCuO_7$	O O O T	Very frequent Very frequent Very frequent Frequent

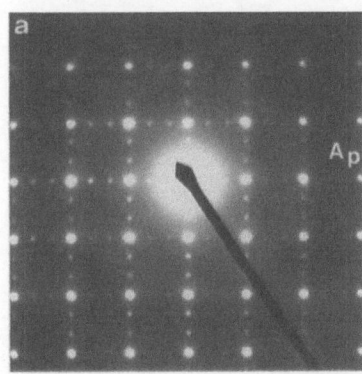

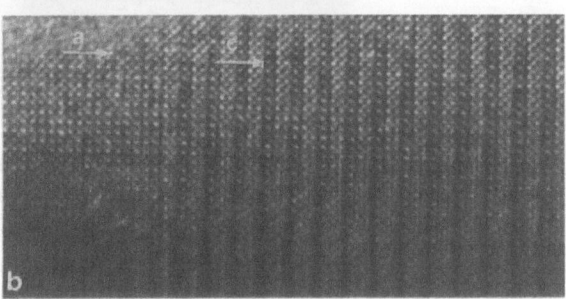

Fig.6.119a,b. Tetragonal phase T_3, $Yba_2Cu_3O_{6.8}$. (a) Typical [001] ED pattern; the $3 \times a_p$ superstructure appears along the equivalent directions of the perovskite cell. (b) Boundary between the [001] and [010] domains

boundaries are not coupled with specific defects. The average width of the domains, some thousands of Ångstroms, is highly sensitive to copper substitution; for instance substitution of iron or zinc for copper produced a sharp decrease of the domain width, up to a tetragonal cell.

On the other hand, the existence of oriented domains with (001) boundaries implies strong disturbances and fractures (Fig.6.10) owing to the mismatch between a and b parameters which must be adjusted through the $[Y]_\infty$ layer.

The domains resulting from cation ordering along one of the equivalent directions of the perovskite subcell are observed in a larger variety of structures. In the "123"-type oxides, the ED patterns which attest to the existence of 90°-oriented domains exhibit $3 \times a_p$ superstructure reflections along two equivalent directions of the perovskite cell (Fig.6.119a). The conditions for the formation of such domains are very simple: $a \simeq b \simeq c/3 = a_p$.

In the tetragonal phase T_2, $YBa_2Cu_3O_{6.5}$, these domains are rare since the a and b parameters differ from $c/3$; few have been reported [6.142]. On the other hand, in the T_3 phase they appear in a systematic way, in agreement with the parameter relationships (Fig.2.37). It can be seen that the boundary between two domains is perfect (Fig.6.119b). This type of domain appears in the same way in the tetragonal phase $La_{1.5}Ba_{1.5}Cu_3O_7$, which exhibits the required parameter relationships.

When the perovskite slices are intergrown with rock-salt-type layers, 90°-oriented domains can be observed, depending on the cations of the

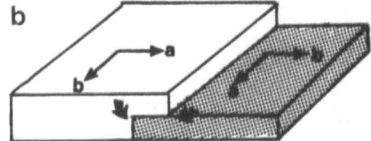

Fig.6.120a,b. $Sr_2NdCu_2O_{5.75}$. (a) Example of two oriented domains: [100] on the left and [010] on the right. The domains correspond to a change in direction of the cation ordering (and also of the oxygens). (b) Idealized drawing of the domains (the a and b axes are those of the perovskite)

$[AO]_\infty$ layers. When mixed $[AO]_\infty$ layers are observed, as in $Sr_2NdCu_2 \cdot O_{5.75}$ or $Tl_{0.5}Pb_{0.5}Sr_2CaCu_2O_7$, the ordering of the cations which occurs along $[100]_p$ or $[010]_p$ is rarely observed (Figs.6.118 and 6.120). In the strontium-neodynium phase, it appears that the two adjacent domains are slightly misoriented (Fig.6.120a) and the broad boundary suggests a partial superposition of the domains (Fig.6.120b).

In the pure thallium oxides, where only one type of cation is located in each $[AO]_\infty$ layer, no domain is observed. The bismuth oxides correspond to a particular case; the 90°-oriented domains are generated by the change in direction of the $[BiO_x]_\infty$ layer modulations (Fig.6.59c).

Two types of tilt boundaries were reported in the high T_c superconductors; they can be described in terms of an array of dislocations arranged along the [100] or [110] axis of the perovskite subcell, implying a small rotation about the axis common to the domains on either side of the boundary. In such a complex structure, a mechanism of dislocation rarely concerns a single atomic plane but rather a row of polyhedra. Thus, it generally requires a classical structural mechanism for a description of the notion of dislocation. The first example (Fig.6.121) was observed in the $Pb_{1.4}Bi_{0.6} \cdot Sr_2Ca_{0.5}Cu_3O_8$ oxide [6.111]: the misorientation of the domains is about 6° and the image shows a translation of $a_p/2$ between the cells of the two domains (black arrows). Such a dislocation is easily explained by a shearing phenomenon taking place in the perovskite layers (Fig.6.121b). The shift can be explained by the formation of double rows of edge sharing CuO_5 pyraminds running along a very similar to the {001} crystallographic shear planes observed in oxides with an octahedral structure, derived from the ReO_3-type structure (Fig.6.121c). The formation of this boundary is com-

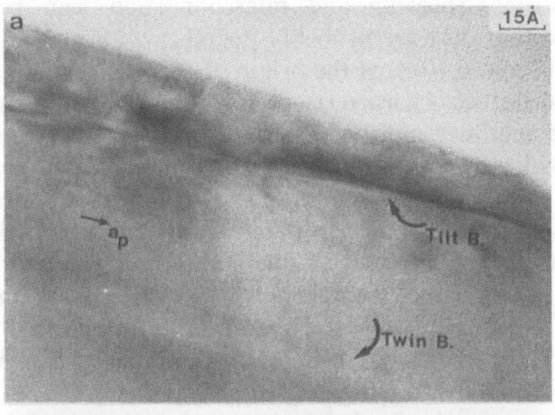

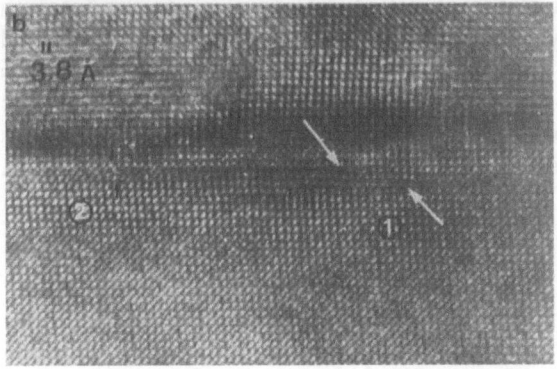

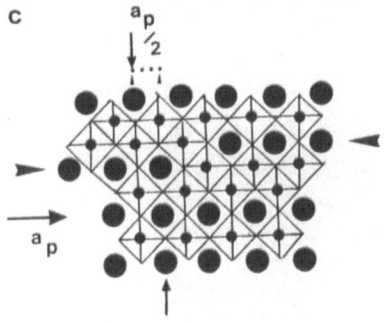

Fig.6.121a-c. $Pb_{1.4}Bi_{0.6}Sr_2Ca_{0.5}Y_{0.5}Cu_3O_8$. (a) [001] image of a crystal where two types of boundaries are observed: a twin boundary, resulting in a classical way from an orthorhombic distortion of the cell, and a tilt boundary. (b) Enlargement of the tilt boundary showing the dislocation–like contrast (labelled 1 and 2) and the translation of the cell through the boundary (*small black arrows*). (c) Model of the crystallographic shear mechanism

patible with the existence of the SrO rock salt layers which exhibit a two-dimensional agreement with the pyramidal copper layers. The second example was observed in $Bi_2Sr_2CaCu_2O_8$ (Fig.6.101); the mechanisms of the dislocations involve the stopping of both types of layers, perovskite and bismuth layers, which is ensured through layer interconnections. It should be noted that both mechanisms, shearing and interconnection, can be regarded as single defects and that they are probably, together with partially intercalated layers (Figs.6.17 and 6.70b), at the origin of the "dislocations" which are sometimes reported but not characterized. The involvement of at least one row of polyhedra, sometimes more (Fig.6.101), in the mechanism leads to domain boundaries which are often disturbed.

Twist boundaries were exclusively observed, up to now, in bismuth and lead oxides. The typical electron diffraction patterns are characterized by the existence of a two-dimensional systems of satellites around the basic spots (Figs.6.82d, 106, 107b). They were frequently observed in bismuth oxides [6.108, 113-116, 126] and can be easily explained by the formation of low angle twist boundaries. The corresponding images exhibit the classical moiré pattern contrast (Fig.6.122). In $Pb_{1.4}Bi_{0.6}Sr_2Ca_{0.5}Y_{0.5}Cu_3O_8$ (n = 1, m = 1.5), some similar features are observed, but contrary to the bismuth oxides (n = 3 families) they are often localized to a part of the crystal (Figs.6.122b, 123a). In every crystal exhibiting such phenomena, it appears that the twist boundaries are only observed in areas where incommensurate modulations are set up, generally along $[100]_p$ or $[110]_p$. These modulations are very similar to those observed in the bismuth oxides but they differ in

Fig.6.122. $Pb_{1.4}Bi_{0.6}Sr_2Ca_{0.5}Y_{0.5}Cu_3O_8$. Examples of moiré patterns observed in some crystals

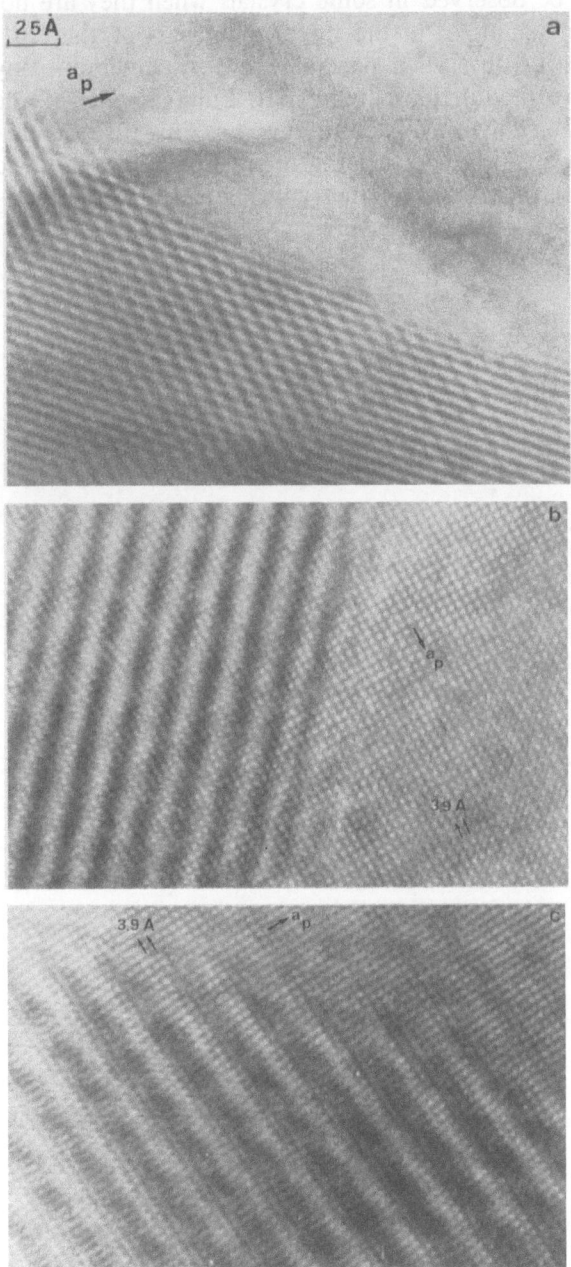

Fig.6.123. (a) [001] image showing that the modulations are not established all over the crystal. **(b, c)** Modulations along two directions are observed. Their superpositions leads to the moiré contrast in the lower part of **(a)**

their orientation; they can be observed in some crystals when they are not superposed on another. For example, in the upper right part of the [001] image (Fig.6.122a) we clearly observe a part of the crystal without any modulations while there are modulations at the left. Enlargements of the two types of modulations are shown in Figs.6.123b and c. This suggests that the twist boundaries which are produced by the interaction of at least two sets of screw dislocations are directly correlated with the existence of modulated distortions of the bismuth and lead layers.

7. Irradiation Effects in the High-T_c Superconducting Oxides

As has been discussed elsewhere (Chap.3), the superconducting properties of the new high-T_c copper oxides are strongly affected by chemical and structural disorder. The factors that may generate defects in the oxygen framework are numerous and have several origins, including oxygen content, size of the grain, nature of the precursors and thermal treatments. They can result either in strong inhomogeneities, for the most stoichiometric $YBa_2Cu_3O_{7-\delta}$ superconductors ($\delta \simeq 0$), or in various microstructures, for the reduced ones ($\delta \geq 0.3$). All these features can be considered a consequence of the disproportionation of Cu(II) into Cu(III) and Cu(I) to take into account the copper coordination states with respect to the oxygen deficiency δ.

In this field of research, irradiation provides a way of varying the state of disorder in a wide damage concentration, from the simplest degrees of defect formation, i.e. vacancies and interstitials, up to levels where phase transformations, e.g. amorphization, can occur. Further, from irradiation studies one can obtain information on the intrinsic defect structures and their influence on the superconducting properties. Radiation effects have thus been extensively studied in recent decades in the "old" high-T_c superconductors of A15 type and the Chevrel phases (for a recent survey of radiation effects in superconductors, see [7.1]). It has been clearly demonstrated that the critical temperature of the A15 compounds depends upon the composition and degree of long-range order, which is easily destroyed by irradiation, producing antisite defects and small static displacements. For $PbMo_6S_8$ it has been found that one displaced atom per formula unit is enough to depress the T_c. In the case of the new high-T_c oxides, it has been established in the previous chapters that oxygen content and oxygen vacancy distribution among the corresponding lattice sites are important parameters for superconductivity. As a consequence, it may be assumed that the oxygen displacements during irradiation can markedly change the electronic structure of these oxides, which exhibit, as is well known, a rather low density of states at the Fermi level.

Furthermore, one prospect of these exciting materials is high-current applications, both in the magnetic technology for fusion reactors or accelerators and in power transmission lines. In the former, the superconductor will be exposed to radiation, for instance to a fast neutron bombardment of about $5 \cdot 10^{18}\,n/cm^2$ during the mean lifetime of a tokamak fusion reactor. For the latter, it is well known that superconducting materials can show a considerable magnetic hysteresis and trapped flux if a high density of defects is introduced [7.2]. Therefore, one should not be surprised by the large number of irradiation experiments undertaken by several research groups, mainly on $La_{2-x}M_xCuO_4$ (M = Sr, Ba) and $YBa_2Cu_3O_{7-\delta}$-type oxides.

The results reviewed in this chapter have been organized as follows. First, we outline briefly the general physical aspects of the fate of electrons, neutrons and heavy ions in matter. Then the effects of electrons and fast neutron radiation, which produce defects by elastic collisions (nuclear stopping) are discussed. Lastly, new results obtained at the Grand Accelerateur National d'Ions Lourds (GANIL) in Caen with highly energetic heavy ions (E > 1 GeV) are presented and compared with those of implantation effects of ions with different masses and energies in superconducting thin films.

7.1 Radiation Damage in Solids

The extensive analysis of the numerous parameters that must be taken into account to describe the interaction of various types of radiation with solids can be found in *Lehmann's* remarkable book [7.3]. Here, we will just recall some general features of the fate of the radiation particles and the subsequent defect production in order to clearly distinguish the effects induced by electron and fast neutron irradiations from those induced by fast heavy ions, since these particles have mainly been used to study the sensitivity to irradiation of the new high-T_c superconducting oxides.

7.1.1 Electronic and Nuclear Stopping Powers

The slowing down of an energetic particle in matter comes from energy losses due to interactions with the constituents of the target material, namely electrons and atomic nuclei. For sufficiently high energies (or velocities) the bombarding particle interacts predominantly with the electrons via a Coulomb potential and a Rutherford-type scattering. According to *Bethe's* theory [7.4] the electronic stopping power, which is a measure of the energy loss per electronic collision, can be written

$$-\left(\frac{dE}{dx}\right)_e = \frac{4\pi Z_1^2 e^4}{m_e v_1^2} n_2 Z_2 \ln\left(\frac{2m_e v_1^2}{I}\right),$$

where I is the mean excitation energy of the atom, Z_1 and v_1 refer to the charge and the velocity of the incident particle, and $n_2 Z_2$ is the electron density of the target.

Bethe's formula should be valid only if v_1 is large compared with the velocities of the fastest electrons of the atom, i.e., for $Z_1 v_0/v_1 \ll 1$ where $v_0 = e^2/\hbar$ is the orbital velocity of the electron in the hydrogen atom.

In the intermediate range of energies ($v_1 \simeq Z_1 v_0$), where the effective charge of the projectile becomes smaller (due to capture of electrons), the stopping power is thus drastically reduced and no longer follows the v_1^{-2} law of Bethe's theory. At low velocities ($v_1 \ll Z_1 v_0$) only less strongly bound electrons can participate in the energy loss process. According to *Lindhard* and *Scharff* [7.5], the electronic stopping power is proportional to v_1

$$-\left(\frac{dE}{dx}\right)_e = \frac{8\pi e^2 a_B Z_1^{7/6} Z_2}{(Z_1^{2/3} + Z_2^{2/3})^{3/2}} \frac{v_1}{v_0},$$

where a_B is the Bohr radius.

In the range of low energies where the screening of the atomic nucleus of the incident particle due to captured electrons begins to become important, the elastic interaction between the atoms of the projectile and the target (nuclear stopping) begins to compete with inelastic collisions (electronic stopping) and finally predominates as the incident particle comes to rest ($v_1 \ll v_0$).

Several types of potentials have been considered to describe the two-body collision. Among them the screened Coulomb potential is the most widely used:

$$V(r) = \frac{Z_1 Z_2 e^2}{r} \phi\left(\frac{r}{a_F}\right),$$

where $\phi(r/a_F)$ is the Thomas–Fermi screening function, r the internuclear distance and a_F the characteristic screening length,

$$a_F = \frac{0.47}{(Z_1^{2/3} + Z_2^{2/3})^{1/2}} \quad [\text{Å}].$$

According to the LSS theory [7.6], the analytical representation of the screening function $\phi(x)$ is

$$\phi(x) = 1 - \frac{x}{(x^2+3)^{1/2}} \quad \text{with} \quad x = \frac{r}{a_F}.$$

The energy loss per elastic collision can then be obtained using *Biersack's* formula [7.7]:

$$-\left(\frac{dE}{dx}\right)_n = \frac{n_2 \, 4\pi a_F \, m_1}{m_1 + m_2} Z_1 Z_2 e^2 \frac{\ln\epsilon}{2\epsilon(1 - \epsilon^{-1.49})} \quad \text{where}$$

$$\epsilon = \frac{m_2}{m_1 + m_2} a_F \frac{E_1}{Z_1 Z_2 e^2}.$$

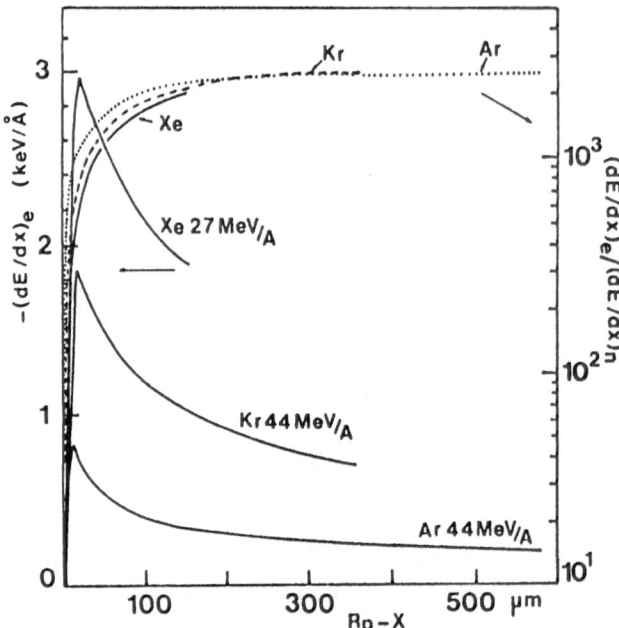

Fig.7.1. Electronic energy losses and the ratio of the electronic stopping power to the nuclear stopping power calculated for Ar, Kr and Xe ions as a function of the distance from the stopping ion zone in the case of $YBa_2Cu_3O_7$

Although the theories that have been proposed are not satisfactory over the whole energy range of the bombarding particle between its initial value E_0 and 0, the existence of correlated experimental data on stopping powers and ranges of radiation in matter allows one to estimate both the electronic and nuclear energy losses as functions of the energy and the penetration depth of the incident particle in the target using tables due to *Hubert* et al. [7.8], *Northcliffe* and *Schilling* [7.9], and *Ziegler* [7.10].

Figure 7.1 depicts an example of the energy losses determined for fast heavy ions (E > 1GeV) in the high-T_c oxide $YBa_2Cu_3O_{7-\delta}$. It is worth noticing here that the electronic stopping power is very large (a few keV/Å) and varies only slowly along the path (in the high-velocity regime). Moreover, it should be pointed out that the ratio of the electronic stopping power to the nuclear stopping power remains constant and of the order of 2000 over 90% of the ion path. The consequence is that the use of high-energy heavy ions provides an ideal situation in which to study the effects induced by electronic stopping in a pure manner in bulk material.

7.1.2 Material Modifications

Both interaction processes, electronic stopping and nuclear stopping, can lead in a crystal to the production of point defects or extended defects. The process that will be predominant in the defect-creation mechanism is, however, strongly dependent on the electronic structure of the target material and on the nature of the bombarding particle. It is thus well known that inelastic energy loss to electrons can cause structural damage in insulators and in semiconductors or, more generally, in materials where electronic excitation may remain localized for a sufficient period of time. In the case of metals this energy appears to be rapidly evacuated as heat without creating structural defects. Therefore the main process that can lead to radiation damage in metals is the elastic energy transfer to atoms. However, it should be mentioned here that recent results obtained at GANIL with very fast heavy ions in an $Fe_{85}B_{15}$ amorphous alloy revealed unambiguously that changes in electrical resistance and dimensions should be attributed to high electronic excitation [7.11].

a) Electronic Energy Loss Effects

Tracks of radiation damage have been described in a variety of dielectric materials and can be considered as a result of high density energy deposition into ionization. Several models of track formation mechanism have been proposed and reviewed by *Fleischer* et al. [7.12]. From small-angle X-ray scattering experiments *Dartyge* et al. [7.13, 14] demonstrated that latent tracks of heavy ions in silicates consisted of extended defects due to successive ionizations separated by gap zones loaded with point defects. In the ion explosion spike model of *Fleischer* et al. [7.12], latent tracks are described as resulting from the positively charged region created by ionization along the path of the incoming ion. The positive ions strongly repel one another and are ejected into interstitial positions surrounding a depleted core region. The application of lattice-imaging to radiation-damage investigations in magnetic insulators has clearly established the dependence of the structure of the defects on the energy deposited by electronic stopping [7.15-17]. The tracks are continuous with a cylindrical shape above a threshold value of the electronic stopping power, while they appear as trails of localized elongated defects below the threshold.

The existence of such threshold effects for the structural transformation of metallic alloys [7.11] and for the formation of continuous tracks in insulators provides an indication of highly nonlinear effects that cannot be neglected in the radiation damage of high-T_c superconducting oxides by fast heavy ions.

b) Nuclear Energy Loss Effects

In problems of the irradiation of solids with electrons and fast neutrons we are mainly concerned with the interaction between these particles and the atomic nuclei of the target, i.e., with elastic nuclear scattering. In the case of heavy ions this nuclear stopping competes with the electronic stopping at very low velocities and finally dominates in the last micrometer of the range where the incident ion comes to rest. As a consequence, it will be considered for ion implantation experiments in thin superconducting films.

Energy transfers of a few tens of electron volts are sufficient to displace atoms from their lattice sites. In typical reactors, for instance, neutrons have energies of a few hundred keV and are thus able to transfer energies in excess of the displacement threshold ($T_d \sim 20\,eV$) to the lattice atoms. Then these displaced atoms, called Primary Knock-On atoms (PKO), can leave their regular sites, leading to the production of Frenkel pairs where vacancies and interstitials are present. However, many of these primary energy transfers can be sufficient to favor a secondary atomic displacement cascade in a region around the path of the PKO. Interstitial atoms are then arranged in a diffuse shell around the edge of the cascade, while the core, often called the depleted zone, is vacancy rich.

Of course, the size and shape of the cascade is strongly dependent on the energy of the PKO and on the atomic number of the target material. This energy loss through atomic collisions can be used to estimate the number of displaced atoms produced in the cascade according to *Kinchin* and *Pease's* formula [7.18] modified by *Norgett* et al. [7.19]:

$$N_d(T) = \begin{cases} 0 & \text{if} \quad T < T_d \\ 1 & \text{if} \quad T_d \le T \le T_{DS} \\ 0.4E_D(T)/T_D & \text{if} \quad T > T_{DS} \end{cases}$$

where T is the energy transferred to the PKO, T_d is the displacement threshold, $E_D(T)$ the damage energy of a PKO and T_{DS} is defined as $E_D(T_{DS}) = 2.5T_d$.

Since it is of great interest to compare the changes induced by different types of bombarding particles on the superconducting (T_c, J_c) and normal (R) properties of the superconducting materials known at present, a convenient measure of damage production is given by the number of displacements per atom (dpa), which represents only the elastic interaction and does not take into account any electronic excitation effect; dpa(E) is calculated using the LSS theory [7.6] as

$$dpa = \phi t \int_{T_d}^{T_M} N_d(T)\, d\sigma(E, T) \; ,$$

where ϕt is the fluence of the particles, $d\sigma(E, T)$ is the differential cross section for an energy transfer by elastic collision between a particle of energy E and a target atom that receives the energy T. The upper limit T_M is the maximum energy that can be transmitted in a head-on collision:

$$T_M = \frac{4m_1 m_2}{(m_1 + m_2)^2} E \; .$$

7.2 Radiation Damage by Electrons and Fast Neutrons in Copper Oxide Superconductors

7.2.1 Defect Structures Produced by Electron Irradiation

As previously reported, the critical temperature of the copper oxides depends on the Cu(III)/Cu(II) ratio, which is strongly sensitive to the amount of oxygen in $YBa_2Cu_3O_{7-\delta}$ and the alkaline earth content of $La_{2-x}Sr_xCuO_4$. In other words, it should be interesting to study the influence of the simplest point defects that can be formed by electron irradiation, i.e., vacancies and interstitials of oxygen and copper atoms, for a better understanding of the superconducting behavior of these materials. The effects of electron irradiation have thus been investigated using either electron accelerators [7.20-23] ($E_e \geq 2MeV$) or the electron microscope ($E \geq 200keV$) [7.24].

Figures 7.2 and 3 show the superconducting transition temperature T_c and the resistance R in the normal state versus total electron fluence for $YBa_2Cu_3O_{7-\delta}$ and $La_{2-x}Sr_xCuO_4$ (x = 0.15,0.2) irradiated below 20 K. The variations of T_c and R are linear with the electron dose in the range $0 \leq \phi t \leq 4 \cdot 10^{19}$ e/cm^2. They correspond to a damage rate expressed as $\Delta T_c / \Delta(\phi t) \simeq -4.0 \cdot 10^{-19}$ K/(e/cm^2) for $YBa_2Cu_3O_{7-\delta}$ [7.21], $-3.30 \cdot 10^{-19}$ K/(e/cm^2) for $La_{1.85}Sr_{0.15}CuO_4$, and $-4.9 \cdot 10^{-19}$ K/(e/cm^2) for $La_{1.8}Sr_{0.2}CuO_4$ [7.20]. The decreases in T_c are very large, exceeding that of electron-irradiated A15 superconductors such as Nb_3Ge [7.25] or Nb_3Sn [7.26] by a factor of ten. The reduced values of T_c (T_c/T_{c0}) are plotted in Fig.7.4 as a function of the resistance in the normal state normalized to its value before irradiation according to the data reported in [7.20,21] for $La_{2-x}Sr_xCuO_4$ (x = 0.15-0.2) and $YBa_2Cu_3O_{7-\delta}$. The resistances were measured at 40 K for the former and 100 K for the latter. A linear correlation between these two quantities is observed in both cases. As a consequence, one might assume that the normal electronic transport properties and superconducting ones

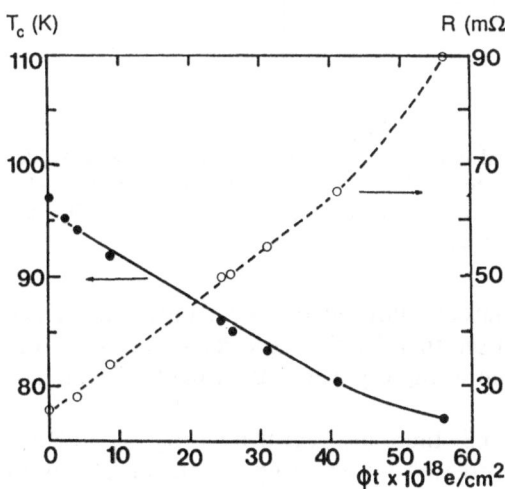

Fig.7.2. Transition temperature and resistance at 100 K as a function of the fluence of 3 MeV electrons in $YBa_2Cu_3O_7$ samples irradiated below 20 K. (From [7.21])

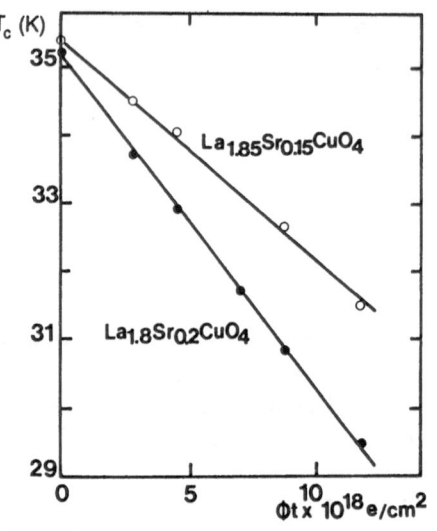

Fig.7.3. Critical temperature T_c versus dose of 2.5 MeV electrons for the two superconductors $La_{1.85}Sr_{0.15}CuO_4$ and $La_{1.8}Sr_{0.2}CuO_4$ irradiated at 20.5 K

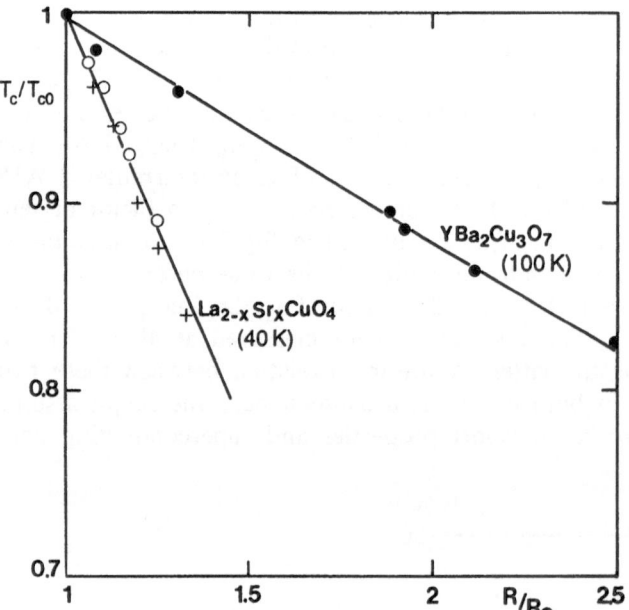

Fig.7.4. Correlation between the critical temperatures normalized to the initial values, T_c/T_{c0}, and the resistances normalized to the initial values, R/R_0, measured at 40 K $[R(40)/R_0(40K)]$ and at 100 K $[R(100)K/R_0(100K)]$ for $La_{2-x}Sr_xCuO_4$ and $YBa_2Cu_3O_7$ after irradiation by 2.5 MeV and 3 MeV electrons, respectively

have been affected by the same defects induced by low-temperature electron irradiation. However, one can see that for the same reduction of T_c the normalized resistance increases more rapidly for $YBa_2Cu_3O_{7-\delta}$ than for $La_{2-x}Sr_xCuO_4$.

As has been discussed in Chap.3, there also exists a strong dependence of T_c and R on oxygen vacancies created by warming the specimens in an inert-gas flow or vacuum. The initial T_c and the metal-like behavior are completely recovered after subsequent heating in oxygen at 723 K or by a low temperature plasma oxidation (353 K), confirming that the changes in T_c and oxygen content are reversible.

According to [7.27-31], for instance, the reduction of the transition temperature ΔT_c related to the vacancy concentration Δx introduced chemically or by heating processes in $YBa_2Cu_3O_{7-\delta}$ ($\delta \leq 0.1$) gives a mean value $\Delta T_c / \Delta x \simeq -5 \cdot 10^{-20}$ K/(vacancy/cm³). Such a result can be compared with the effect of radiation damage by evaluating the number of electron-induced defects per unit volume, as discussed by *Moser* et al. [7.22]. Using the data of *Stritzker* et al. [7.21] for $YBa_2Cu_3O_{7-\delta}$ samples irradiated with 3 MeV electrons, one obtains a ΔT_c related to the number of electron-induced defects, $\Delta T_c / n_F \simeq -10^{-18}$ K/(defect/cm³), implying a higher sensitivity of T_c to the electron irradiation damage. This may be explained by atomic displacements involving not only oxygen atoms but also copper atoms, although electron-microscopy observation of twin-boundary motion under electron irradiation seems to argue, as reported by *Kirk* et al. [7.24], for the direct displacement of oxygen atoms. However, it should be outlined here that the observation was made at room temperature, which implies, as shown by *Stritzker* et al. [7.21], important thermalization effects and/or subsequent rapid diffusion of the oxygen atom defects in Cu(1)-O chains.

A more suprising result is that reported by *Zelenskij* et al. [7.32], which shows that for high-energy electron irradiation ($E_e = 35$ MeV) at 5 K an increase of T_c is obtained for $YBa_2Cu_3O_{7-\delta}$ ceramics after subsequent annealing at 105 K, whereas irradiation at 185 and 375 K leads to a T_c decrease (Fig. 7.5). Similar behavior seems to be observed for $La_{1.8}Ba_{0.2} \cdot CuO_{4-x}$ samples. It is suggested that a thermal mobility of vacancies may be realized, leading to more-perfect atomic chains O-Cu-O and therefore to the T_c increase, the saturation of the T_c value at fluences higher than $5 \cdot 10^{17}$ e/cm² being due to the balance achieved between formed and decaying O-

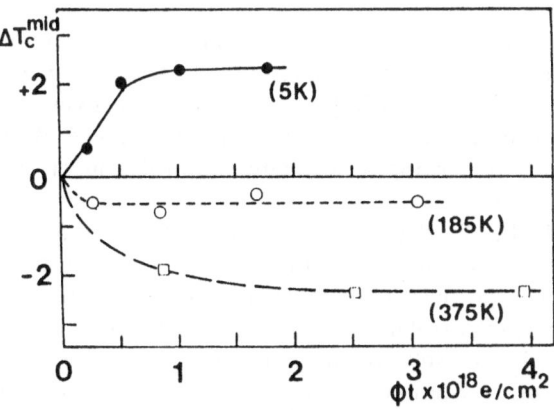

Fig. 7.5. Increase and decrease of T_c measured at the midpoint for Y-Ba-Cu-O samples irradiated with 30 MeV electrons at 5 K, 185 K and 375 K. (From [7.32])

271

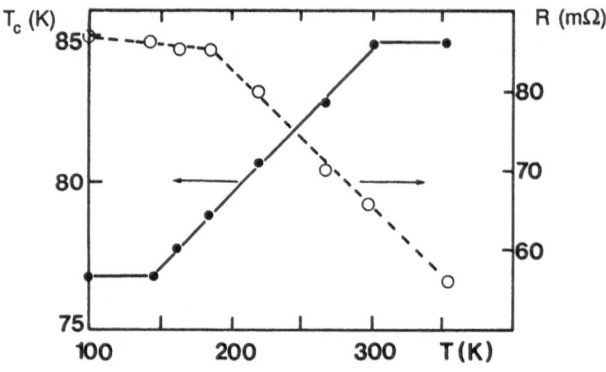

Fig.7.6. Transition temperature and resistance at 100 K of bulk $YBa_2Cu_3O_7$ samples as a function of the annealing temperature. (From [7.21])

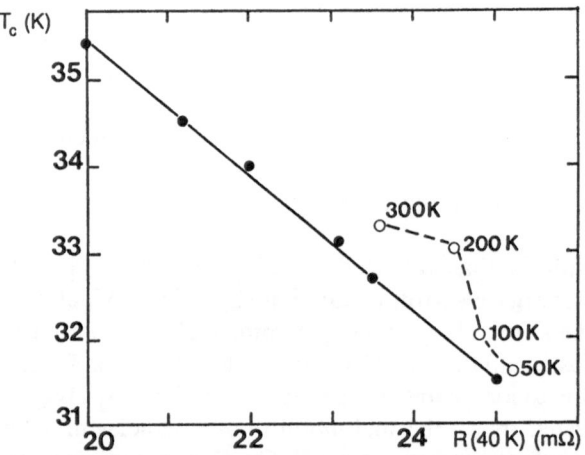

Fig.7.7. Effect of various annealings on the critical temperature and the resistance measured at 40 K for $La_{1.85}Sr_{0.15}CuO_4$ irradiated at 20.5 K by 2.5 MeV electrons at a total dose of $1.2 \cdot 10^{19} e/cm^2$. Observe that the linear relation between T_c and R(40K) under irradiation is no longer verified after annealing

Cu-O chains. Such a result should be compared with the annealing behavior of $YBa_2Cu_3O_{7-\delta}$ samples irradiated by 2MeV electrons at a fluence of 2.5 $\times 10^{18}$ e/cm² (Fig.7.6). It is thus observed [7.22] that a complete recovery of T_c (determined from ac-susceptibility measurements) can be achieved after a moderate annealing at 100 K, suggesting a correlated annihilation of the vacancy- and interstitial-type defects.

In contrast, only half of the T_c depression obtained for $\phi t = 5 \cdot 10^{19}$ e/cm² can be recovered, probably owing to an uncorrelated recovery process, which maintains vacancy-type defects on oxygen and copper sites up to 300 K. Furthermore, the linear correlation of T_c and R observed under irradiation is no longer verified after subsequent annealings, as can be seen in Fig.7.7 for $La_{1.85}Sr_{0.15}CuO_4$. Similar results are also reported for $YBa_2Cu_3O_{7-\delta}$ [7.21]. After the 200 K annealing for $La_{1.85}Sr_{0.15}CuO_4$ and the 400 K annealing for $YBa_2Cu_3O_{7-\delta}$, the transition width δT_c, which remained approximately constant under irradiation, increases suddenly. According to *Moser* et al. [7.22] the irradiation of $YBa_2Cu_3O_{7-\delta}$ by an electron dose of $2.2 \cdot 10^{19}$ e/cm^2 leads indeed to a 10%-90% transition width of 2.6 K, which increases to 5.2 K after an annealing treatment at 400 K for 1 h in He atmosphere. In the case of $La_{1.8}Sr_{0.2}CuO_4$ [7.20], the increase of δT_c is accompanied by an increase of the remaining resistivity at low temperature. This may suggest that some superconducting regions of the samples have been transformed into nonsuperconducting ones, leading to a partial destruction of the Josephson coupling between individual grains and rendering the sample more inhomogeneous.

7.2.2 Changes Induced by Fast Neutron Irradiation

The study of neutron irradiation effects in the new high-T_c superconducting oxides is a point of crucial importance, owing to possible applications in fusion reactors. Low temperature fast neutron irradiation of $YBa_2Cu_3O_{7-\delta}$ and $La_{1.85}Sr_{0.15}CuO_4$ superconductors has thus been performed for fluences ranging between 10^{16} and 10^{19} n/cm^2 [7.33-38].

In general, a rapid decrease of T_c and a considerable broadening of the transition width δT_c are observed, although some differences are noted, which are attributed to less perfect intergrain material. Therefore it is necessary to consider the influence of the material quality in irradiation experiments ranging from polycrystalline to single crystalline with quite different initial values of T_c, δT_c and resistivity.

The dependence on neutron fluence of T_c midpoint and resistivity measured at 100 K for $YBa_2Cu_3O_{7-\delta}$ samples by *Müller* et al. [7.38] is depicted in Fig.7.8. As outlined above, the irradiation-induced decrease of T_c is accompanied by a broadening of the transition width, which is indicated by the vertical bars. T_c decrease corresponding to a certain resistivity increase is observed to be enhanced by an increase of ϕt, but it is shown that a more sensitive behavior occurred for samples exhibiting higher resistivities before irradiation.

For fluences below $3 \cdot 10^{18}$ n/cm^2, *Cost* et al. [7.36, 37] found that the critical temperature deduced from their magnetic measurements decreased

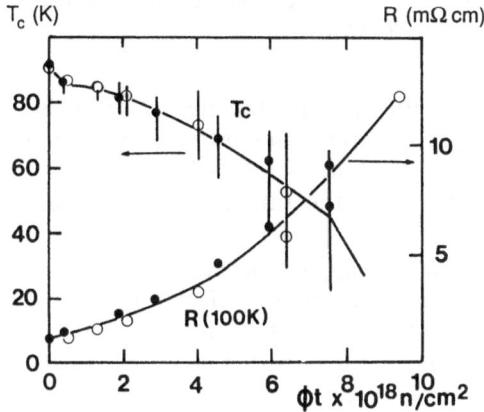

Fig.7.8. T_c and residual resistivity as functions of neutron fluence for two sintered samples of $YBa_2Cu_3O_7$ (from [7.38]). The vertical bars indicate the full transition width (0-100%)

273

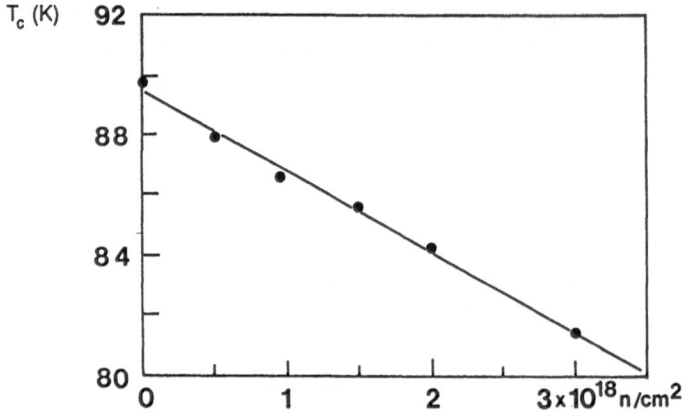

Fig.7.9. Superconducting transition temperature T_c vs neutron fluence (E>0.1 MeV) for $YBa_2Cu_3O_7$. The solid line is a linear least-squares fit to the data (From [7.37])

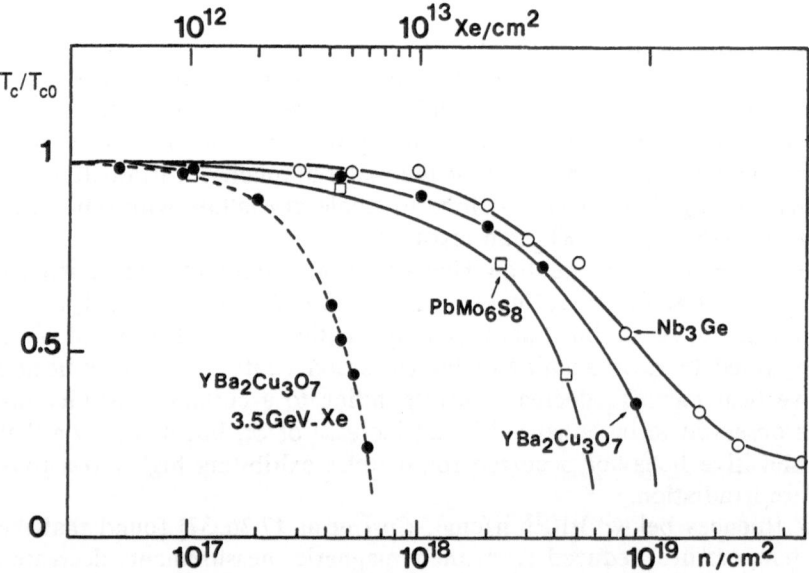

Fig.7.10. Normalized transition temperature T_c/T_{c0} vs neutron fluence (E>0.1 MeV) for $YBa_2Cu_3O_7$ compared with that of Nb_3Ge and $PbMo_6S_8$

roughly linearly with fluence (Fig.7.9) from the unirradiated value of 89.8 ±0.25 K at a rate of 3 K/(10^{18} n/cm²), comparable to neutron irradiation of $La_{1.85}Sr_{0.15}CuO_4$. In this case T_c decreased at a rate of roughly 2 K/ (10^{18} n/cm²) [7.33], but 10 times higher than the rate of A15 compounds [7.1, 39].

A more severe decrease is observed at higher fluences. A comparison with A15 alloys and Chevrel phases (Fig.7.10) showed clearly that the T_c degradation is larger than in A15 compounds and slightly less than in $PbMo_6S_8$. Moreover, the decrease of the T_c onset is considerably less than that of the T_c midpoint and T_c offset, as one can see in Fig.7.11. According

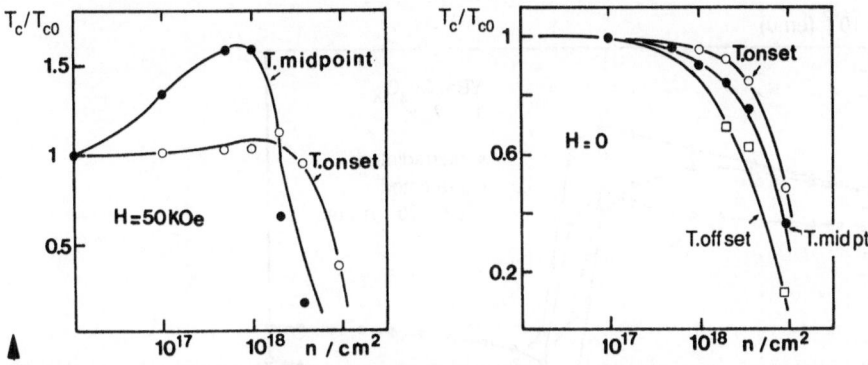

Fig.7.11. Normalized transition temperatures T_c/T_{c0} measured at the onset, the midpoint and the offset in zero field vs neutron fluence for $YBa_2Cu_3O_7$. (From [7.34])

Fig.7.12. Normalized transition temperature T_c/T_{c0} measured at the onset and the midpoint in a field of 50 kOe vs neutron fluence for $YBa_2Cu_3O_7$. (From [7.34])

to *Küpfer* et al. [7.34], such a result could be interpreted by considering the dominant role of the intragrain properties with respect to the onset and that of the intergranular links with respect to the offset.

As mentioned before, the primary defects due to fast neutron irradiation are displacement cascades. However, although the atoms are constrained to the cascade volume, it should be pointed out that some of their vibrational energy may be transmitted to regions significantly outside this volume by focusing collisions [7.40]. Then the rapid decrease of T_c between 10^{18} and 10^{19} n/cm^2 should be attributed to a high concentration of these disordered regions. *Küpfer* et al. pointed out that the size of such regions should be much larger than in A15 superconductors because of thermal spike effects, which would enhance the oxygen diffusion in Cu-O chains and thereby the correlated T_c depresssion.

The most gradual decrease of T_c at lower fluence could be caused by a homogeneous disorder due to replacement collisions of secondary knock-on atoms. The T_c onset in an applied magnetic field of 50 kOe shows a similar fluence dependence to that in zero field (Fig.7.12), whereas the T_c midpoint exhibits a broad maximum at about $\phi t = 10^{18}$ n/cm^2. It was discussed in detail by *Küpfer* et al. [7.41] that such a χ' transition in a magnetic field reflects not only structural inhomogeneities responsible for the transition width but also the H_{c2} anisotropy. Figure 7.12 could thus be considered to indicate a decreasing anisotropy of H_{c2} up to a fluence of 10^{18} n/cm^2.

In accordance with conventional superconducting materials, the critical current density J_c is expected to increase upon irradiation, owing to a flux pinning effect that manifests itself as a magnetization current. This behavior was observed in $La_{1.85}Sr_{0.15}CuO_4$ [7.33] and $YBa_2Cu_3O_{7-\delta}$ [7.36, 37] superconductors. Figure 7.13 shows the magnetic moment at 75 K plotted versus applied magnetic field H_a for a $YBa_2Cu_3O_{7-\delta}$ polycrystalline sample irradiated by fast neutrons at $\phi t = 9.6 \cdot 10^{17}$ n/cm^2. The curve is very hysteretic after irradiation, while it is nearly reversible above 35 kOe for the un-

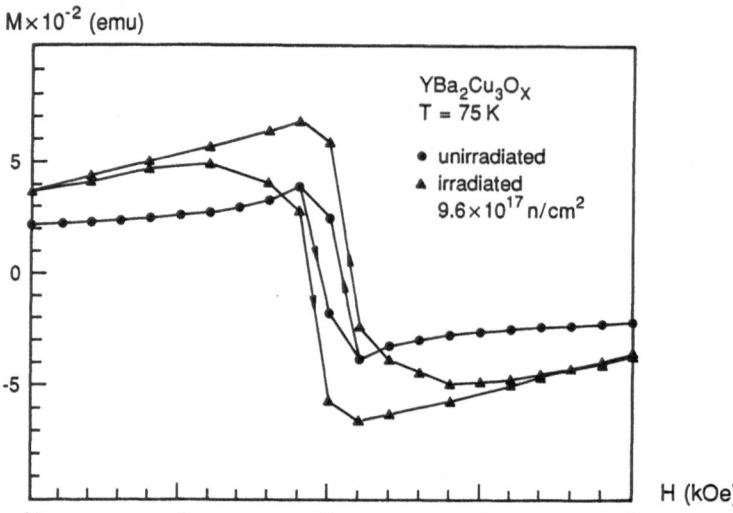

Fig.7.13. Magnetic moment of $YBa_2Cu_3O_x$ sample vs applied magnetic field at 75 K for the neutron fluence indicated. (From [7.36])

irradiated sample. In terms of the critical-state model of *Bean* [7.42] a critical current J_c will flow in a superconductor in response to an applied field. For a long cylinder oriented parallel to the field direction, J_c can be calculated from

$$J_c = \frac{15}{R}(M_+ - M_-) \quad [A/cm^2]$$

where R is the mean sample radius [cm], and M_+ and M_- are the sample magnetization [e.m.u./cm³] in increasing and decreasing fields, respectively.

According to *Cost* et al. [7.36,37], the critical current shows a strong field dependence at all fluences (Fig.7.14), dropping by a factor of 8 between 0 and 40 kOe. In single-crystal samples, J_c has been found to be highly anisotropic, with the ab plane supporting currents 10 times greater than those in the c direction [7.43]. The effects of fast neutron irradiation have been investigated particularly by *Umezawa* et al. [7.35]. For fields along the a direction, both the magnitude and the field dependence of the critical magnetization current are improved by neutron bombardment, as pictured in Fig.7.15. This increase is much smaller for fields along the c direction (Fig.7.16).

Such improvements of J_c deduced from magnetization measurements appear, however, to contradict the results reported by *Küpfer* et al. [7.34] from inductive measurements, as shown in Fig.7.17 for $\phi t = 4 \cdot 10^{18} n/cm^2$. It has been argued that the difference could come from the nature of the critical current density, which is effectively measured by means of both techniques. Magnetization measurements would give the intragrain J_c, which is expected to increase because of strong pinning effects inside the grain, whereas inductive measurements would lead to the intergrain J_c, which is more sensitive to the weak link structure of the intergranular material.

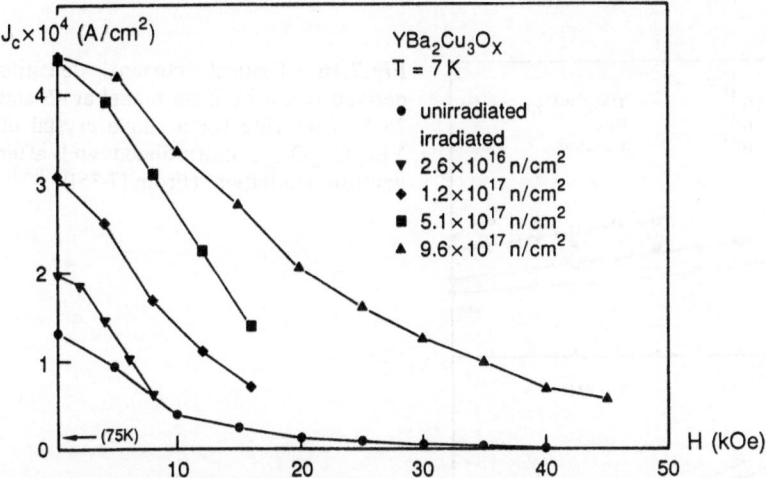

Fig.7.14. Critical current density vs applied magnetic field at 7 K for the neutron fluences indicated. (From [7.36])

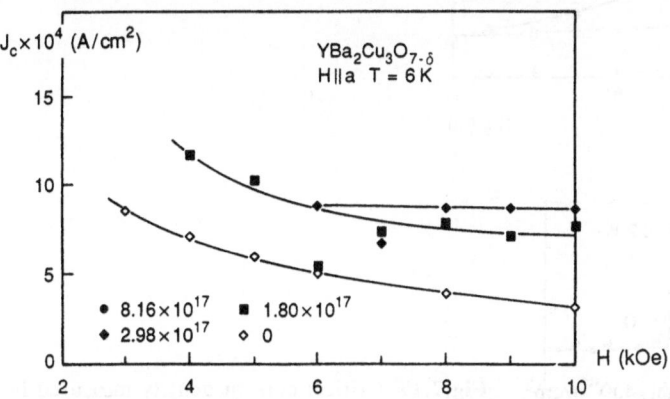

Fig.7.15. Critical current densities derived from the Bean model at 6 K and applied magnetic field H parallel to the a direction for a single crystal of $YBa_2Cu_3O_{7-\delta}$, unirradiated and after neutron irradiation. (From [7.35])

7.3 Phase Transformations Induced by Fast Heavy Ions in the High-T_c Copper Oxides

In contrast to electrons and fast neutrons, which produce defects by elastic collisions (nuclear stopping) with the target nuclei, fast heavy ions ($E \geq$ 1 GeV) allow one to investigate radiation damage induced by inelastic collisions with the target electrons, because the electronic stopping power is about 2000 times greater than the nuclear stopping power over 90% of the ion path. In the following we will describe recent experimental results obtained for sintered high-T_c copper oxides irradiated with 2.9 GeV Kr and 3.4 GeV Xe ions. The results are then compared with those obtained in thin

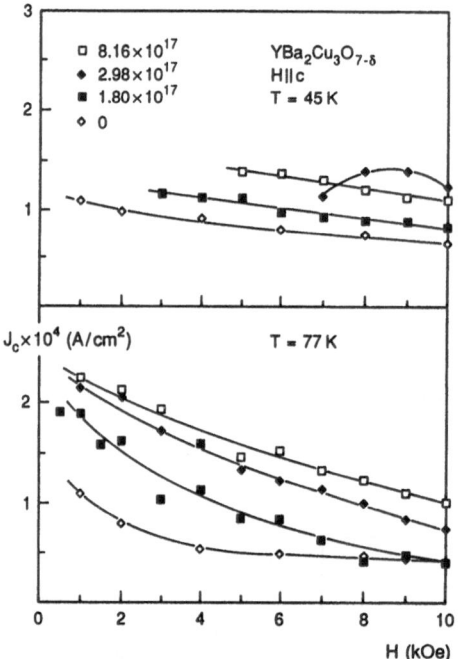

$J_c \times 10^5$ (A/cm^2)

$\square$ 8.16×10^{17} YBa$_2$Cu$_3$O$_{7-\delta}$
$\blacklozenge$ 2.98×10^{17} H$\|$c
$\blacksquare$ 1.80×10^{17} T = 45 K
$\diamond$ 0

$J_c \times 10^4$ (A/cm^2)   T = 77 K

H (kOe)

Fig.7.16. Critical current densities derived from the Bean model at 45 and 75 K with H$\|$c for a single crystal of YBa$_2$Cu$_3$O$_{7-\delta}$, unirradiated and after neutron irradiation. (From [7.35])

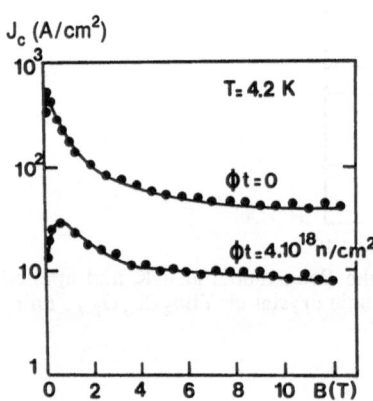

J_c (A/cm^2)

T = 4.2 K

$\phi t = 0$

$\phi t = 4 \cdot 10^{18}$ n/cm^2

B(T)

Fig.7.17. Critical current density measured inductively as a function of field at 4.2 K for a polycrystalline sample of YBa$_2$Cu$_3$O$_{7-\delta}$, unirradiated and after neutron irradiation. (From [7.34])

superconducting films by ion implantation, which has been used for film patterning and the fabrication of Superconducting QUantum Interference Devices (SQUIDs).

7.3.1 Improvement of T_c in the Grain Surface Superconductor La$_2$CuO$_4$

As described in Chap.3, the pristine oxide La$_2$CuO$_4$ can exhibit semimetallic behavior at room temperature [7.44] and semiconducting behavior at low temperatures [7.45]. A metallic state can be reached in the Sr-doped compounds La$_{2-x}$Sr$_x$CuO$_4$. Moreover, for x > 0.06 a superconducting transition appears, T_c increasing to 37 K for x = 0.12 [7.46]. A similar superconducting transition has also been demonstrated for La_2CuO_4; it appears to depend upon the annealing conditions of synthesis under oxygen. A bulk superconductivity with a T_c at 37 K can thus be realized after an anneal at 773 K under a high oxygen pressure (500 bars) [7.47] while a grain surface superconducting behavior is achieved at 32 K for La_2CuO_4 specimens annealed at 673 K in an oxygen flow [7.48, 49]. In the latter case a spin density wave phase has been shown to appear at 220 K inside the grain [7.50-52].

Contrary to the Sr-doped compounds $La_{2-x}Sr_xCuO_4$, the radiation effects in undoped La_2CuO_4 had not been studied, to our knowledge. Thus La_2CuO_4 samples were irradiated at GANIL with 2.9 GeV krypton ions. The irradiation was carried out at 50 K in a He-flow cryostat equipped with a CLTS resistance for the temperature regulation. The resistance of the samples was measured in situ during beam-stops by means of four probe contacts. Each time a predetermined fluence was reached in the range $5 \cdot 10^{11} - 4 \cdot 10^{12}$ Kr/cm^2, a full resistance versus temperature curve was recorded between 10 and 60 K. The samples were irradiated along their smallest direction, which was less than the projected range of 2.9 GeV Kr ions in La_2CuO_4. As a result, the energy deposited by inelastic collisions could be estimated from the electronic stopping power as having a mean value of 10 MeV/μm.

Figure 7.18 shows the variation of the resistance as a function of the temperature for different Kr fluences for a typical sample [7.53]. As the fluence increases, the resistance in the normal state decreases and the peak just above the superconducting transition gradually disappears. This decrease of resistance is accompanied by a distinct increase of the T_c mid-

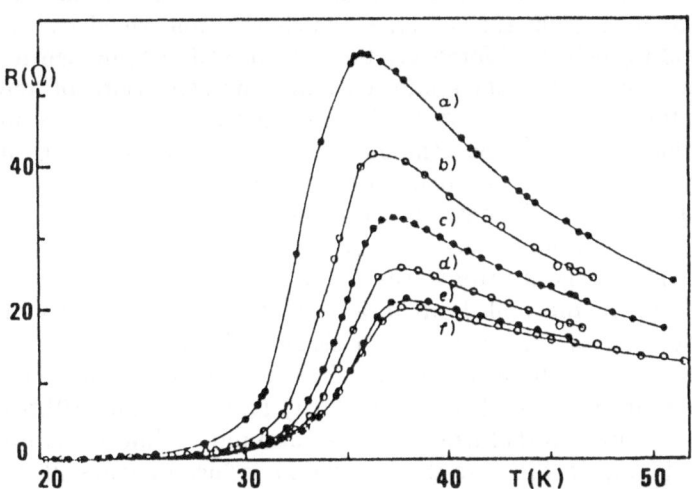

Fig.7.18. The resistance as a function of temperature for a sample of La_2CuO_4 irradiated by 2.9 GeV krypton ions. The different curves correspond to the following fluences: (a) $\phi t = 0$, (b) $4.8 \cdot 10^{11}$ cm^{-2}, (c) 10^{12} cm^{-2}, (d) $1.9 \cdot 10^{12}$ cm^{-2}, (e) $3.1 \cdot 10^{12}$ cm^{-2} and (f) $3.9 \cdot 10^{12}$ cm^{-2}

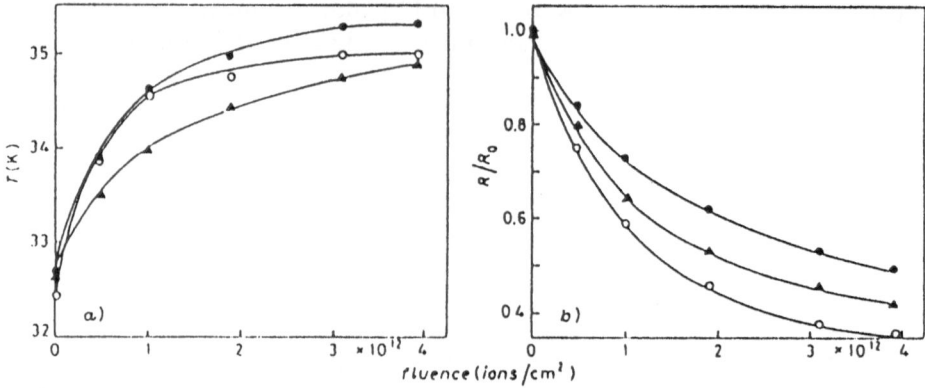

Fig.7.19. (a) Transition temperature measured at the midpoint as a function of the fluence for three different samples of La_2CuO_4. (b) Resistance at the onset point normalized to the initial value as a function of the fluence for three samples of La_2CuO_4

point while the transition width remains approximately constant. Figure 7.19 gives the evolution of the resistance at the T_c onset and of the T_c midpoint with the fluence for three measured samples. The correlation between the variation of the resistance and the transition temperature is evident. Both properties saturate for a fluence of about $2 \cdot 10^{12}$ ions/cm^2.

Contrary to the results described before for $La_{2-x}Sr_xCuO_4$ superconductors, the undoped compound La_2CuO_4 thus reveals an increase of the transition temperature together with a decrease of the normal resistance with increasing fluence. One can only speculate on how this increase of T_c occurs, since the nature of the superconductivity in La_2CuO_4 (and in other copper oxides) is not yet well understood. As the La_2CuO_4 specimens that have been irradiated were prepared under normal O_2 pressure, the superconductivity was of a grain surface type, which was confirmed by the Meissner and shielding effects. Moreover, the presence of a Spin density Wave (SDW) phase should be taken into account. The sensitivity of the SDW to the irradiation-induced defects has been demonstrated elsewhere in the case of low-dimensional organic compounds such as $(TMTSF)_2ClO_4$ in which a metallic state was stabilized under irradiation [7.54]. If SDW competes with superconductivity, the destruction of the former should enhance the superconducting transition, as has been shown when superconductivity develops together with charge density waves [7.55]. The loss of coherence of SDW under irradiation in the bulk leads to a decrease of the resistivity of the grains, i.e., a decrease of the macroscopy resistance above T_c.

We can also compare such an irradiation effect to that of the pressure, which leads to a strong increase of T_c (0.5K/kbar) between 7 and 30 kbar [7.47]. It is indeed well known that irradiation can also induce internal stress around the defects [7.56]. In this case the irradiation-induced change of T_c ($\Delta T_c \sim 3K$) would correspond to a pressure of 6 kbar. In both cases, the $R(T)$ curves are modified in a similar manner.

7.3.2 Heavy-Ion-Induced Changes of Superconducting and Normal Properties of Polycrystalline Ceramics $YBa_2Cu_3O_{7-\delta}$ and $Bi_2Sr_2CaCu_2O_8$

Sintered bars ($5\times1\times0.1\,mm^3$) of "123" and "2212" superconducting phases have been irradiated by 2.9GeV Kr and 3.5GeV Xe ions, following a procedure similar to that mentioned above for La_2CuO_4. Irradiation was carried out at 100 K for in situ resistance measurements and at 300 K for magnetic and electron microscopy investigations.

Figures 7.20 and 21 show the temperature dependence of the resistance between 100 K and 4.2 K for $YBa_2Cu_3O_{6.9}$ and $Bi_2Sr_2CaCu_2O_8$ samples prior to and after irradiation with 3.5 GeV xenon ions at fluences ranging from $5\cdot10^{11}$ to $9.1\cdot10^{12}$ cm^{-2} [7.57]. With increasing Xe fluence, the following main features can be recognized.

i) The superconductivity is destroyed for $YBa_2Cu_3O_{6.9}$ at higher fluences ($\phi t > 6\cdot10^{12}$ cm^{-2}) than for $Bi_2Sr_2CaCu_2O_8$ ($\phi t > 2\cdot10^{12}$ cm^{-2}), which exhibits an extreme sensitivity to Xe-induced defects.

ii) The T_c offset decreases more rapidly than the T_c onset, leading to a transition width δT_c that increases, in the case of $YBa_2Cu_3O_{6.9}$, from 1 K before irradiation to 58 K at $\phi t = 6\cdot10^{12}$ Xe/cm^2.

iii) The resistance at 100 K is drastically increased during irradiation, emphasizing once again the highest sensitivity of $Bi_2Sr_2CaCu_2O_8$.

The rate of depression of T_c with fluence ($dT_c/d\phi t$) and the increase of δT_c clearly depend on the oxygen deficiency of the "123" phases, as can be seen in Fig.7.22. The increase of both parameters is thus larger for samples of low oxygen content, i.e., δ = 0.46 and 0.62, which show a T_c depression below 4.2 K at smaller fluences than $YBa_2Cu_3O_{6.9}$ does. Further, such

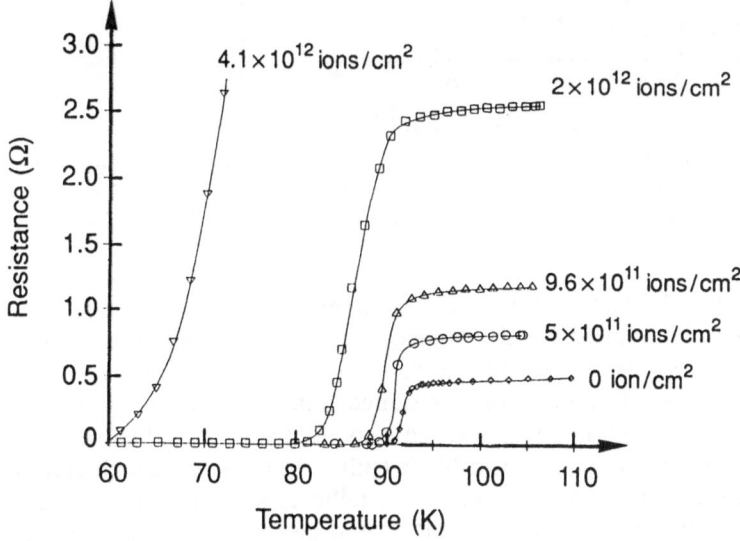

Fig.7.20. In situ measurements of the R-T curves for $YBa_2Cu_3O_{7-\delta}$ irradiated with 3.5GeV Xe ions

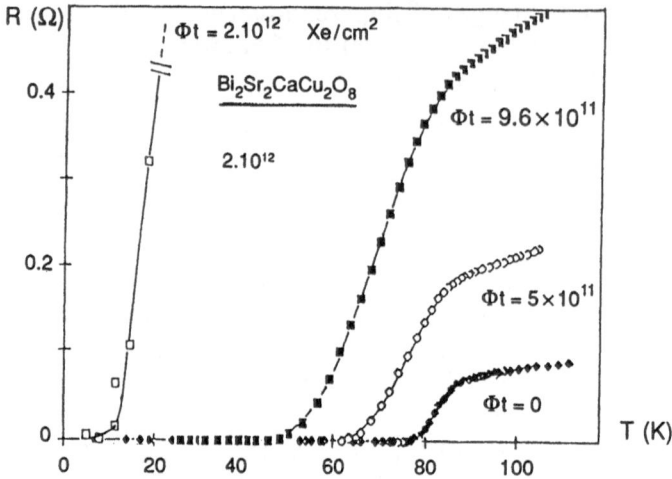

Fig.7.21. In situ measurements of the R–T curves for $Bi_2Sr_2CaCu_2O_8$ irradiated with 3.5GeV Xe ions

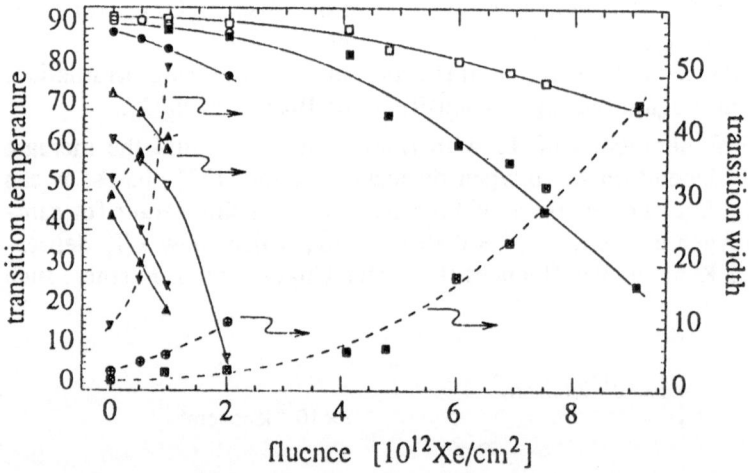

Fig.7.22. The reduced transition temperature T_c/T_{c0} (black symbols correspond to 10% and white symbols to 90%) and the transition width δT_c versus fluence for $YBa_2Cu_3O_{7-\delta}$ ($\delta = 0.10\ \square$, $0.28\ \circ$, $0.46\ \triangledown$, $0.62\ \triangle$) irradiated with 3.5GeV Xe ions at 100 K

phases exhibit a still larger increase of ΔR than the highest-T_c superconductor. This is accompanied by the appearance of a semiconducting phase mixed with the superconducting phase, as indicated by the change of sign of the temperature coefficient of the resistance (Fig.7.23). As shown in Fig. 7.24, in this fluence region ΔR obeys an exponential law, indicating that the irradiation-induced depression of T_c is truly a bulk effect which can be attributed to the transformation from the metallic to a semiconducting and then to an insulating phase. In all cases it should be pointed out that the T_c offset decreases more steeply than the T_c onset, which stays nearly constant at low fluences. Such a behavior is the result of a broadening of the transi-

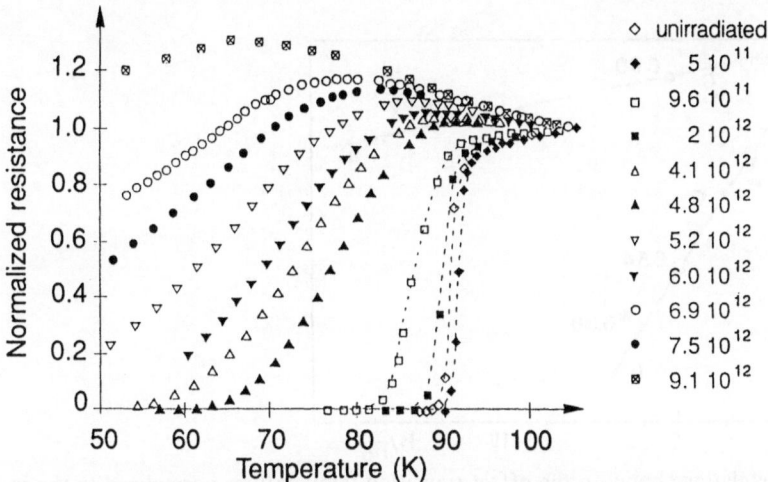

Fig. 7.23. Normalized transition curves of $YBa_2Cu_3O_{6.9}$ irradiated with 3.5 GeV Xe ions. Note the continuous transition from metallic to semiconducting behavior

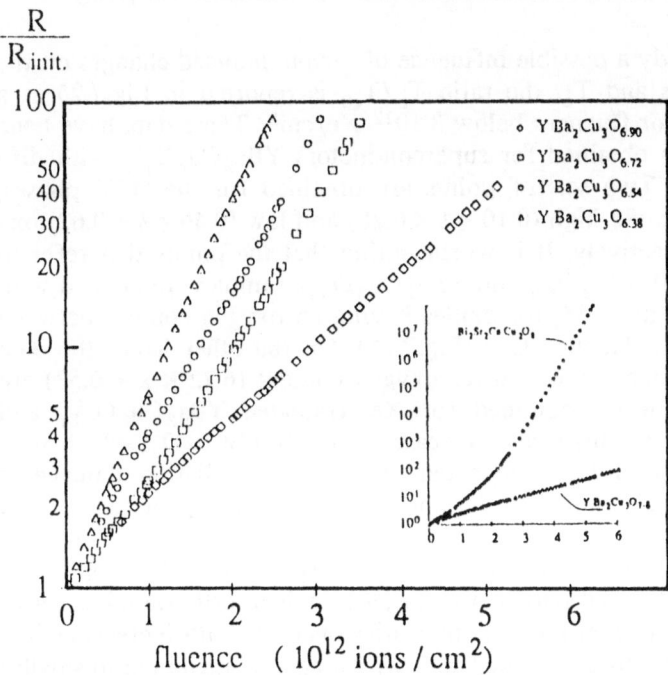

Fig. 7.24. Exponential variations (with fluence) of the resistance normalized to the initial value at 100 K for $YBa_2Cu_3O_{7-\delta}$ ($\delta = 0.10, 0.28, 0.46, 0.62$) and $Bi_2Sr_2CaCu_2O_8$ oxides irradiated with 3.5 GeV Xe ions

tion width, which suggests, similarly to neutron irradiation that the induced defects have been inhomogeneously distributed, leaving an undamaged network of high-T_c material.

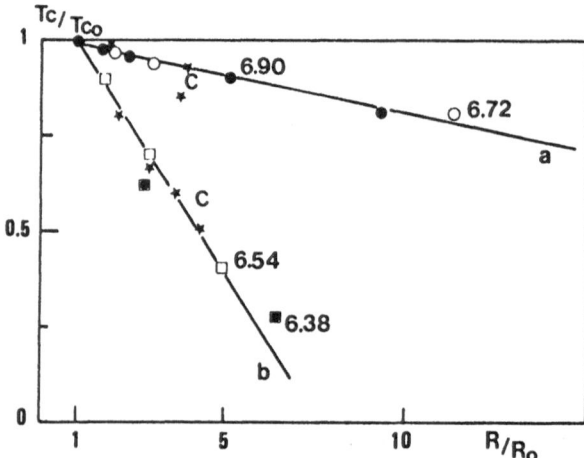

Fig.7.25. Correlations between the offset transition temperatures normalized to the initial values T_c/T_{c0} and the resistance normalized to the initial value at 100 K for $YBa_2Cu_3O_{7-\delta}$ irradiated with 3.5GeV Xe ions. Curve (a) corresponds to δ = o.1, 0.25 and (b) to δ = 0.46 and 0.62. C indicates results obtained from *Cava* et al. [7.55]

In order to study a possible influence of xenon-induced changes on the transport properties and T_c, the ratio T_c/T_{c0} is reported in Fig.7.25 as a function of R/R_0 for fluences below $3 \cdot 10^{12}$ Xe/cm^2. These data have been compared to results obtained for superconductors $YBa_2Cu_3O_{7-\delta}$ with different values of δ. Two sets of points are obtained for the "123" phases, which correspond to the high ($0.10 \le \delta \le 0.28$) and low ($0.46 \le \delta \le 0.62$) oxygen amounts, respectively. It is worth noting that the points that refer to xenon-irradiated $YBa_2Cu_3O_{6.9}$ and $YBa_2Cu_3O_{6.72}$ samples are aligned with the data of *Cava* et al. [7.58] for oxides having an oxygen content between 6.76 and 6.98 (points labelled C in Fig.7.25). On the other hand, the data that relate to specimens with a lower oxygen content ($6.72 \le x \le 6.53$) are grouped with the results obtained for Xe-irradiated $YBa_2Cu_3O_{6.54}$ and $YBa_2Cu_3O_{6.38}$ sinters. This is also the case for $Bi_2Sr_2CaCu_2O_8$, which suggests that the sensitivity of the superconductor to radiation damage is strongly correlated to atomic disorder and chemical inhomogeneities present before irradiation.

X-ray and electron diffraction analysis of irradiated "123" samples revealed that the undamaged part of the target remained orthorhombic. Nevertheless, one observes that the line intensities decrease with increasing fluence between $5 \cdot 10^{11}$ and $5 \cdot 10^{12}$ Xe/cm^2, which can be attributed to smaller amounts of superconducting phase due to the extension of disordered zones and displacements of atoms that no longer contribute to the coherent scattering. Moreover, one can see no significant changes, either in the angular position of the diffraction lines or in the intensity ratio of the lines (020) and (006) to the line (200), which would indicate a orthorhombic to tetragonal transformation during irradiation. However, it should be pointed out that ED patterns showed, besides the orthorhombic arrangement of white dots, streaks that appeared along the [100], [010] and [110] directions

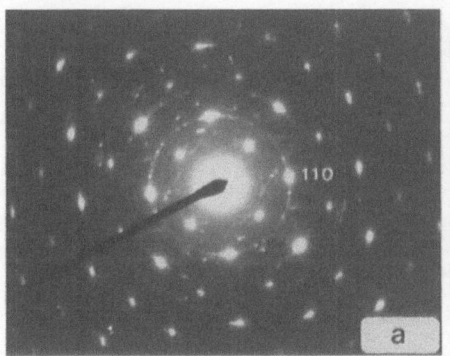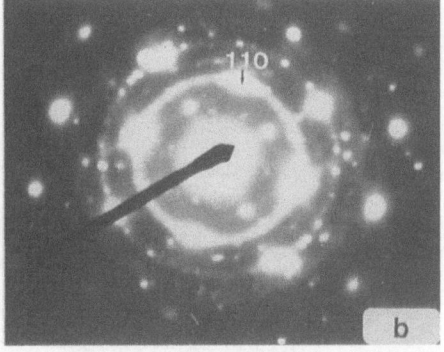

Fig.7.26a,b. Typical [001] electron diffraction patterns from different samples of $YBa_2Cu_3O_{6.90}$ microcrystals irradiated with 3.5GeV Xe ions ($\phi t = 10^{12} cm^{-2}$) showing arcing of the spots and streaks appearing parallel to $\langle 110 \rangle$ and $\langle 100 \rangle$ directions

(Fig.7.26). A similar feature was previously described for annealed $YBa_2 \cdot Cu_3O_{6.55}$ [7.59] and substituted $YBa_2Cu_{2.85}Fe_{0.15}O_7$ phases [7.60]. This suggests that it is the consequence of a similar structural mechanism: local variations of oxygen stoichiometry linked to short-range ordering of the vacancies in various ways which could be assigned, in the case of Xe-irradiated samples, to statistical fluctuations of the energy deposited by electronic stopping.

In order to elucidate the real nature of the atomic transformation responsible for the T_c decrease and δT_c increase, a High-Resolution Electron Microscopy (HREM) investigation has been undertaken in addition to the electron diffraction study [7.61]. After taking account of the random orientation of the grains with respect to the xenon ion beam, we noted that the nature of the defects appeared to be strongly dependent upon the direction of the ion beam relative to the layers of the $YBa_2Cu_3O_7$ structure. This is to be compared with previous results obtained for barium hexaferrite, $BaFe_{12}O_{19}$, which is also anisotropic [7.62].

Typical induced damaged zones can be seen in Fig.7.27a for grains corresponding to the layers normal to the Xe ion beam. They appear as amorphous white lines parallel to the layers. A comparison with the calculated through-focus series shows that these disturbed slices are located at the levels of Cu(1) and BaO layers. If one assumes that the electrons are more localized in the BaO layers than in the Cu(1)-O chains, it would be expected that the former will be much more sensitive to ionization processes than the latter. As a consequence, ionization spikes could be easily initiated in the BaO layers and then could develop within the Cu(1)-O chains, leading to local disorder and amorphization of the layers. Amorphous domains ($\simeq 50$Å width), probably due to high-density energy deposition, can thus be observed in Fig.7.27b, in which they have been arrowed. According to *Thompson* [7.63], it can be assumed that high-density cascades could have produced local transients of both temperature and pressure, which might be at the origin of the nucleation and growth of defects through the surrounding perovskite triple layers that can be seen in Fig.7.27b.

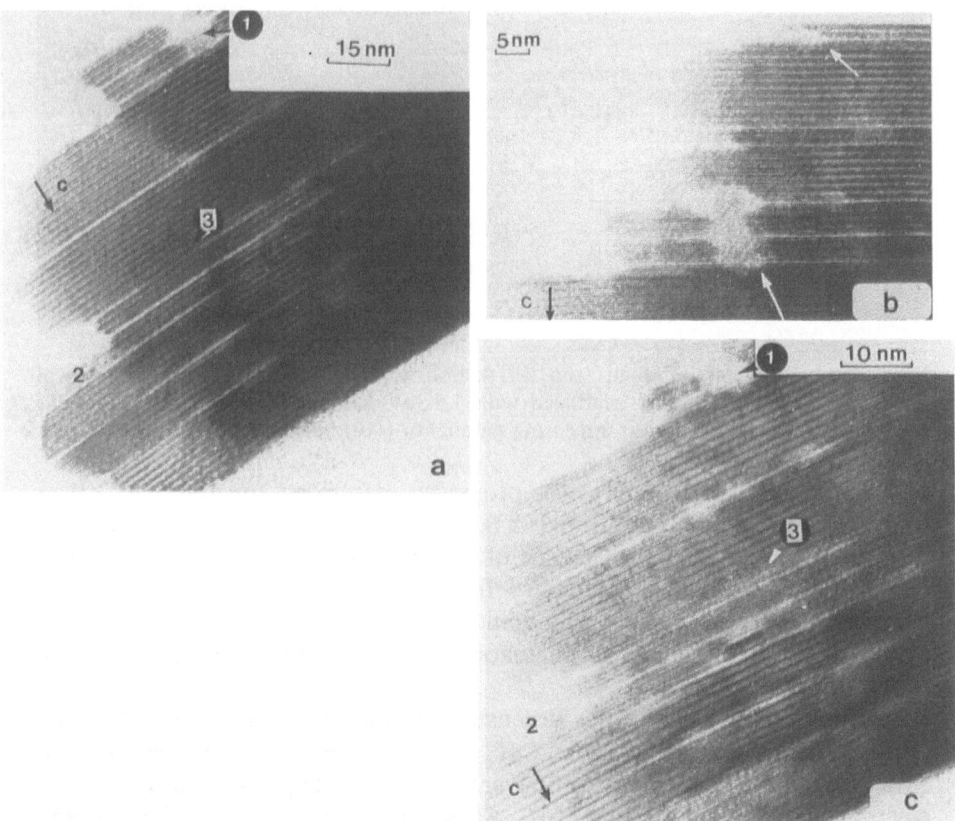

Fig.7.27a-c. Images of a crystal of $YBa_2Cu_3O_{6.90}$ corresponding to the layers normal to the Xe ion beam. (a) Damaged zones parallel to the layers are observed as white lines. The area labelled 3 corresponds to a local superstructure with a $2a_p$ periodicity (b) Microcrystals within which amorphous zones are observed (white arrows) probably due to high-density energy deposition. (c) The same microcrystal as (a) after prolonged electron irradiation

A high concentration of these disordered regions could explain the rapid decrease of T_c at a fluence between $3 \cdot 10^{12}$ and $7 \cdot 10^{12}$ Xe/cm². Furthermore, it should be pointed out that such damaged crystals showed a higher sensitivity to the electron irradiation in the microscope than the precursor crystal. This effect is illustrated in Fig.7.27c with the same area of the sample as in Fig.7.27a but after prolonged observation in the electron microscope.

Attention should be drawn to two features: (i) the areas labelled *1* and *2* in Fig.7.27a became amorphous in Fig.7.27c; (ii) the area labelled *3*, which corresponds in Fig.7.27a to a contrast modulation implying a doubling of the lattice parameter a, shows a crystalline-to-amorphous transformation after electron irradiation (Fig.7.27c). Such modulations were previously observed in unirradiated samples [7.60]; they were interpreted as the intergrowth of $[CuO_2]_\infty$ and $[CuO]_\infty$ chains along the a-axis, leading to

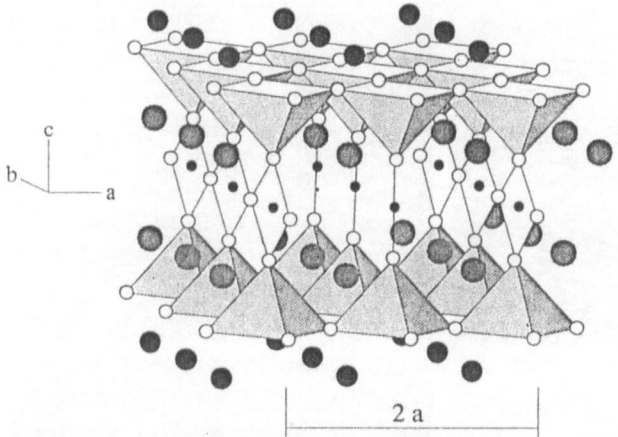

2 a

Fig.7.28. Schematic drawing of the intergrowth of $[CuO_2]_\infty$ and $[CuO]_\infty$ chains appearing as local contrast modulations in $YBa_2Cu_3O_{6.90}$ microcrystals irradiated with 3.5 GeV Xe ions at $\phi t = 10^{12} cm^{-2}$

the formulation $YBa_2Cu_3O_{6.55}$, as sketched in Fig.7.28. This value of the oxygen content could involve a possible loss of oxygen under irradiation, as has been suggested by *Clark* et al. [7.64] for 500 keV O^+ implanted thin films of $YBa_2Cu_3O_x$. As the superstructure "2a" disappears after prolonged irradiation in the electron microscope, it may be assumed that the crystalline-to-amorphous transition occurs when a critical density of point defects, i.e., oxygen vacancies and interstitials, is exceeded, which supposes local changes of the oxygen content between O_7 and O_6 in the Cu(1)-O chains.

The data provide a new insight into the lamellar morphology and amorphous transformations that have been described earlier for the reduced phases [7.60]. It is well known that $YBa_2Cu_3O_6$ is tetragonal with Cu(I) in a twofold coordination state. However, ED and neutron-diffraction studies revealed a partial amorphization of the samples. In the case of the reduced phases [7.60], corresponding to oxygen stoichiometries between 7 and 6, the HREM investigations showed that the crystals exhibit a lamellar morphology with preferential orientation normal to the c axis. Such features could then be attributed to an excessive density of chemical disorder or point defects, which can increase the free energy of the crystal above that of the liquid, i.e., the amorphous phase. The growth of this amorphous phase in the Cu(1) and BaO layers could affect the intergrain regions more strongly than the intragrain ones and hence could explain why the T_c offset decreases more rapidly than the T_c onset at the highest fluences.

For grains corresponding to BaO and CuO layers parallel to the ion beam, [001] lattice images show modulation of both contrast and strain (Fig. 7.29a, b). The main directions of the modulations are $[100]_p$, $[110]_p$, $[210]_p$ and $[310]_p$, where p refers to the perovskite subcell. They are established on domains whose widths range from several tens to several thousands of Ångstroms. These images are similar to those observed in oxygen-deficient

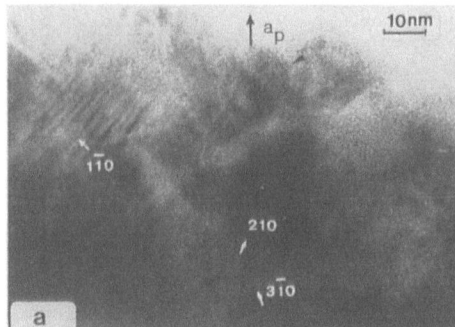
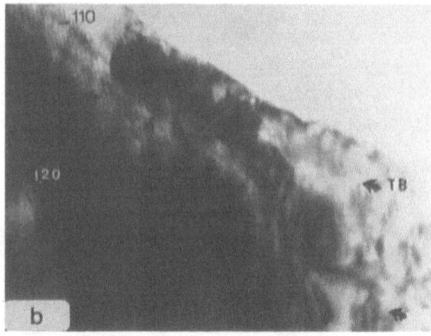
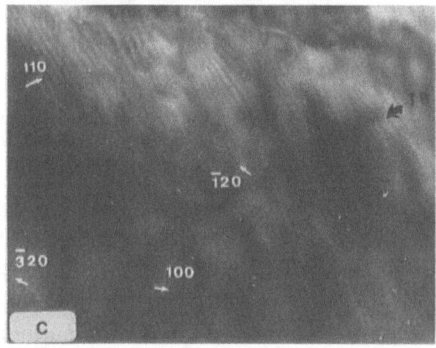

Fig.7.29a–c. Examples of strains and modulations of the contrast in (001) images of $YBa_2Cu_3O_{6.90}$ microcrystals irradiated with 3.5GeV Xe ions ($\phi t = 10^{12} cm^{-2}$). The main directions are referred to the perovskite subcell. (a) The modulations or superstructures (black arrowed) arise in distinct areas. (b, c) They are established throughout the crystal. The twin boundaries (TB) are arrowed.

$YBa_2Cu_3O_{6.67}$ samples [7.59], in which the modulation phenomena were assumed to arise from local ordering of oxygen atoms and vacancies in the Cu(1)-O plane. Several models have been proposed to explain such local superstructures, which correspond to oxygen contents varying from 6.5 to 6.7. They can be considered as a way to accommodate local variations of oxygen stoichiometry and to ensure a favorable coordination state of the copper atoms. As they arise in orthorhombic twinned crystals, it may be assumed that the majority of the oxygen atoms are ordered in the Cu(1)-O chains. Figure 7.29c shows a large area of the crystal where the twin boundaries are clearly observable. In the lattice images presented in Fig.7.30, one can see the appearance of dotted latent tracks similar to those observed in magnetic insulators [7.65] and probably due to higher electronic-energy-loss effects. Two features should be emphasized from this image: (i) the modulations have completely disappeared but not the strain; (ii) the twin boundaries are obviously more disturbed (see enlargement, Fig.7.30b). This suggests larger atomic displacements that no longer allow local ordering of oxygen atoms and vacancies.

The existence of trails of damage formed of successive extended defects, which should act, as is well known, as pinning centers for flux lines, can now explain the reduction in magnitude of the Meissner and shielding effects for the irradiated samples with respect to the unirradiated ones. Further, such disordered microregions are responsible for the hysteresis observed in the M(H,T) curves and therefore for the enhancement of the

288

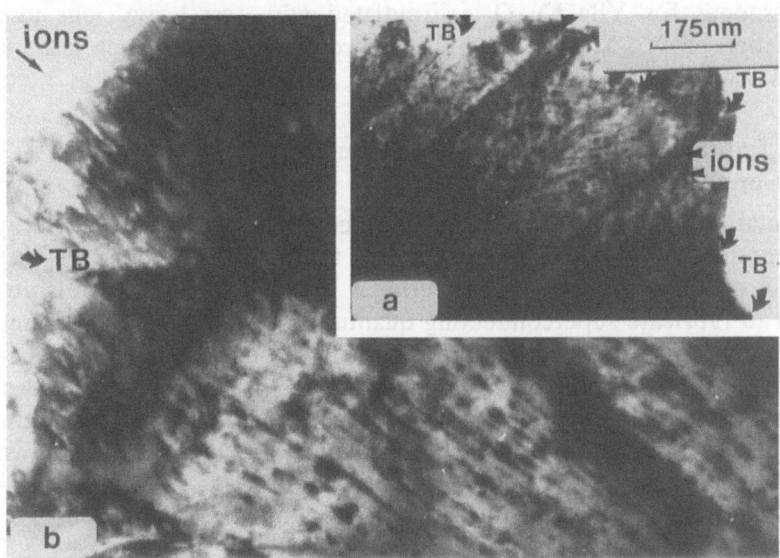

Fig.7.30. Latent tracks of 3.5 GeV Xe ions visualized by HREM in "1.2.3" microcrystals: (001) images. (b is an enlargement of a)

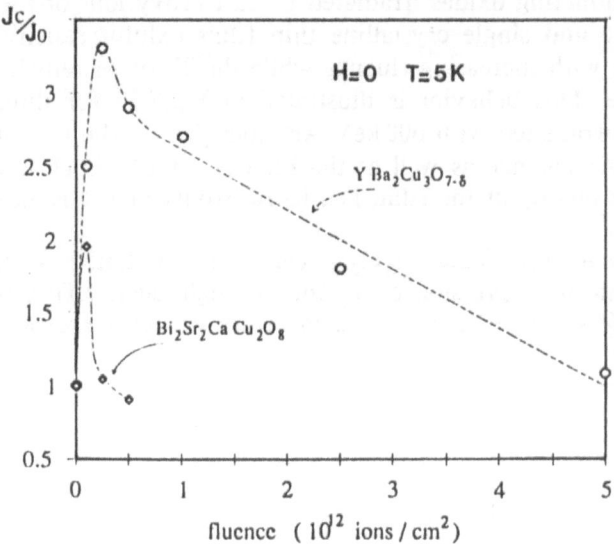

Fig.7.31. Critical magnetization current density J_c/J_{c0} versus fluence at 5 K and zero field for $YBa_2Cu_3O_{6.90}$ and $Bi_2Sr_2CaCu_2O_8$ superconductors irradiated with 3.5GeV Xe ions at 300 K. The dashed lines are only guides to the eye

critical magnetization current density, which increases if the defect density remains below a threshold value corresponding to a fluence of $\phi t = 3 \cdot 10^{11}$ Xe/cm^2 for $YBa_2Cu_3O_{6.90}$ and $\phi t = 10^{11}$ Xe/cm^2 for $Bi_2Sr_2CaCu_2O_8$ (Fig.7.31). It should be pointed out that the maximum value of J_c (J_c/J_{c0} =

3.5) corresponds for $YBa_2Cu_3O_{6.90}$ irradiated with fast heavy ions to a small variation of T_c ($\Delta T_c = 1 K$), unlike neutron irradiation, which leads to $\Delta T_c = 5 K$.

7.3.3 Ion Implantation Effects in Thin Films of Copper Oxide Superconductors

Ion implantation has been used in thin-film technology [7.66,67] to destroy the superconductivity in the film surrounding the device, without removing material, in order to produce narrow bridges for critical current measurements and to fabricate superconducting quantum interferometers. Although the film thicknesses were smaller than the mean projected range of the incident ions, it was unambiguously demonstrated that for light particles, nuclear collisions rather than electronic excitations were responsible for the induced changes in electronic properties and microstructure. This is shown in Fig.7.32, where the conductivity normalized to its value prior to irradiation has been plotted versus the energy deposited into elastic collisions, i.e., displacement per atom (dpa) for $YBa_2Cu_3O_7$ thin films irradiated by 500 keV O^+ and 2 MeV As^+ [7.68]. The drop in conductivity is thus exponential with dose and therefore does not obey Matthiessen's rule for metallic materials.

Like bulk superconducting oxides irradiated by fast heavy ions or fast neutrons, polycrystalline and single crystalline thin films exhibit resistive transitions that broaden with increasing fluence while the T_c onset remains approximately the same. This behavior is illustrated in Fig.7.33 for three $YBa_2Cu_3O_7$ thin films irradiated with 600 keV Ar^+ ions [7.69]. The rate of depression of T_c with Ar fluence as well as the increase of δT_c is clearly shown to depend on the quality of the film, i.e., its resistivity at 100 K before irradiation.

As mentioned above for $YBa_2Cu_3O_{7-\delta}$ ceramics irradiated with 3.5 GeV Xe ions, the rate of depression of T_c for the high-quality film is small at low fluences ($\phi t \leq 2 \cdot 10^{13} Ar/cm^2$) and increases suddenly for $\phi t >$

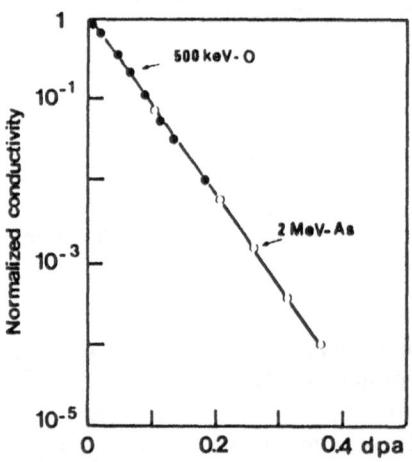

Fig.7.32. Conductivity normalized to unity at zero dose as a function of the energy deposited into elastic collisions (dpa) for $YBa_2Cu_3O_7$ superconducting film irradiated by 500 keV O^+ and 2 MeV As^+ ions. (From [7.68])

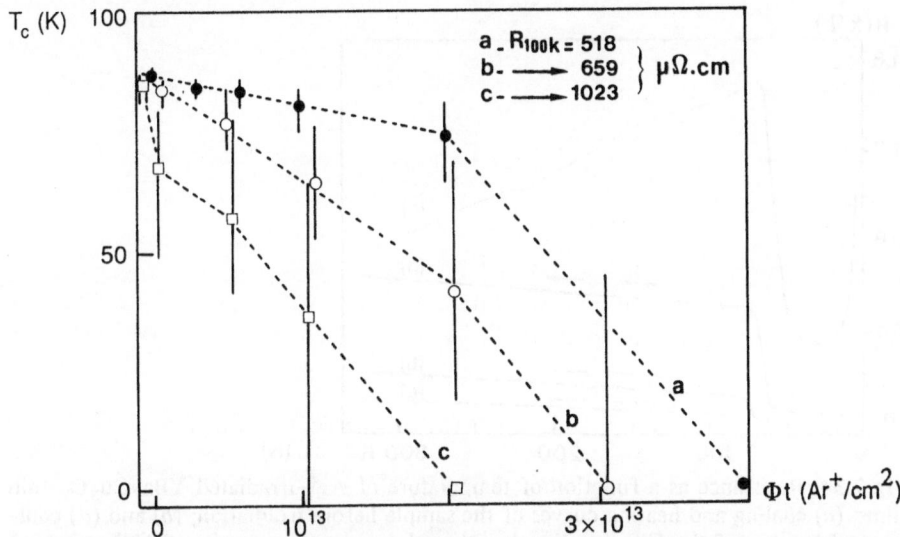

Fig.7.33. Transition temperatures and transition widths versus fluence of 600 keV Ar+ ions for three $YBa_2Cu_3O_7$ thin films of different quality. (From [7.69])

$2 \cdot 10^{13}$ Ar/cm^2. This behavior is still more pronounced after irradiation with 300 keV protons [7.70]. It is accompanied by a rapid increase of both δT_c and resistivity, correlated to the appearance of a semi-conducting phase. It was demonstrated that, independent of the mass and the energy of the incident ions, such a metallic-to-semiconducting transition required a deposited energy between 0.01 and 0.04 dpa [7.69-72]. From a crystallo-graphic point of view, the extent of the semiconducting phase seems to be indicated by a continuous c-axis expansion [7.68], which suggests an ortho-rhombic-to-tetragonal phase transformation [7.70]. This has recently been shown by electron microscopy [7.73].

During annealing at room temperature, the semiconducting phase part-ly retransformed to the superconducting one, as illustrated in Fig.7.34. Such a recovery stage could be due to displaced oxygen atoms, which may re-combine with vacancies at low temperatures.

Changes of the electron properties of $YBa_2Cu_3O_7$ films above 0.04 dpa have been ascribed to a crystalline-to-amorphous transition, which is complete at about 0.1 dpa as determined by X-ray diffraction. According to *Clark* et al. [7.64], TEM investigation of $YBa_2Cu_3O_7$ films formed of large "pancake"-shaped grains revealed the accumulation of amorphous ma-terial at the grain boundaries after irradiation with 500 keV O+ ions. With increasing O+ fluence ($\phi t \geq 2 \cdot 10^{14}$ cm^{-2}) the amorphous layer became con-tinuous on all boundaries to a thickness of about 150Å. For $\phi t \geq 6 \cdot 10^{14}$ O+/cm^2 all the grains were amorphous. The gradual formation of the amorphous layer on the grain boundaries at low doses accounts for the T_c reduction by loss of phase coherence between the grains. As discussed by *Clark* et al. [7.68], the energy required to amorphize $YBa_2Cu_3O_7$ is rather low if it is compared to that reported for rare-earth perovskites [7.74-75].

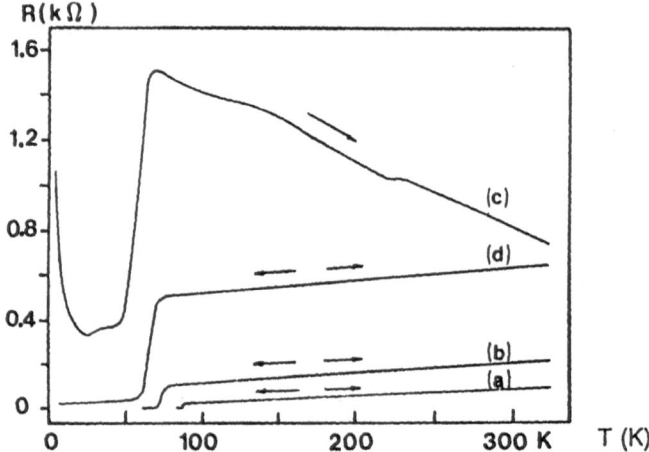

Fig.7.34. Resistance as a function of temperature of Ar^+-irradiated $YBa_2Cu_3O_x$ thin films: (a) cooling and heating curves of the sample before irradiation; (b) and (d) cooling and heating of the film irradiated with Ar^+ ions at fluences $\phi t = 3 \cdot 10^{13}$ Ar^+/cm^2 and $6 \cdot 10^{13}$ Ar^+/cm^2, respectively, at 77 K and then annealed to RT; (c) annealing curve after irradiation with $\phi t = 6 \cdot 10^{13}$ Ar^+/cm^2 at 77 K. (From [7.69])

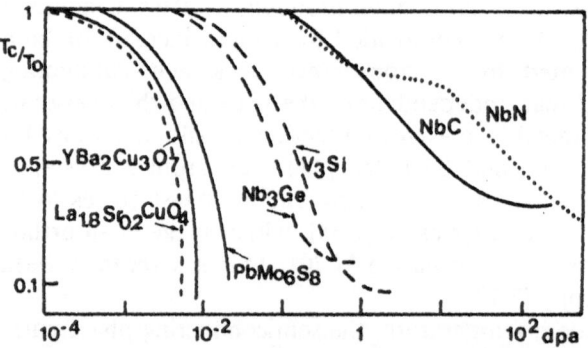

Fig.7.35. Influence of irradiation-induced defects on the transition temperatures, normalized to the initial values, expressed as a function of the energy deposited into elastic collisions (dpa) for the old and new superconductors. (From [7.76])

$CmAl_2O_3$ was found to become amorphous for a deposited energy of about 15 eV/atom, i.e., 0.3 dpa [7.74]. This could be the result of the presence of vacancies in the oxygen network and hence of a displacement threshold far smaller than 20 eV. Moreover, it is evident that the numbers of displaced atoms that caused the T_c reduction and change in microstructure are small with respect to those reported for conventional superconductors. This is clearly illustrated in Fig.7.35, which shows a plot of the normalized T_c decreases as a function of dpa for old and new superconductors [7.76]. The motion of oxygen atoms easily displaced from their lattice sites could be the cause of this extreme sensitivity of the high T_c copper oxides to irradiation.

However, the detailed mechanism of the changes both of the superconducting and normal properties remains for the most part still obscure.

According to *Valles* et al. [7.77] the resistivity in the insulating regime be-haves like $\exp(T^{-1/2})$ but with a Hall coefficient which changes very little. This could imply that the metal-insulator transition would be the result of a reduction in mobility, rather than a drop in carrier density.

Furthermore, contrary to the hasty conclusion of *Summers* et al. [7.78], the role of the energy deposited into electronic excitations and/or ioniza-tions of the target atoms must not be neglected in the damage mechanism, as recently demonstrated by *Roas* et al. [7.79] for high-quality epitaxial $YBa_2Cu_3O_{7-\delta}$ films irradiated by 173 MeV-Xe ions. In agreement with the results described here above for $YBa_2Cu_3O_{7-\delta}$ ceramic irradiated with 3.5 GeV-Xe ions, it is clearly established that a much stronger sensitivity to irradiation occurs with fast heavy ions because of the formation of latent tracks of amorphized material. This is illustrated in Fig.7.36 where the de-pendences of the resistance normalized to its initial value at 100K on the number of displaced atoms (dpa) have been depicted for a large variety of projectiles which correspond to nuclear and electronic stopping powers dif-fering by several orders of magnitude. It is evident from such a plot that besides the mere energy loss in elastic collisions, an additional damage mechanism must be taken into account for the heavier projectiles, i.e. fast heavy ions having an electronic stopping power several orders of magnitude higher than the nuclear stopping power. Moreover, it is obvious that the electronic stopping power has to exceed a threshold value as found for fer-

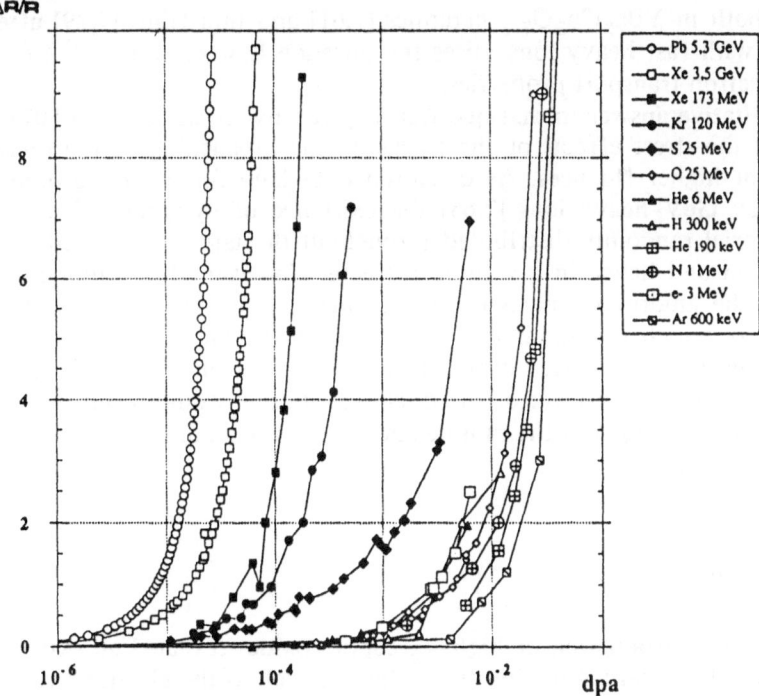

Fig.7.36. Variation of the resistance normalized to its initial value at 100 K with the number of displacements per atom (dpa) for a large variety of projectiles

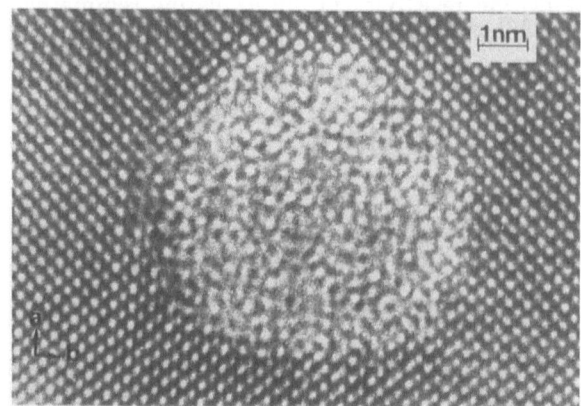

Fig.7.37. HREM micrograph of a 5.3 GeV-Pb ion latent track in a $YBa_2Cu_3O_{7-\delta}$ microcrystal, (001) plane. Note that the disordered region appears to be entirely confined to the core, having a diameter of approximately 70 Å

rites [7.15] in order to generate latent tracks. This value has been estimated by *Hansel* et al. [7.80] as ranging between 18 and 24 MeV/μm in YBCO films. It has more recently been shown that the electronic energy loss begins to compete with nuclear stopping when a threshold value of 8 MeV/μm is exceeded [7.81].

The HREM evidence of the existence of amorphous latent tracks (Fig.7.37) both in $YBa_2Cu_3O_{7-\delta}$ ceramics [7.61] and thin films [7.79] after irradiation with fast heavy ions raises the question how these tracks influence the electron transport properties.

A qualitative answer to that question is given by *Hensel* et al. [7.80] by considering the stress effects of the tracks on the broadening of the phase transitions at higher fluences. As discussed elsewhere for magnetic oxides irradiated by GeV-heavy ions [7.65] the existence of cylinders of amorphized material randomly distributed throughout the target will result in a density defect and hence in a volume increase. However, this volumic expansion of the amorphous parts of the sample is hindered by the unperturbed matrix which surrounds the tracks, thus involving the development of highly anisotropic stresses normal to the tracks. As shown by *Hensel* et al. [7.80] the random character of the track distribution would result in very different strains throughout the sample and so to a transition width that will broaden with dose.

7.4 Conclusions

Low-temperature irradiation of high-T_c copper oxides revealed an extreme sensitivity of the superconducting properties, as well as the electrical resistance, to lattice defects. The T_c value is reduced below 4.2 K for a number of displacements per atom 10 at 20 times smaller than for A15 superconductors. Moreover, this sensitivity appears to be enhanced for samples that exhibit the greatest chemical disorder prior to irradiation, i.e., large oxygen deficiency for $YBa_2Cu_3O_{7-\delta}$ oxides, great atomic disorder for the $Bi_2Sr_2 \cdot CaCu_2O_8$ phase.

Compared to fast neutron and fast heavy ion irradiation, electron irradiation led to a general shift of the R(T) curves towards low temperatures, implying that intragrain and intergrain properties had been degraded in the same manner. On the other hand, a more rapid decrease of the T_c-offset than the T_c-onset, together with a strong broadening of the transition width and a sudden increase of the resistance, has been observed above a threshold value of the energy deposited by elastic collisions (nuclear stopping) in the case of fast neutrons and by inelastic collisions (electronic stopping) in the case of fast heavy ions. Such behavior has been attributed to the destruction of the links that ensure the intergrain coupling. Moreover it should be pointed out that the damage rate ΔT_c/dpa was strongly enhanced by the electronic energy loss effect, its value being increased by two orders of magnitude with respect to that obtained for elastic collisions only. Similar results have been observed in the ion implantation of thin superconductor films, which showed a crystalline-to-amorphous transformation between 0.02 and 0.08 dpa.

Radiation damage investigation of fast heavy ions irradiation effects both in HTSC ceramics and films gives evidence of the formation of latent tracks of amorphized material whereever the electronic energy loss exceeds a threshold value closed to 20 MeV/μm. In the case of epitaxial YBCO thin films, a defect-induced enhancement of the critical current density is only observed when the magnetic field is aligned along the tracks.

8. Concluding Remarks

This survey of the solid-state chemistry of the copper oxides shows that they exhibit very complex nonstoichiometry phenomena, due to the Jahn-Teller effect of copper and to its ability to present several valence states - Cu(I), Cu(II), Cu(III) - and consequently multiple coordinations. Although the relationship between the crystal chemistry of these materials and their superconducting properties is so far not understood, we would like to discuss here very briefly some keywords which characterize these oxides and will probably play a role in superconductivity.

8.1 Low Dimensionality of the Structure

An examination of all the copper oxides that exhibit superconducting properties (Table 5.1) clearly shows that they all exhibit a structure formed of insulating layers (i.e., rock-salt-type layers, or calcium or yttrium layers) that alternate with superconducting copper layers. The anisotropy of the conductivity in single crystals of these materials has also been demonstrated. Moreover, investigations of many mixed-valence copper oxides that exhibit a three-dimensional framework of corner-sharing CuO_5, CuO_4 and CuO_6 polyhedra have shown that a great number of them are excellent metallic conductors, but have never shown a sign of superconductivity. Thus, the two-dimensionality of the structure appears to be a necessary condition for the existence of superconductivity at high temperatures in these oxides, in agreement with the theory developed by *Labbé* and *Bock* [8.1], and *Friedel* [8.2] at the very beginning of the rush to study copper oxide superconductors. Nevertheless, it should also be pointed out that this condition is not sufficient for the appearance of superconductivity, since two-dimensional oxides like $TlSr_2NdCu_2O_7$ [8.3] or $PbBaYSrCu_3O_{8-\delta}$ [8.4] are not superconductors.

8.2 Mixed Valence of Copper and Hole Delocalization

The language of physicists and solid-state chemists is generally different, which can lead to misunderstandings. The valence state, or the oxidation state, for the solid-state chemist represents a formal state of valence. In the case of the trivalent state of copper, this does not mean that there are actually Cu^{3+} ions in the structure, but only that there are additional holes with respect to the normal divalent copper oxides. Thus the mixed valence of copper, Cu(II)/Cu(III), which is necessary for the existence of the metal-like properties and further for high-T_c superconductivity, corresponds for the solid-state chemist to a delocalization of the holes in the copper-oxygen layers. It is of course not possible to determine from the crystal chemistry whether the holes are distributed in the copper band or in the oxygen band. Nevertheless, recent X-ray absorption results [8.5] and electron spectroscopy [8.6] suggest a tendency to a distribution of the holes in the oxygen band, leading to an intermediate configuration for copper called Cu $3d^9$ $\underline{L}$, where $\underline{L}$ represents the ligand hole.

Thus, it is clear from the examination of the La_2CuO_4 family and of the $YBa_2Cu_3O_7$ oxides that the mixed valence Cu(II)/Cu(III), taken in the above sense is necessary for superconductivity. This is also true for the Tl_1 cuprates $TlA_2Ca_{m-1}Cu_mO_{2m+3}$ (A = Ca, Sr), whose structure determination shows definitely that the oxygen sites are all occupied and consequently, from the charge balance, that Tl(III), Cu(II) and Cu(III) coexist. In contrast, examination of the charge balance of Tl_2 cuprates $Tl_2Ba_2Ca_{m-1}Cu_m \cdot O_{2m+4}$, leads to the conclusion that only Tl(III) and Cu(II) coexist, in spite of the high T_c observed for these materials. This suggests that the thallium layers play the role of a reservoir of holes for the copper oxygen layers, leading to the configuration $\langle Cu-O \rangle$, i.e., Cu $3d^9$ $\underline{L}$ [8.7]. The behavior of the bismuth cuprates is also complex. The ideal formulation of those oxides, $Bi_2Sr_2Ca_{m-1}Cu_mO_{2m+4}$, would suggest, from the charge balance, only Cu (II). However, the existence for bismuth of an oxidation state less than 3, which was observed by X-ray absorption [8.8] implies that the Bi-O layers be characterized by an excess of electrons coming from the Cu-O layers. Thus it appears that the rock-salt layers in both bismuth and thallium layers are not insulating, but exhibit a metallic or semi-metallic character [8.9], and become superconductive at low temperature by a proximity effect suggested by *Tournier* and co-workers [8.10].

8.3 The Model of Copper Disproportionation

The oxygen nonstoichiometry observed in the "123" oxide $YBa_2Cu_3O_{7-\delta}$ raises the issue of coordination, and of the valence state of copper for intermediate compositions (Chaps.2 and 4). The resolution of the average structures of the orthorhombic form, as well as of the tetragonal form, for $\delta \neq 1$ and 0, has shown that the oxygen vacancies appear only at the same level as the Cu(1) level (Fig.8.1). A statistical distribution of such vacancies would lead to a threefold coordination for copper, which is not likely. Moreover, consideration of the critical temperature versus the oxygen deficiency δ, as observed by different researchers, shows that superconductivity exists not only for $\delta < 0.50$, but is also obtained for $\delta \geq 0.50$ (Fig.8.2). This latter domain would correspond (from the classical charge balance) to a Cu(II)/Cu(I) system, which is not compatible with an electronic delocaliza-

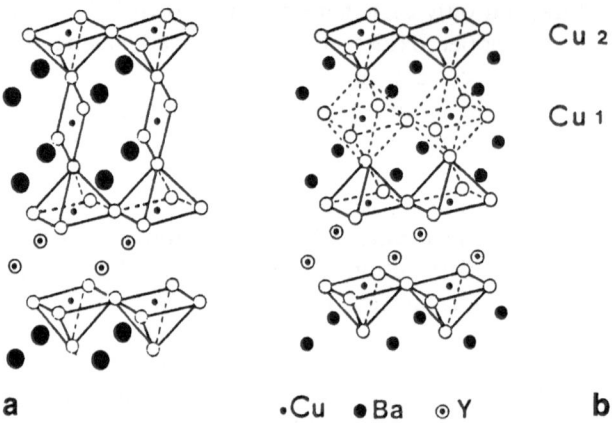

Fig.8.1a,b. Structure of $YBa_2Cu_3O_{7-\delta}$: (a) orthorhombic $YBa_2Cu_3O_7$, (b) tetragonal $YBa_2Cu_3O_{7-\delta}$

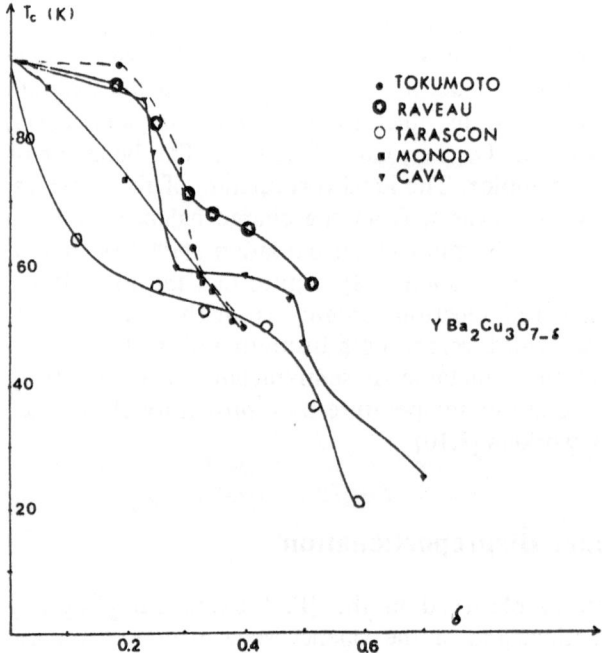

Fig.8.2. Critical temperature versus δ in $YBa_2Cu_3O_{7-\delta}$

tion, and consequently not with superconductivity, either. In order to explain this particular behavior of $YBa_2Cu_3O_{7-\delta}$, a model was proposed [8.11] involving a static disproportionation of copper according to the equilibrium

$$2Cu(II) \rightleftarrows Cu(I) + Cu(III) .$$

In these conditions, a correct coordination of copper, i.e., 4-5 or 6-fold coordination for Cu(II)/Cu(III) and twofold coordination for Cu(I), can

298

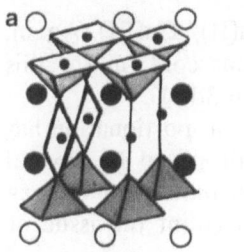 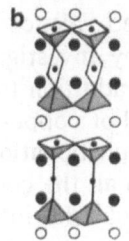

Fig.8.3a,b. Two structural models taking into consideration the copper disproportionation in $YBa_2Cu_3O_{7-\delta}$, leading to $YBa_2 \cdot Cu_3O_6$ insulating regions alternating with $YBa_2Cu_3O_7$ superconducting regions in the form of **(a)** chains or **(b)** layers

be obtained. It results that the formula of $YBa_2Cu_3O_{7-\delta}$ should not be written $YBa_2Cu^{II}_{2+2\delta}Cu^{III}_{1-2\delta}O_{7-\delta}$, but should take the form $(YBa_2Cu^{II}_2, Cu^{III}O_7)_{1-\delta}(YBa_2Cu^{II}_2Cu^IO_6)_{\delta}$, leading to a $[Cu(III)]/Cu_{total}$ ratio of $(1-\delta)/3$, i.e., greater than that obtained from the charge balance [8.11]. Such a model implies the coexistence in the structure of insulating microdomains $YBa_2Cu_3O_6$ and superconducting microdomains $YBa_2Cu_3O_7$. Such domains can take the form of "O_6" and "O_7" chains alternating along a or b (Fig.8.3a), or "O_6" and "O_7" layers alternating along c (Fig.8.3b) or more complicated arrangements involving particular orderings, as shown in Chap.6. The high-resolution electron microscopy observations are quite in agreement with such a model, showing that microdomains can be very tiny, leading to modulations, but can also correspond to larger regions, leading to true superstructures. But this model is confirmed, above all, by the X-ray absorption spectroscopy observations. With this latter technique, one observes besides $Cu3d^9$ L the existence of Cu(I) for the whole range $0 \le \delta < 1$. Moreover, it appears that the Cu(I) content increases as δ increases. It should also be pointed out that the other superconductors, La_2CuO_4-type oxides, bismuth and thallium cuprates, for which no copper disproportionation is expected, do not exhibit any Cu(I) from this spectroscopy observation. Finally, the recent discovery of the lead superconductor $Pb_2Sr_2Ca_{0.5} \cdot Y_{0.5}Cu_3O_8$ [8.12] definitely confirms this model of copper disproportionation. The structure is built up of layers of Cu^IO_2 sticks, which alternate along c with double pyramidal copper layers $[Cu_2O_5]_\infty$ and $[(Pb,Sr)O]_\infty$ layers (Fig.8.4). The layers of Cu^IO_2 sticks in which copper exhibits two-fold coordination can be considered as containing only Cu(I), and will be insulating. On the other hand, the pyramidal copper layers can be assumed

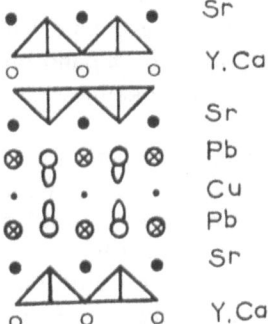

Sr
Y,Ca
Sr
Pb
Cu
Pb
Sr
Y,Ca

Fig.8.4. Structure of the oxide $Pb_2Sr_2Ca_{0.5}Y_{0.5}Cu_3O_8$

to be superconducting, involving the mixed valence Cu(II)/ Cu(III). Again, preliminary X-ray absorption spectroscopy investigations confirm for this phase the existence of univalent copper in addition to Cu $3d^9$ $\underline{L}$.

It should also be noted that this model of copper disproportionation has nothing to do with the dynamical disproportionation proposed by several investigators, which was in fact applied to all the copper oxides for the explanation of superconductivity and did not take into account the issue of copper coordination.

This possibility of Cu(II) disproportionation is of central importance, since it will be greatly influenced by the experimental conditions, and especially the redox conditions. The synthesis of the lead cuprates confirms this point of view since a low oxygen pressure is necessary to allow such a disproportionation to be realised.

8.4 Role of the Lone Pair Cations and of the Alkaline Earth Elements

The role of the alkaline earth ions has already been pointed out. It is indeed remarkable that, if one disregards La_2CuO_4, all the high-T_c superconductors require the presence of alkaline earth ions, Sr^{2+} or Ba^{2+}. The explanation has already been given and was a motive for the investigation of the mixed-valence copper oxides (Chap.2): the basic properties of these cations allow a partial oxidation of Cu(II) into Cu(III) to be realised. But attention should also be drawn to the fact that the oxidizing power increases with the basic character of the cation, i.e., a better oxidation of copper is obtained for barium cuprates than for strontium cuprates, while pure calcium cuprates do not exhibit a mixed valence of copper. This "oxidizing power" of the alkaline earth ions explains why the thallium barium cuprates are more easily synthesized than the thallium strontium cuprates. Nevertheless, it is not always possible to use the most oxidizing cation, as seen for instance for bismuth cuprates, which cannot be obtained in the case of barium, due to the fact that Bi(III) is oxidized into Bi(V) before copper oxidation. In the same way, the lead cuprates are so far the only known superconductors for strontium and they can be obtained only provided the gaseous atmosphere is not oxidizing, in order to avoid the oxidation of Pb(II) into Pb(IV). Thus it is now sure that the synthesis of these materials involves complex redox reactions and that this optimization will depend on the oxygen pressure but also on the nature and content of the alkaline earth cations associated with the copper.

Concerning the role of the lone pair cations in superconductivity, it can be seen from comparison of the bismuth and lead superconductors with the thallium superconductors that the first ones, which are characterized by a $6s^2$ lone pair of Bi(III) and Pb(II), do not exhibit higher critical temperatures than the thallium family, in which Tl(III) does not exhibit any lone pair character. Thus it appears that there is no relationship between superconductivity and the presence of a lone pair cation. However, lone pair cations such as Pb(II) or Bi(III) must be considered for their ability to form layered structures, and in this respect they are interesting, since they favor the two-dimensional character which seems to be necessary for the production of superconductivity at high temperatures. Finally, it is worth pointing out that the $6s^2$ lone pair may play a role in the incommensurability phenomena observed in several superconductors. The fact that all the bismuth cuprates exhibit incommensurate superstructures, whereas such a phenomenon is not observed for the Tl(III) cuprates, is in agreement with this point of view. Moreover, the existence of such satellites in isotypic iron oxides $Bi_2Sr_{3-x}Ca_xFe_2O_9$ [8.13] and $Bi_2Sr_4Fe_3O_{12}$ [8.14], which do not superconduct, clearly demonstrates that such a phenomenon is not unique to copper and has no connection with superconductivity. Contrary to bismuth cuprates, the superconducting cuprates $Pb_2Sr_2Ca_{1-x}Y_xCu_3O_8$ [8.12] and $Pb_{2-x}Bi_xSr_2Ca_{1-y}Y_yCu_3O_8$ [8.15] do not present any incommensurability, in spite of the presence of lone pair cations Pb(II) and Bi(III). This latter phenomenon is easily explained by the very lacunary structure of these compounds, allowing the $6s^2$ lone pair to spread towards the oxygen vacancies, i.e., in the Cu(I) layers (Fig.8.4); this avoids then the distortion of the layers, i.e. their undulation, so that no incommensurability is observed. Since the $6s^2$ lone pair plays the role of an anion, there is no doubt that an excess oxygen should also lead to a displacive incommensurability in such structures even in absence of a lone pair cation.

Thus, incommensurability is a consequence of the presence of both lone pair cation and/or excess oxygen in rather close-packed layers, which leads to a distortion and a puckering of those layers.

There is now no doubt that recently many advances have been made and will be made in the future in high-temperature superconductivity, not only in copper oxides but in a more general sense in oxygen bronzes. Nevertheless it should be kept in mind that, contrary to appearances, mixed valence oxides are complex materials and we must first focus our attention on the crystal chemistry of these systems and avoid drawing hasty conclusions from physical experimental results that are obtained on sometimes poorly characterized materials.

References

Chapter 1

1.1 M. Sergent: La Vie Sci. **3**, 195 (1986)
1.2 J. Friedel, F. Gallais: La Vie Sci. **3**, 181 (1986)
1.3 D. Jérome, H.T. Schulz: Adv. Phys. **31**, 299 (1982)
1.4 T. Ishiguro, K. Yamaji: *Organic Superconductors*, Springer Ser. Solid-State Sci., Vol.88 (Springer, Berlin, Heidelberg 1989)
1.5 G. Saito, S. Kagoshima (eds.): *The Physics and Chemistry of Organic Superconductors*, Springer Proc. Phys., Vol.51 (Springer, Berlin, Heidelberg 1990)
1.6 W.A. Little: Phys. Rev. A **134**, 1416 (1964)
1.7 J.G. Bednorz, K.A. Müller: Z. Phys. B **64**, 189 (1986)
1.8 C. Michel, B. Raveau: Rev. Chim. Miner. **21**, 407 (1984)
1.9 G. Hägg: Z. Phys. Chem. B **29**, 192 (1935)
1.10 A. Magnéli: Ark. Kemi. **1**, 269 (1949)
1.11 A. Magnéli, B. Blomberg: Acta Chem. Scand. **5**, 372 (1951)
1.12 J.B. Goodenough: *Les Oxydes des Métaux de Transition* (Gauthier-Villars, Paris 1973)
1.13 R.K. Stanley, R.C. Morris, W.G. Moulton: Phys. Rev. B **20**, 1903 (1979)
1.14 M.R. Skokan, W.G. Moulton, R.C. Morris: Phys. Rev. B **20**, 3670 (1979)
1.15 A.W. Sleight, J.L. Gillson, P.E. Bierstedt: Solid State Commun. **17**, 27 (1975)
1.16 R.D. Shannon, P.E. Bierstedt: J. Am. Ceram. Soc. **53**, 635 (1970)
1.17 D.C. Johnston, H. Prakash, W.H. Zachariasen, R. Viswanathan: Mater. Res. Bull. **8**, 777 (1973)
1.18 A. Deschanvres, B. Raveau, Z. Sekkal: Mater. Res. Bull. **6**, 699 (1971)

Chapter 2

2.1 J.B. Goodenough, N.F. Mott, M. Pouchard, G. Demazeau, P. Hagenmuller: Mater. Res. Bull. **8**, 647 (1973)
2.2 L. Orgel: *An Introduction to Transition Metal Chemistry*, 2nd ed. (Wiley, New York 1966)
2.3 G. Demazeau, C. Parent, M. Pouchard, P. Hagenmuller: Mater. Res. Bull. **7**, 913 (1972)
2.4 B. Raveau: Proc. Indian Natl. Sci. Acad. **52A**, 67 (1986)
2.5 A. Reller, D.A. Jefferson, J.M. Thomas, M. Kuppal: Proc. R. Soc. London A **394**, 224 (1984); J. Phys. Chem. **87**, 913 (1983)
2.6 S.R. Ruddelesden, P. Popper: Acta Crystallogr. **10**, 538 (1957); ibid. **11**, 54 (1958)
2.7 J.M. Longo, P.M. Raccah: J. Solid State Chem. **6**, 526 (1973)
2.8 B. Grande, H.K. Müller-Buschbaum, M. Schweizer: Z. Anorg. Allg. Chem. **428**, 120 (1977)
2.9 H.K. Müller-Buschbaum, W. Wollschlager: Z. Anorg. Allg. Chem. **414**, 76 (1975); ibid. **428**, 120 (1977)

2.10 Y. Tokura, H. Takagi, S. Uchida: Nature 337, 345 (1989)
2.11 H. Takagi, S. Uchida, Y. Tokura: Phys. Rev. Lett. 62, 1197 (1989)
2.12 J.T. Markert, M.B. Maple: Solid State Commun. 70, 145 (1989)
2.13 C.L. Teske, H.K. Müller-Buschbaum: Z. Anorg. Allg. Chem. 371, 325 (1969)
2.14 C.L. Teske, H.K. Müller-Buschbaum: Z. Anorg. Allg. Chem. 379, 234 (1970); ibid. 377, 144 (1970)
2.15 C.L. Teske, H.K. Müller-Buschbaum: Z. Anorg. Allg. Chem. 370, 134 (1969)
2.16 D.M. de Leeuw, C. Mutsaers, C. Langereis, H. Smoorenburg, P. Rommers: Physica C 152, 39 (1988)
2.17 T. Rouillon, M. Hervieu, B. Raveau: Private communication
2.18 R. Kipka, H.K. Müller-Buschbaum: Z. Naturforsch. B 32, 121 (1977)
2.19 M. Arjomand, D. Machin: J. Chem. Soc., Dalton. Trans. 1061 (1975)
2.20 L. Teske, H. Müller-Buschbaum: Z. Naturforsch. B 27, 296 (1972)
2.21 C. Michel, L. Er-Rakho, N. Hervieu, J. Pannetier, B. Raveau: J. Solid State Chem. 68, 143 (1987)
2.22 J.G. Bednorz, K.A. Müller: Z. Phys. B 64, 189 (1986)
2.23 L. Er-Rakho, C. Michel, B. Raveau: J. Solid State Chem. 73, 514 (1988); J. Phys. Chem. Solids 49, 451 (1988)
2.24 L. Er-Rakho, C. Michel, J. Provost, B. Raveau: J. Solid State Chem. 37, 151 (1981)
2.25 W. David, W. Harrison, M. Ibberson, M. Weller, J.R. Grasmeder, P. Lanchester: Nature 328, 328 (1987)
2.26 M. Hervieu, B. Domenges, F. Deslandes, C. Michel, B. Raveau: C.R. Acad. Sci. 307, serie II, 1444 (1988); Angew. Chem. 27, 440 (1988)
2.27 C. Michel, B. Raveau: Rev. Chim. Miner. 21, 407 (1984)
2.28 T. Fujita, Y. Aoki, Y. Maeno, J. Sakurai, H. Fukuba, H. Fujii: Jpn. J. Appl. Phys. 26, 202 (1987)
2.29 N. Nguyen, J. Choisnet, M. Hervieu, B. Raveau: J. Solid State Chem. 39, 120 (1981)
2.30 J.B. Goodenough, G. Demazeau, M. Pouchard, P. Hagenmuller: J. Solid State Chem. 8, 325 (1973)
2.31 N. Nguyen, L. Er-Rakho, C. Michel, J. Choisnet, B. Raveau: Mater. Res. Bull. 15, 891 (1980)
2.32 R.J. Cava, B. Battlog, R.B. van Dover, J.J. Krajewski, J.W. Waszczak, R.M. Fleming, W.F. Peck, L.W. Rupp, P. Marsh, A.C. James, L.F. Schneemeyer: Nature 345, 602 (1990)
2.33 C. Michel, L. Er-Rakho, B. Raveau: J. Solid State Chem. 39, 161 (1981)
2.34 R.S. Roth, K.L. Davis, J.R. Dennis: Adv. Ceram. Mater. 2(3B), 303 (1987)
2.35 R. Famery, F. Queyroux: Mat. Res. Bull. 24, 275 (1989)
2.36 M.R. Freund, H.Z. Müller-Buschbaum: Z. Naturforsch. B 32, 609 (1977)
2.37 T. Ishiguro, N. Ishizana, N. Mizutani, H. Kato: J. Solid State Chem. 49, 232 (1983)
2.38 M.K. Wu, J.R. Ashburn, C.J. Torng, P.H. Hor, R.L. Meng, L. Gao, Z.J. Huang, Y.Z. Wang, C.W. Chu: Phys. Rev. Lett. 58, 908 (1987)
2.39 R.J. Cava, B. Battlog, R.B. Vandover, D.N. Murphy, S. Sunshine, T. Siegrist, J.P. Remeika, E.A. Rietman, S. Zahurak, G.P. Espinosa: Phys. Rev. Lett. 58, 1676 (1987)
2.40 C. Michel, F. Deslandes, J. Provost, P. Lejay, R. Tournier, M. Hervieu, B. Raveau: C.R. Acad. Sci. 304, 1050 (1987)
2.41 Y. Lepage, W.R. McKinnon, J.M. Tarascon, L.H. Greene, G.N. Hull, D.M. Hwang: Phys. Rev. B 35, 7245 (1987)
2.42 G. Roth, D. Ewert, G. Heger, C. Michel, M. Hervieu, B. Raveau, F. D'Yvoire, A. Revcolevschi: Z. Phys. B 69, 21 (1987)
2.43 M.A. Beno, L. Soderholm, D.W. Capone, D. Hinks, J.O. Jorgensen, I.K. Schuller,

C.U. Segre, K. Zhang, J.D. Grace: Appl. Phys. Lett. **51**, 57 (1987)

2.44 J. Capponi, C. Chaillout, A.W. Hewat, P. Lejay, M. Marezio, N. Nguyen, B. Raveau, J.L. Soubeyroux, J.L. Tholence, R. Tournier: Europhys. Lett. **12**, 1301 (1987)

2.45 F. Izumi, H. Azano, T. Ishigaki, E. Takayama, Y. Uchida, N. Watanabe, T. Nishikawa: Jpn. J. Appl. Phys. **26**, 649 (1987)

2.46 F. Izumi, H. Azano, T. Ishigaki, E. Takayama, Y. Uchida, N. Watanabe: Jpn. J. Appl. Phys. **26**, 1214 (1987)

2.47 J.D. Jorgensen, M. Beno, D.G. Hinks, L. Soderholm, K.J. Volkin, R.L. Hitterman, J.D. Grace, I.K. Schuller, C.U. Segre, K. Zhang, M.S. Kleefisch: Phys. Rev. B **36**, 3608 (1987)

2.48 B. Domenges, M. Hervieu, V. Caignaert, B. Raveau, J.L. Tholence, R. Tournier: J. Microsc. Spectrosc. Electron. **13**, 75 (1988)

2.49 G. Roth, B. Renker, G. Heger, M. Hervieu, B. Domenges, B. Raveau: Z. Phys. B **69**, 53 (1987)

2.50 A. Santoro, S. Miraglia, F. Beech, S.A. Sunshine, D.W. Murphy, L.F. Schneemayer, J.Y. Waszcak: Mater. Res. Bull. **22**, 1007 (1987)

2.51 P. Bordet, C. Chaillout, J.J. Capponi, J. Chenavas, M. Marezio: Nature **327**, 687 (1987)

2.52 F. Boterel, J. Wang, J.M. Haussonne, G. Desgardin, B. Raveau: Industrie chimique SEE Meeting (September 1988); see also V. Caignaert, M. Hervieu, J. Wang, G. Desgardin, B. Raveau, F. Bosterel, J.M. Haussonne: Physica C **170**, 139 (1990)

2.53 C. Michel, B. Raveau: J. Solid State Chem. **43**, 73 (1982)

2.54 B. Chevalier, C. Michel, V. Caignaert, M. Hervieu, G. Demazeau, J. Etourneau, B. Raveau: Private communication

2.55 N. Nguyen, J. Choisnet, B. Raveau: Mat Res. Bull. **17**, 567 (1982)

2.56 D.M. de Leeuw, C.A. Mutsaers, G.P. Geelen, C. Langeyers: J. Solid State Chem. **80**, 276 (1989)

2.57 N. Nguyen: Dissertation. Caen (1982) pp.126-131

2.58 V. Caignaert, R. Retoux, M. Hervieu, C. Michel, B. Raveau: J. Solid State Chem. **91**, 41 (1991)

2.59 V. Caignaert, R. Retoux, C. Michel, M. Hervieu, B. Raveau: Physica C **167**, 483 (1990)

2.60 R.A. Steadman, D.M. de Leeuw, G.P. Geelen, E. Fikkee: Physica C **162–164**, 542 (1989)

2.61 J.R. Grasmeder, M.T. Weller: J. Solid State Chem. **85**, 88 (1990)

Chapter 3

3.1 J.B. Goodenough: *Metallic Oxides*, Progress in Solid State Chemistry, Vol.5 (Pergamon, Oxford 1971)

3.2 G. Hagg: Nature **135**, 874 (1935)

3.3 G. Hagg: Z. Phys. Chem. B **29**, 192 (1935)

3.4 B.W. Brown, E. Banks: Phys. Rev. **84**, 609 (1951)

3.5 B.W. Brown, E. Banks: J. Am. Chem. Soc. **76**, 963 (1954)

3.6 J. Hubbard: Proc. R. Soc. London A **276**, 238 (1963); ibid A **277**, 237 (1964); ibid. A **281**, 401 (1964); ibid. A **285**, 542 (1965); ibid. A **296**, 82,100 (1966)

3.7 J. Labbé, J. Bok: Europhys. Lett. **3**, 1225 (1988)

3.8 J. Friedel: J. Physique **48**, 1787 (1987)

3.9 L.F. Mattheiss: Phys. Rev. Lett. **58**, 1028 (1987)

3.10 A. Bianconi, J. Budnick, A.M. Flank, A. Fontaine, P. Lagarde, A. Marcelli, H. Tolentino, B. Chamberland, C. Michel, B. Raveau, G. Demazeau: Phys. Lett. A **127**, 285 (1988)

3.11 A. Bianconi, J. Budnick, B. Chamberland, A. Clozza, E. Dartyge, G. Demazeau, M. De Santis, A.M. Flank, A. Fontaine, J. Jegoudez, P. Lagarde, L.L. Lynds, C. Michel, F.A. Otter, H. Tolentino, B. Raveau, A. Revcolevschi: Physica C 153-155, 115 (1988)

3.12 A. Bianconi, J. Budnick, G. Demazeau, A.M. Flank, A. Fontaine, P. Lagarde, J. Jegoudez, A. Revcolevschi, A. Marcelli, M. Verdaguer: Physica C 153-155, 117 (1988)

3.13 G. Kaindl, D.D. Sarma, O. Strebel, C.T. Simmons, V. Neukirch, R. Hoppe, H.P. Müller: Physica C 153-155, 139 (1988)

3.14 T. Takahashi: Proc. 34th Jpn. Phys. Soc. Symp. 31 PA-4 (1987)

3.15 P. Ganguly, C.N.R. Rao: Mater. Res. Bull. 8, 408 (1973)

3.16 M. Foex: Bull. Soc. Chim. Fr. 109 (1961)

3.17 T. Kenjo, S. Yajima: Bull. Chem. Soc. Jpn. 46, 1329 (1973)

3.18 N. Nguyen, F. Studer, B. Raveau: J. Phys. Chem. Solids 44, 389 (1983)

3.19 K.K. Singh, P. Ganguly, J.B. Goodenough: J. Solid State Chem. 52, 254 (1984)

3.20 S. Uchida, H. Takagi, H. Yanagisawa, K. Kishio, K. Kitazawa, K. Fueki, S. Tanaka: Jpn. J. Appl. Phys. 26, L 445 (1987)

3.21 J.D. Jorgensen, H.B. Schüttler, D.G. Hinks, D.W. Capone, K. Zhang, M.B. Brodsky, D.J. Scalapino: Phys. Rev. Lett. 58, 1024 (1987)

3.22 E.F. Skelton, W.T. Elam, D.U. Gubser, R.A. Hein, V. Letourneau, M.S. Osofsky, S.B. Quadri, L.E. Toth, S.A. Wolf: In *Novel Superconductivity*, ed. by A. Wolf, V. Kresin (Plenum, New York 1987) p.825

3.23 S. Uchida, H. Takagi, H. Ishi, H. Eisaki, T. Yabe, S. Tajima, S. Tanaka: Jpn. J. Appl. Phys. 26, L 440 (1987)

3.24 J.B. Goodenough: Mater. Res. Bull. 8, 423 (1973)

3.25 A. Fujimori, E. Takayama-Muromachi, Y. Uchida, B. Okai: Phys. Rev. B 35, 8814 (1987)

3.26 H. Nakayama, H. Fujita, T. Nogami, Y. Shirota: Physica C 153-155, 149 (1988)

3.27 N. Nocker, J. Fink, J.C. Fuggle, P.J. Durham, W.M. Temmerman: Physica C 153-155, 119 (1988)

3.28 Z.S. Shen, J.W. Allen, J.J. Yeh, J.S. Kang, W. Ellis, W. Spicer, I. Lindau, M.B. Maple, Y.D. Dalichouach, M.S. Torikachvili, J.Z. Sun, T.H. Geballe: Phys. Rev. B 36, 8414 (1987)

3.29 N. Nguyen: Dissertation, University of Caen (1982)

3.30 J. Beille, R. Cabanel, C. Chaillout, B. Chevalier, G. Demazeau, F. Deslandes, J. Etourneau, P. Lejay, C. Michel, J. Provost, B. Raveau, A. Sulpice, J.L. Tholence, R. Tournier: C.R. Acad. Sci. 304, Ser.II, 1097 (1987)

3.31 W. Kang, G. Collin, M. Ribault, J. Friedel, D. Jerome, J.M. Bassat, J.P. Coutures, Ph. Odier: J. de Phys. 48, 1181 (1987)

3.32 P.M. Grant, S.S.P. Parkin, V.Y. Lee, E.M. Engler, M.L. Ramirez, J.E. Vasquez, G. Lim, R.D. Jacowitz, G.L. Greene: Phys. Rev. Lett. 58, 2482 (1987)

3.33 S.M. Fine, M. Greenblatt, S. Simizu, S.A. Friedberg: In *Chemistry of High Temperature Superconductors*, ed. by L. Nelson, ACS Symp. Ser., Vol.351 (American Chemical Society, Washington, DC 1987) p.95; Phys. Rev. B 36, 5716 (1987)

3.34 S.A. Shaheen, N. Jisrawi, Y.H. Lee, Y.Z. Zhang, M. Croft, W.L.H. Zhen, L. Rebelsky, S. Horn: Phys. Rev. B 36, 7214 (1987)

3.35 J.M. Tarascon, L.H. Greene, B.G. Bagley, W.R. McKinnon, P. Barboux, G.W. Hull: In *Novel Superconductivity*, ed. by A. Wolf, V. Kresin (Plenum, New York 1987) p.705

3.36 J. Beille, B. Chevalier, G. Demazeau, F. Deslandes, J. Etourneau, O. Laborde, C. Michel, P. Lejay, J. Provost, B. Raveau, A. Sulpice, J.L. Tholence, R. Tournier: Physica B 46, 307 (1987)

3.37 G. Demazeau, F. Tresse, Th. Plante, B. Chevalier, J. Etourneau, C. Michel, M.

Hervieu, B. Raveau, P. Lejay, A. Sulpice, R. Tournier: Physica C **153–155**, 824 (1988)

3.38   J.D. Jorgensen, B. Dabrowsky, S. Pei, D.G. Hinks, L. Soderholm, B. Morosin, J.E. Shirber, E.L. Venturini, D.S. Ginley: Phys. Rev. B**38**, 11337 (1988)

3.39   J.D. Jorgensen, B. Dabrowsky, S. Pei, D.R. Richards, D.G. Hinks: Phys. Rev. B**40**, 2187 (1989)

3.40   C. Chaillout, S.W. Cheong, Z. Fisk, M.S. Lehmann, M. Marezio, B. Morosin, J.E. Shirber: Physica C **158**, 183 (1989)

3.41   T. Kenjo, S. Yajima: Bull. Chem. Soc. Jpn. **50**, 2847 (1977)

3.42   I.S. Shapygin, B.G. Kalhan, V.B. Lazarev: J. Inorg. Chem. **24**, 620 (1979)

3.43   L. Er-Rakho: Dissertation, University of Caen (1981)

3.44   J.G. Bednorz, K.A. Müller: Z. Phys. B **64**, 189 (1986)

3.45   N.F. Mott: Philos. Mag. **24**, 935 (1971)

3.46   N.F. Mott: *Metal-Insulator Transitions* (Taylor and Francis, London 1974)

3.47   M.W. Shafer, T. Penney, B.L. Olson: Phys. Rev. B **36**, 4047 (1987)

3.48   T. Penney, M.W. Shafer, B.L. Olson, T.S. Plaskett: Adv. Ceram. Mater., special issue on superconducting materials (July 1987)

3.49   M. Petravic, A. Hamzic, B. Leontic, L. Forro: Proc. *Adriatico Research Conference on High Temperature Superconductors*, ed. by S. Lundqvist, E. Tosatti, M. Tosi, Y. Lu (World Scientific, Singapore 1987) p.419

3.50   N.P. Ong, Z.Z. Wang, J. Clayhold, J.M. Tarascon, L.H. Greene, W.R. McKinnon: Phys. Rev. B **35**, 8807 (1987)

3.51   J.B. Goodenough, G. Demazeau, M. Pouchard, P. Hagenmüller: J. Solid State Chem. **8**, 325 (1973)

3.52   Y. Hidaka, Y. Enomoto, M. Suzuki, M. Oda, T. Murakami: Jpn. J. Appl. Phys. **26**, L 377 (1987)

3.53   H. Takagi, S. Uchida, K. Kitazawa, S. Tanaka: Jpn. J. Appl. Phys. **26**, L 123 (1987)

3.54   J.G. Bednorz, M. Takashige, K.A. Müller: Europhys. Lett. **3**, 379 (1987)

3.55   T. Fujita, Y. Aoki, Y. Maeno, J. Sakurai, H. Fukuba, M. Fujii: Jpn. J. Appl. Phys. **26**, L 368 (1987)

3.56   C.W. Chu, P.H. Hor, R.L. Meng, L. Gao, Z.J. Huang, Y.Q. Wang: Phys. Rev. Lett. **58**, 405 (1987)

3.57   C.N.R. Rao, P. Ganguly: Curr. Sci. **56**, 47 (1987)

3.58   K. Kitazawa, M. Sakai, S. Uchida, H. Takagi, K. Kishio, S. Kanbe, S. Tanaka, K. Fueki: Jpn. J. Appl. Phys. **26**, L 342 (1987)

3.59   H.C. Ku, S.W. Hau, S.Y. Tsaur: Physica C, **153–155**, 922 (1988)

3.60   R.J. Cava, R.B. Van Dover, B. Batlogg, E.A. Rietman: Phys. Rev. Lett. **58**, 408 (1987)

3.61   J.M. Tarascon, L.H. Greene, W.R. McKinnon, G.W. Hull, T.H. Geballe: Science **235**, 1373 (1987)

3.62   T.P. Orlando, K.A. Delin, S. Foner, E.J. McNiff, Jr., J.M. Tarascon, L.H. Green, W.R. McKinnon, G.W. Hull: Preprint.

3.63   D.W. Capone, D.G. Hinks, J.D. Jorgensen, K. Zhang: Appl. Phys. Lett. **50**, 543 (1987)

3.64   A.J. Panson, G.R. Wagner, A.I. Braginski, J.R. Gavaler, M.A. Janocko, H.C. Pohl, J. Talvacchio: Appl. Phys. Lett. **50**, 1104 (1987)

3.65   K. Fueki, K. Kitazawa, K. Kishio, T. Hasegawa, S. Uchida, H. Takagi, S. Tanaka: In *Chemistry of High Temperature Superconductors*, ed. by L. Nelson, ACS Symp. Ser., Vol.351 (Americal Chemical Society, Washington, DC 1987) p.38

3.66   C. Politis, J. Geerk, M. Dietrich, B. Obst: Z. Phys. B **66**, 141 (1987)

3.67   M. Hervieu, A. Maignan, C. Martin, C. Michel, J. Provost, B. Raveau: J. Solid State Chem. **75**, 212 (1988)

3.68 S. Nakajima, M. Kikuchi, N. Kobayashi, H. Iwasaki, D. Shindo, Y. Syono, Y. Muto: In *Advances in Superconductivity II*, ed. by T. Ishiguro, K. Kajimura (Springer-Verlag, Tokyo 1990) p.219

3.69 W.A. Green, D.M. de Leeuw, L.F. Feiner: Physica C **165**, 55, 305 (1990)

3.70 Y. Kubo, Y. Shimakawa, T. Manako, T. Satoh, S. Ijima, T. Ichihashi, H. Igarashi: Physica C **162-164**, 991 (1989)

3.71 K.V. Ramanujachary, S. Li, M. Greenblatt: Physica C **165**, 377 (1990)

3.72 C. Martin, A. Maignan, J. Provost, C. Michel, M. Hervieu, R. Tournier, B. Raveau: Physica C **168**, 8 (1990)

3.73 R. Yoshizaki, H. Kurahashi, N. Ishikawa, M. Akamatsu, J. Fujikami, H. Ikeda: In Proc. of the Tsukuba Seminar on High T_c Superconductivity, ed. by K. Masuda, T. Arai, I. Iguehi, R. Yoshizaki (University of Tsukuba, Tsukuba 1989) p.21

3.74 J.B. Torrance, A. Bezinge, A.I. Nazzal, S.S.P. Parkin: Physica C **162-164**, 291 (1989)

3.75 K. Kishio, K. Kitazawa, N. Sugii, S. Kanbe, K. Fueki, H. Takagi, S. Tanaka: Chem. Lett. 635 (1987)

3.76 P. Ganguly, R.A. Ram Mohan, K. Sreedhan, C.N.R. Rao: Solid State Commun. **62**, 807 (1987)

3.77 J.M. Tarascon, L.H. Greene, W.R. McKinnon, G.W. Hull: Solid State Commun. **63**, 499 (1987)

3.78 A.A. Abrikosov, L.P. Gor'kov: Sov. Phys.-JETP **12**, 1243 (1961)

3.79 M.A. Subramanian, J. Gopalakrishnan, A.W. Sleight: J. Solid State Chem. **84**, 413 (1990)

3.80 C. Michel, J. Provost, F. Deslandes, B. Raveau, J. Beille, R. Cabanel, P. Lejay, A. Sulpice, J.L. Tholence, R. Tournier, B. Chevalier, G. Demazeau, J. Etourneau: Z. Phys. B **68**, 417 (1987)

3.81 A.W. Sleight, J.L. Gilson, P.E. Bierstedt: Solid State Commun. **17**, 27 (1975)

3.82 N. Nguyen, C. Michel, F. Studer, B. Raveau: Mater. Chem. **7**, 413 (1982)

3.83 J.B. Torrance, Y. Tokura, A. Nazzai, S.S.P. Parkin: Phys. Rev. Lett. **60**, 542 (1988)

3.84 R.J. Cava, B. Batlogg, R.B. van Dover, J.J. Krajewski, J.V. Waszczak, R.M. Fleming, W.F. Peck Jr., L.W. Rupp Jr., P. Marsh, A.C.W.P. James, L.F. Schneemeyer: Nature **345**, 602 (1990)

3.85 G. Demazeau, C. Parent, M. Pouchard, P. Hagenmüller: Mater. Res. Bull. **7**, 913 (1972)

3.86 J.B. Goodenough, N.F. Mott, M. Pouchard, G. Demazeau, P. Hagenmüller: Mater. Res. Bull. **8**, 647 (1973)

3.87 C. Michel, L. Er-Rakho, B. Raveau: Mater. Res. Bull. **20**, 667 (1985)

3.88 C. Michel, L. Er-Rakho, B. Raveau: J. Phys. Chem. Solids **49**, 451 (1988)

3.89 J.B. Torrance: Private communication

3.90 S. Li, M. Greenblatt: In *Chemistry of Oxide Superconductors*, ed. by C.N.R. Rao, IUPAC (Blackwell, Oxford 1988) p.35

3.91 F. Herman, R.V. Kasowski, W.Y. Hsu: Phys. Rev. B **37**, 2309 (1988)

3.92 C. Michel, L. Er-Rakho, M. Hervieu, J. Pannetier, B. Raveau: J. Solid State Chem. **68**, 143 (1987)

3.93 L. Er-Rakho, C. Michel, B. Raveau: J. Solid State Chem. **73**, 514 (1988)

3.94 D.W. Murphy, S.A. Sunshine, P.K. Gallagher, H.M. O'Bryan, R.J. Cava, B. Batlogg, R.B. Van Dover, L.F. Schneemeyer, S.M. Zahurak: In *Chemistry of High Temperature Superconductors*, ed. by D.L. Nelson, ACS Symp. Ser., Vol.351 (Americal Chemical Society, Washington, DC 1987) p.181

3.95 R. Vijayaraghavan, R.A. Mohan Ram, C.N.R. Rao: J. Solid State Chem. **78**, 316 (1989)

3.96 J. Provost, C. Michel, F. Studer, B. Raveau: Synth. Met. **4**, 147,157 (1981)

3.97 M. Hervieu, C. Michel, B. Domengès, A. Maignan, B. Raveau: Phys. Status Solidi 106, 645 (1988)

3.98 S. Tsurumi, T. Iwata, Y. Tajima, M. Ikita: Jpn. J. Appl. Phys. 26, L 1865 (1987)

3.99 M.K. Wu, J.R. Ashburn, C.J. Torng, P.H. Hor, R.L. Meng, L. Gao, Z.J. Huang, Y.Q. Wang, C.W. Chu: Phys. Rev. Lett. 58, 908 (1987)

3.100 R.J. Cava, B. Batlogg, R.B. Van Dover, D.W. Murphy, S. Sunshine, T. Siegrist, J.P. Remeika, E.A. Rietman, S. Zahurak, G.P. Espinosa: Phys. Rev. Lett. 58, 1676 (1987)

3.101 Y. Lepage, W.R. McKinnon, J.M. Tarascon, L.H. Greene, G.W. Hull, D.M. Hwang: Phys. Rev. B 35, 7245 (1987)

3.102 C. Michel, F. Deslandes, J. Provost, P. Lejay, R. Tournier, M. Hervieu, B. Raveau: C.R. Acad. Sci. 304, Ser.II, 1059 (1987)

3.103 J. Karpinski, E. Kaldis, S. Rusieki: Proc. Symp. High Temp. Superconductors, E-MRS Fall Meeting, Strasbourg (1988), J. Less Common. Met. 150, 207 (1989)

3.104 P.H. Hor, L. Gao, R.L. Meng, Z.J. Huang, Y.Q. Wang, K. Forster, J. Vassilous, C.W. Chu, M.K. Wu, J.R. Ashburn, J.C. Torng: Phys. Rev. Lett. 58, 911 (1987)

3.105 A. Driessen, R. Griessen, N. Koeman, E. Salomons, R. Brouwer, D.G. de Groot, K. Heeck, H. Hemmes, J. Rector: Phys. Rev. B 36, 5602 (1987)

3.106 K. Murata, H. Ihara, M. Tokumoto, M. Hirabayashi, N. Terada, K. Senzaki, Y. Kimma: Jpn. J. Appl. Phys. 26, L473 (1987)

3.107 A. Umezawa, G.W. Crabtree, J.Z. Liu, T.J. Moran, S.K. Malik, L.H. Nunez, W.L. Kwok, C.H. Sowers: Phys. Rev. B 38, 2843 (1988)

3.108 A.P. Malozemoff, W.J. Gallagher, R.E. Schwall: In Chemistry of High Temperature Superconductors, ed. by L. Nelson, ACS Symp. Ser., Vol. 351 (American Chemical Society, Washington, DC 1987) p.280

3.109 O. Laborde, J.L. Tholence, P. Lejay, A. Sulpice, R. Tournier, J.J. Capponi, C. Michel, J. Provost: Solid State Commun. 63, 877 (1987)

3.110 S. Jin, R.C. Sherwood, T.H. Tiefel, R.B. Van Dover, D.W. Johnson, Jr., G.S. Grader: Appl. Phys. Lett. 51, 855 (1987)

3.111 G. Paterno, C. Alvani, S. Casadio, U. Gambardella, L. Maritato: Physica C 153-155, 1341 (1988)

3.112 J.C. Golowski, H. Lauvray: In Proc. Annual Meeting, American Ceramic Society, Dallas, April 1990

3.113 R.L. Meng, C.W. Chu: In Proc. Annual Meeting, American Ceramic Society, Dallas, April 1990

3.114 T. Vankatesan, X.D. Wu, B. Dutta, A. Inam, M.S. Hegde, D.M. Huang, C.C. Chang, L. Nazar, B. Wilkens: Appl. Phys. Lett. 54, 581 (1989)

3.115 X.X. Xi, G. Linker, O. Meyer, E. Nold, B. Obst, F. Ratzel, R. Smithey, B. Strehlau, F. Weshenfelder, J. Geerk: Z. Phys. B 74, 13 (1989)

3.116 M. Gurvitch, A.T. Fiory: In Novel Superconductivity, ed. by S.A. Wolf, V. Kresin (Plenum, New York 1987)

3.117 S.W. Tozer, A.W. Kleinsasser, T. Penney, D. Kaiser, F. Holtzberg: Phys. Rev. Lett. 59, 1768 (1987)

3.118 L. Foro, M. Raki, C. Ayache, P.C.E. Stamp, J.Y. Henry, J. Rossat-Mignot: Physica C 153-155, 1357 (1988)

3.119 L.Y. Vinnikov, G.A. Emelchenko, P.A. Konnovich, Y.A. Ossipyan, L.L. Buravov, V.N. Laukhin: Physica C 153-155, 1359 (1988)

3.120 N.P. Ong, Z.Z. Wang, S. Hagen, T.W. Jing, J. Clayhold, J. Horvath: Physica C 153-155, 1072 (1988)

3.121 A. Gupta, G. Koren, R.J. Baseman, A. Segmüller, W. Holber: Physica C 162-164, 127 (1989)

3.122 O. Brunner, J.M. Triscone, M.G. Karkut, L. Antognazza, A.D. Kent, T. Driscoll, T. Boichat, O. Fisher: Physica C 162-164, 647 (1989)

3.123 G.V. Subba Rao, U.V. Varadaraju, K.A. Thomas, R. Vijayashree, R. Srinivasan,

V. Sankaranarayanan, N.P. Raju, T.R. Ravindran, L.S. Vaidhyanathan: In *Progress in High Temperature Superconductivity*, ed. by S. Lundqvist, E. Tosatti, M. Tosi, Y. Lu (World Scientific, Singapore 1987) Vol.1, p.449

3.124 S. Tanaka, S. Uchida, H. Takagi, S. Tajima, T. Hasegawa, K. Kishio, K. Kitazawa, K. Fueki: In *Progress in High Temperature Superconductivity*, ed. by S. Lundqvist, E. Tosatti, M. Tosi, Y. Lu (World Scientific, Singapore 1987) Vol.1, p.107

3.126 N.P. Ong, Z.Z. Wang, J. Clayhold, J.M. Tarascon, L.H. Greene, W.R. McKinnon: In *Novel Superconductivity*, ed. by S.A. Wolf, V. Kresin (Plenum, New York 1987) p.1061

3.126 Y. Iye, T. Tamegai, T. Sakakibara, T. Goto, N. Miura, H. Takeya, H. Takei: Physica C 153-155, 26 (1988)

3.127 P.B. Allen, W.E. Pickett, H. Krakauer: Phys. Rev. B 36, 3926 (1987)

3.128 P.B. Allen, W.E. Pickett, H. Krakauer: In *Novel Superconductivity*, ed. by S.A. Wolf, V. Kresin (Plenum, New York 1987) p.489

3.129 A. Bianconi, A. Congiu Castellano, M. De Santis, P. Delogu, A. Gargano, R. Giorgi: Solid State Commun. 63, 1135 (1987)

3.130 A. Bianconi, M. De Santis, A.M. Flank, A. Fontaine, P. Lagarde, A. Marcelli, H. Katayama-Yoshida, A. Kotani: Physica C 153-155, 1760 (1988)

3.131 R.J. Cava, B. Batlogg, C.H. Chen, E.A. Rietman, S.M. Zahurak, D. Werder: Nature 329, 423 (1987)

3.132 P. Monod, M. Ribault, F. D'Yvoire, J. Jegoudez, G. Collin, A. Revcolevschi: J. de Phys. 48, 1369 (1987)

3.133 M. Tokumoto, H. Ihara, T. Matsubara, M. Hirabayashi, N. Terada, H. Oyanagi, K. Murata, Y. Kimura: Jpn. J. Appl. Phys. 26, L1566 (1987)

3.134 D.C. Johnston, A.J. Jacobson, J.M. Newsam, J.T. Lewandowski, D.P. Goshorn, D. Xie, W.B. Yelon: In *Chemistry of High Temperature Superconductors* ed. by D.L. Nelson, ACS Symp. Ser., Vol.351 (American Chemical Society, Washington, DC 1987) p.136

3.135 C.N.R. Rao: In *Chemistry of Oxide Superconductors*, ed. by C.N.R. Rao, IUPAC (Blackwell, Oxford 1988) p.1

3.136 H. Oyanagi, H. Ihara, T. Matsubara, M. Tokumoto, T. Matsushita, M. Hirabayashi, K. Murata, N. Terada, T. Yao, H. Iwasaki, Y. Kimura: Jpn. J. Appl. Phys. 26, L1561 (1987)

3.137 S. Uchida, H. Takagi, T. Hasegawa, K. Kishio, S. Tajima, K. Kitazawa, K. Fuck, S. Tanaka: In *Novel Superconductivity*, ed. by A. Wolf, V. Kresin (Plenum, New York 1987) p.855

3.138 B. Raveau, F. Deslandes, C. Michel, M. Hervieu: In Proc. Int'l Discussion Meeting, Mauterndorf, Austria, February 1988 (Plenum, New York 1988)

3.139 R.C. Budhani, S.M.H. Tzeng, R.F. Bunshab: Phys. Rev. B 36, 8873 (1987)

3.140 Z.Z. Wang, J. Clayhold, N.P. Ong, J.M. Tarascon, L.H. Greene, W.R. McKinnon, G.W. Hull: Phys. Rev. B 36, 7222 (1987)

3.141 B. Raveau, C. Michel, M. Hervieu, J. Provost: Physica C 153-155, 5 (1988)

3.142 F. Baudelet, G. Collin, E. Dartyge, A. Fontaine, J.P. Kappler, G. Krill, J.P. Itie, J. Jegoudez, M. Maurer, Ph. Monod, A. Revcolevschi, H. Tolentino, G. Tourillon, M. Verdaguer: Z. Phys. B 69, 141 (1988)

3.143 M. Lengeler, M. Wilhelm, B. Jobst, W. Schwaen, B. Seebacher, U. Hillebrecht: Physica C 153-155, 143 (1988)

3.144 M. Maurer, M.F. Ravet, J.C. Ousset, T. Gourieux, A. Fontaine, H. Tolentino, E. Dartyge, G. Tourillon, G. Krill, P.H. Kes: Physica C 153-155, 145 (1988)

3.145 C.N.R. Rao, P. Ganguly, J. Gopalakrishnan, D.D. Sarma: Mater. Res. Bull. 22, 1159 (1987)

3.146 M. Hervieu, B. Domengés, B. Raveau, J.M. Tarascon, M. Post, W.R. McKinnon: Mater. Lett. 8, 73 (1989)

3.147 C.P. Burmester, M.E. Mann, G. Ceder, L.T. Wille, D. de Fontaine: Physica C **162-164**, 225 (1989)
3.148 Y. Bruynseraede, J.P. Locquet, J. Vanacken, B. Wuyts, C. Van Haesendonck, I.K. Schuller: Physica C **162-164**, 23 (1989)
3.149 A.J. Jacobson, J.M. Newsam, D.G. Johnston, J.P. Stokes, S. Bahttacharya, J.T. Lewandowski, D.P. Goshorn, M.J. Higgins, M.S. Alvarez: In *Chemistry of Oxide Superconductors*, ed. by C.N.R. Rao, IUPAC (Blackwell, Oxford 1988) p.43
3.150 H. Tolentino, E. Dartyge, A. Fontaine, G. Tourillon, T. Gourieux, G. Krill, M. Maurer, M.F. Ravet: *Proc. of the Int'l Symp. on the Electronic Structure of High T_c Superconductors*, Rome (1988), ed. by A.B. Bianconi (Pergamon, Oxford 1989)
3.151 S.W. Keller, K.J. Leary, T.A. Faltens, J.N. Michaels, A.M. Stacy: In *Chemistry of High Temperature Superconductors* ed. by D.L. Nelson, ACS Symp. Ser., Vol.351 (American Chemical Society, Washington, DC 1987) p.115
3.152 A.M. Umarji, P. Somasundaram, C.N.R. Rao: Physica C **153-155**, 497 (1988)
3.153 F. Deslandes: Private communication
3.154 A.S. Bhalla, R. Ross, L.E. Cross: In *Chemistry of Oxide Superconductors*, ed. by C.N.R. Rao, IUPAC (Blackwell, Oxford 1988) p.71
3.155 P.H. Hor, R.L. Meng, Y.Q. Wang, L. Gao, Z.J. Huang, J. Bechtold, K. Foster, C.W. Chu: Phys. Rev. Lett. **58**, 1891 (1987)
3.156 H. Takagi: Jpn. J. Appl. Phys. Suppl. **26-3**, 1029 (1987)
3.157 L. Soderholm, Z. Zhang, G.G. Hinks, M.A. Beno, J.D. Jorgensen, C.V. Segre, I.K. Schuller: Nature **328**, 604 (1987)
3.158 G.V. Subba Rao: In *Chemistry of Oxide Superconductors*, ed. by C.N.R. Rao, IUPAC (Blackwell, Oxford 1988) p.63
3.159 D.W. Murphy, S. Sunshine, R.B. Van Dover, R.J. Cava, B. Batlogg, S.M. Zahurak, L.F. Schneemeyer: Phys. Rev. Lett. **58**, 1088 (1987)
3.160 K. Ikeda: Jpn. J. Appl. Phys. Superconduct. Mater. Ser. **1**, 85 (1988)
3.161 A. Manthiram, S.J. Lee, J.B. Goodenough: J. Solid State Chem. **73**, 278 (1988)
3.162 D.M. de Leeuw, C.A.H.A. Mutsaers, H.A.M. Van Hal, H. Verwey, A.H. Carim, H.C.A. Smoorenburg: Physica C **156**, 126 (1988)
3.163 W.T. Fu, H.W. Zandbergen, C.J. Van der Beck, L.J. de Jongh: Physica C **156**, 133 (1988)
3.164 H.W. Zandbergen, W.T. Fu, L.J. de Jongh: Physica C **156**, 307 (1988)
3.165 F. Keller-Berest, S. Megtert, G. Collin, P. Monod, M. Ribault: Physica C **161**, 150 (1989)
3.166 B.W. Veal, A.P. Paulikas, J.W. Downey, H. Claus, K. Vanderwoort, G. Toulins, H. Shi, M. Jensen, L. Morss: Physica C **162-164**, 97 (1989)
3.167 C. Michel, F. Deslandes, J. Provost, P. Lejay, R. Tournier, M. Hervieu, B. Raveau: C.R. Acad. Sci. **304**, Ser.II, 1169 (1987)
3.168 Y. Syono, A. Tokiwa, M. Kikuchi, K. Kusaba, R. Suzuki, T. Kajitani, D. Shindo, N. Kobayashi, O. Nakatsu, Y. Muto: Jpn. J. Appl. Phys. Supercond. Mater. Ser. **1**, 42 (1988)
3.169 L. Govea, R. Escudero, R. Dios-Jara, C. Pina, F. Morales, C. Wang, R.A. Barrio: Physica C **153-155**, 940 (1988)
3.170 K. Mori, M. Sasakawa, Y. Isikawa, K. Kobayashi, K. Sato: Jpn. J. Appl. Phys. Supercond. Mater. Ser. **1**, 81 (1988)
3.171 A. Maeda, T. Yabe, K. Uchinokura, S. Tanaka: Jpn. J. Appl. Phys. **26**, L 1368 (1987)
3.172 A. Maeda, T. Yabe, K. Uchinokura, M. Izumi, S. Tanaka: Jpn. J. Appl. Phys. Lett. **26**, L 1550 (1987)
3.173 B. Chevalier, B. Buffat, G. Demazeau, B. Lloret, J. Etourneau, M. Hervieu, C. Michel, B. Raveau, R. Tournier: J. de Phys. **48**, 1619 (1987)

3.174 T. Wada, N. Suzuki, T. Maeda, S. Uchinokura, S. Tanaka: Appl. Phys. Lett. **52**, 1989 (1988)

3.175 K. Zhang, B. Dabrowsky, C.U. Segre, D.G. Hinks, I.K. Schuller, J.D. Jorgensen, M. Slashi: J. Phys. C **20**, L935 (1987)

3.176 B.W. Veal, W.K. Kwok, A. Umezawa, G.W. Crabtree, J.D. Jorgensen, J.W. Downey, L.J. Nowicki, A.W. Mitchell, A.P. Paulikas, C.H. Sowers: Appl. Phys. Lett. **51**, 279 (1987)

3.177 A. Manthiram, X.X. Tang, J.B. Goodenough: Phys. Rev. B **37**, 3734 (1988)

3.178 M.R. Harrison, I.C.S.T. Hegedus, W.G. Freeman, R. Jones, P.P. Edwards, W.I.P. David, C.C. Wilson: In *Chemistry of Oxide Superconductors*, ed. by C.N.R. Rao, IUPAC (Blackwell, Oxford 1988) p.131

3.179 K. Takita: Prog. High Temp. Supercond. **5**, 73 (1988)

3.180 B. Okai, K. Takahashi, H. Nozaki, M. Sacki, M. Kosuge, M. Ohita: Jpn. J. Appl. Phys. **26**, L1648 (1987)

3.181 A.P. Gonçalvez, I.C. Santos, E.B. Lopes, R.T. Henriques, M. Almeida, M.O. Figueiredo: Phys. Rev. B **37**, 7476 (1988)

3.182 I.K. Schuller, D.G. Hinks, J.D. Jorgensen, L. Soderholm, M. Beno, K. Zhang, C.U. Segree, Y. Bruynseraede, J.P. Locquet: In *Novel Superconductivity*, ed. by S.A. Wolf, V. Kresin (Plenum, New York 1987) p.647

3.183 U. Neukirch, C.T. Simmons, P.S. Sladeczek, C. Laubschat, O. Strebel, G. Kaindl, D.D. Sarma: Europhys. Lett. **5**, 567 (1988)

3.184 E.E. Alp, G.K. Shenoy, L. Soderholm, G.L. Goodman, D.G. Hinks, B.W. Veal: Mater. Res. Soc. Symp. Proc. **99**, 177 (1988)

3.185 R. Retoux, F. Studer, C. Michel, B. Raveau, A. Fontaine, E. Dartyge: Phys. Rev. B **41**, 193 (1990)

3.186 J.J. Neumeier, T. Bjornholm, M.B. Maple, I.K. Schuller: Phys. Rev. Lett. **63**, 2516 (1989)

Chapter 4

4.1 J.M. Tarascon, L.H. Greene, B.G. Agley, W.R. McKinnon, P. Barboux, G.N. Hull: In *Novel Superconductivity*, ed. by S.A. Wolf, V. Kresin (Plenum, New York 1987) p.705

4.2 M.A. Subramanian, J. Gopalakrishnan, C.C. Torardi, T.R. Askew, R.B. Flippen, W. Sleight: Science **240**, 495 (1988)

4.3 C. Michel, J. Provost, F. Deslandes, B. Raveau, J. Beille, R. Cabanel, P. Lejay, A. Sulpice, J.L. Tholence, R. Tournier, B. Chevalier, G. Demazeau, J. Etourneau: Z. Phys. B **68**, 417 (1987)

4.4 S. Boo Nam, S. Woo Nam, J. Ok Nam: In *Novel Superconductivity*, ed by S.A. Wolf, V. Kresin (Plenum, New York 1987) p.993

4.5 K. Kitazawa, H. Takagi, K. Kishio, T. Hasegawa, S. Uchida, S. Tajima, S. Tanaka, K. Fueki: Physica C **153–155**, 9 (1988)

4.6 C.V. Tomy, R. Prasad, N.C. Sony, S.D. Malik: Physica C **153–155**, 174 (1988)

4.7 F. Nakamura, T. Fukuda, Y. Ochiai, A. Tominaga, Y. Narahara: Physica C **153– 155**, 178 (1988)

4.8 S.M. Greene, C. Jiang, H.L. Luo, Y. Mei, C. Politis: Physica C **153–155**, 182 (1988)

4.9 E. Lin, G. Lu, C. Wei, Z. Liu, Y. Sun, J. Lan, X. Zhu, G. Li, S. Feng, Y. Dai, Z. Gan: Physica C **153–155**, 190 (1988)

4.10 Z. Chen, Y. Qian, D. Sun, M. Fang, J. Xia, Z. Zhao, Y. Zhao, W. Kuan: Physica C **153–155**, 409 (1988)

4.11 F. Moran, U. Amador, M. Barahona, M.A. Alario-Franco, A. Vergas, J. Rodriguez Carvajal: Physica C **153–155**, 423 (1988)

4.12 M.B. Maple, Y. Dalichaouch, E. Early, B. Lee, J. Markert, J. Neumeier, C. Seaman, K. Yang, H. Zhou: Physica C 153-155, 858 (1988)
4.13 L. Gouea, R. Escudero, D. Rios-Jara, C. Pina, F. Murales, C. Wang, R. Barrio: Physica C 153-155, 940 (1988)
4.14 D.W. Murphy, S. Sunshine, R.B. Van Dover, R.J. Cava, B. Batlogg, S.M. Zahurak, L.F. Schneemeyer: Phys. Rev. Lett. 58, 1888 (1987)
4.15 F. Zuo, B.R. Patton, D.L. Cox, S.I. Lee, Y. Song, J.P. Golben, X.D. Chen, S.Y. Lee, Y. Cao, Y. Lu, J.R. Gaines, J.C. Garland, A.J. Epstein: Phys. Rev. B 36, 3603 (1987)
4.16 K. Kitazawa, K. Kishio, H. Takagi, T. Hasegawa, S. Kanbe, S. Uchida, S. Tanaka, K. Fueki: Jpn. J. Appl. Phys. 26, L 339 (1987)
4.17 U.V. Varadaraju, S. Natarajan, T.S. Sampath Kumar, M. Parantha Man, G.V. Subba Rau, N.P. Raju, R. Srivasan: Physica B 148, 417 (1987)
4.18 G.V. Subba Rao, U.V. Varadaju, K.A. Thomas, R. Vijaya Shree, R. Srivasan, V. Sankaranrayanan, N.P. Raju, T.R. Ravindran, L.S. Vaidhyathan: Int'l J. Mod. Phys. B 1, 1097 (1987)
4.19 G.V. Subba Rao: In Chemistry of Oxide Superconductors, ed. by C.N.R. Rao (Blackwell, Oxford 1988) p.63
4.20 G. Svensson, Z. Hegedus, L. Wang, O. Rapp: Physica C 153-155, 864 (1988)
4.21 Y. Dalichaouch, M.S. Torikachvili, E.A. Early, B.W. Lee, C.L. Seaman, K.N. Yang, H. Zhou, M.B. Maple: Solid State Commun. 65, 999 (1988)
4.22 B. Okai, M. Kosuge, H. Nozaki, K. Takahashi, M. Ohta: Jpn. J. Appl. Phys. 27, L41 (1988)
4.23 J. Amador, C. Cascales, I. Rasines: Mat. Res. Soc. Symp. Proc. 99, 249 (1988)
4.24 M.E. Lopez-Morales, D. Rios-Jara, J. Taguena-Martinez, R. Escudero: Physica C 153-155, 942 (1988)
4.25 F. Zuo, X. Chen, A. Chakraborty, B.R. Patton, J.R. Gaines, A.J. Epstein: Solid State Commun. 68, 239 (1988)
4.26 E.M. McCarron, M.K. Crawford, J.B. Parise: J. Solid State Chem. 78, 192 (1989)
4.27 J.B. Parise, E. McCarron: J. Solid State Chem. 83, 188 (1989)
4.28 K. Remsching, R. Rogl, E. Bauer, R. Eibler, G. Hilscher, H. Kircmayr, N. Pillmayr: In High T_C Superconductors, ed. by H.A. Weber (Plenum, New York 1988) p.99
4.29 G. Hilscher, N. Pillmayr, R. Eibler, E. Bauer, K. Remsching, P. Rogl: Z. Phys. B 72, 461 (1988)
4.30 K. Remsching, P. Rogl, R. Eibler, G. Hilscher, N. Pillmayr, H. Kircmayr, E. Bauer: Physica C 153-155, 906 (1988)
4.31 Y. Maeno, T. Fujita: In Novel Superconductivity, ed. by S.A. Wolf, V. Kresin (Plenum, New York 1987) p.1073
4.32 H. Adrian, O. Baver, H. Niederhofer, G. Adrian, S. Nielsen: Physica C 153-155, 928 (1988)
4.33 K. Westerholt, M. Arndt, H.T. Wuller, H. Bach, P. Stauche: Physica C 153-155, 862 (1988)
4.34 B. Ullmann, R. Wordenweber, K. Heinemann, H. Krebs, H.C. Freyhardt: Physica C 153-155, 872 (1988)
4.35 M. Ishikawa, T. Takabatake, A. Tohdake, Y. Nakazawa, T. Shibuya, K. Koga: Physica C 153-155, 890 (1988)
4.36 J. Bieg, J. Jing, H. Engelmann, Y. Hsia, U. Gonser, P. Gutlich, R. Jakubi: Physica C 153-155, 952 (1988)
4.37 B. Raveau, F. Deslandes, C. Michel, M. Hervieu, G. Heger, G. Roth: In High T_c Superconductors, ed. by H.A. Weber (Plenum, New York 1988) p.3
4.38 F. Deslandes: Private communication

4.39 G. Roth, G. Heger, B. Renker, J. Pannetier, V. Caignaert, M. Hervieu, B. Raveau: Z. Phys. **74**, 43 (1988); Physica C **153-155**, 972 (1988)

4.40 Z.Q. Qiu, Y.W. Du, H. Tang, J.C. Walker, W.A. Bryden, K. Mourjani: J. Magn. Magn. Mater. **69**, L221 (1987)

4.41 P. Chaudouet, S. Garcon, J.P. Senateur, F. Weiss, P. Dalmas, P. Vuillet, A. Yaouanc: Booklet of publications on hight-Tc superconductors, Laboratoires de Physique et de Chimie de Grenoble (May 1988)

4.42 P. Bordet, J.L. Hodeau, P. Strobel, M. Marezio, A. Santoro, C. Chaillout, J.J. Capponni: Physica C **153-155**, 582 (1988)

4.43 B. Raveau, C. Michel, M. Hervieu: In *Novel Superconductivity*, ed. by S.A. Wolf, V. Kresin (Plenum, New York 1987) p.599

4.44 B. Raveau, C. Michel, M. Hervieu: In *Chemistry of Oxide Superconductors*, ed. by C.N.R. Rao, IUPAC (Blackwell, Oxford 1988) p.15

4.45 M. Veit, J. Langen, M. Galffy, H.D. Josarnt, A. Erle, S. Blumenroder, H. Schmidt, E. Zierngiebl, G. Guntherodt: Physica C **153-155**, 900 (1988)

4.46 G. Ferey, A. Lebail, Y. Lalignant, M. Hervieu, B. Raveau, A. Sulpice, R. Tournier: J. Solid State Chem. **73**, 619 (1988)

4.47 S.B. Oseroff, D.C. Vier, J.F. Smyth, C.T. Salling, S. Schultz, Y. Dalichaouch, B.W. Lef, M.B. Maple, Z. Fisk, J.O. Thompson, J.L. Smith, E. Zierngiebl: In *Novel Superconductivity*, ed. by S.A. Wolf, V. Kresin (Plenum, New York 1987) p.679

4.48 R. Survanaravanan, O. Gorochov, M. Rateau, H. Pankovska: Physica C **153-155**, 874 (1988)

4.49 A. Matsushita, H. Aoki, Y. Asada, T. Hatano, K. Kimura, T. Matsumoto, K. Nakamura, K. Ogawa: Physica B&C **148**, 342 (1987)

4.50 T. Siegrist, L.F. Schneemeyer, J.V. Waszczak, N.P. Singh, R.L. Opila, B. Battlogg, L.N. Rupp, D.W. Murphy: Phys. Rev. B **36**, 8365 (1987)

4.51 A.K. Rajarajan, P. Guptasarma, V.R. Palkar, M.S. Multani, L.G. Guppa, R. Vijayaraghavan: Physica C **153-155**, 894 (1988)

4.52 I. Felner, M. Kowitt, Y. Lehavi, L. Ben-Dor, Y. Wolfus, R. Barbara, I. Nowik: Physica C **153-155**, 898 (1988)

4.53 A. Gupta, H.F. Braun: Physica C **153-155**, 904 (1988)

4.54 Y. Saito, J. Noji, A. Endo, H. Higuchi, K. Fujimojo, T. Oikawa, A. Hattori, K. Furuse: Physica B **148**, 336 (1987)

4.55 H.M. Sung, J.H. Kung, J.M. Liang, R.S. Liu, Y.C. Chen, P.J. Wu, L.J. Chen: Physica C **153-155**, 868 (1988)

4.56 S.G. Eriksson, L.G. Johansson, C. Olsson: Physica C **153-155**, 902 (1988)

4.57 D. Morin, J. Schneck, L. Pierre, J. Primot, J.C. Toledano, F. Glas, J. Etrillard: Physica C **153-155**, 932 (1988)

4.58 C. Perrin, O. Pena, M. Sergent, P. Christensen, G. Fonteneau, J. Lucas: Physica C **153-155**, 934 (1988)

4.59 R. Herrmann, N. Kubicki, H.U. Muller, H. Dwelk, N. Pruss, A. Krapk, L. Rothkirch: Physica C **153-155**, 938 (1988)

4.60 B. Chevalier, J. Etourneau, G. Demazeau: Private communication

Chapter 5

5.1 B. Raveau: Workshop on Superconductivity, Bordeaux (1987)

5.2 B. Aurivillius: Arkiv Kemi **1**, 463 (1949)

5.3 J. Labbé, J. Bok: Europhys. Lett. **3**, 1225 (1987)

5.4 J. Friedel: J. Physique **48**, 1787 (1987)

5.5 C. Michel, M. Hervieu, M.M. Borel, A. Grandin, F. Deslandes, J. Provost, B. Raveau: Z. Phys. B **68**, 421 (1987)

5.6   H. Maeda, J. Tanaka, M. Fukutomi, T. Asano: Jpn. J. Appl. Phys. 27, L209, L548 (1988)
5.7   J.M. Tarascon, Y. Lepage, P. Barboux, B.G. Bagley, L.M. Greene, W.R. McKinnon, G.W. Hull, M.G. Giroud, B.W. Hwang: Phys. Rev. B 37, 9382 (1988)
5.8   M.A. Subramanian, C.C. Torardi, J.C. Calabrese, J. Gopalakrishnan, K.J. Morrissey, T.R. Askew, R.B. Flippen, U. Chowdry, A.N. Sleight: Science 239, 1015 (1988)
5.9   M. Hervieu, B. Domenges, C. Michel, B. Raveau: Mod. Phys. Lett. 2, 835 (1988)
5.10 M. Hervieu, C. Michel, B. Domenges, Y. Laligant, A. Lebail, G. Ferey, B. Raveau: Mod. Phys. Lett. 2, 491 (1988)
5.11 P. Bordet, J.J. Capponi, C. Chaillout, J. Chenavas, A.W. Hewat, E.A. Hewat, J.L. Hodeau, M. Marezio, J. Tholence, D. Tranqui: Physica C 153-155, 623 (1988)
5.12 H.G. von Schnering, L. Walz, M. Schwarz, W. Becker, M. Hartweg, T. Popp, B. Hettich, P. Muller, G. Kampf: Angew. Chem. Int'l Ed. Engl. 27, 574 (1988)
5.13 T. Kajitani, K. Kusuba, M. Kikuchi, N. Kobayashi, Y. Syono, T. Williams, M. Irabayashi: Jpn. J. Appl. Phys. 27, L587 (1988)
5.14 S.A. Sunshine, T. Siegrist, L.F. Schneemeyer, D.W. Murphy, R.J. Cava, B. Batlogg, R.B. Van Dover, R.M. Fleming, S.H. Glarum, S. Nakahara, R. Farrow, J.J. Krajewski, S.M. Zahurak, J.V. Waszcazak, J.H. Marshal, P. March, L.W. Rupp, W.F. Peck: Phys. Rev. B 38, 893 (1988)
5.15 J.M. Tarascon, W.R. McKinnon, P. Barboux, D.M. Hwang, B.G. Bagley, L.H. Greene, G. Hull, Y. Lepage, N. Stoffel, M. Giroud: Phys. Rev. B 38, 8885 (1988)
5.16 N. Kijima, H. Endo, J. Tsuchiya, A. Sumiyama, M. Mizuno, Y. Oguri: Jpn. J. Appl. Phys. 27, L821 (1988)
5.17 C.C. Torardi, M.A. Subramanian, J.C. Calabrese, J. Gopalakrishnan, K.J. Morrissey, T.R. Askew, R.B. Flippen, U. Chowdry, A.W. Sleight: Science 240, 631 (1988)
5.18 H.W. Zandbergen, Y.K. Huang, M.J.V. Menken, J.N. Li, K. Kadowaki, A.A. Menovsky, G. Van Tendeloo, S. Amelinckx: Nature 332, 620 (1988)
5.19 U. Endo, S. Koyama, T. Kawai: Jpn. J. Appl. Phys. 27, L1476 (1988)
5.20 N. Mizumo, H. Endo, J. Tshuchiya, N. Kijima, A. Sumiyama, Y. Oguri: Jpn. J. Appl. Phys. 27, L1225 (1988)
5.21 S. Koyama, U. Endo, T. Kawai: Jpn. J. Appl. Phys. 27, L1861 (1988)
5.22 P. Lejay, P. de Rango, A. Sulpice, B. Giordanengo, R. Tournier, R. Retoux, F. Deslandes, C. Michel, M. Hervieu, B. Raveau: Rev. Phys. Appl. 211, 485 (1989)
5.23 R. Retoux, V. Caignaert, J. Provost, C. Michel, M. Hervieu, B. Raveau: J. Solid State Chem. 79, 157 (1989)
5.24 C.C. Torardi, M.A. Subramanian, J.C. Calabrese, J. Gopalakrishnan, E.M. McCarron, K.J. Morrissey, T.R. Askew, R.B. Flippen, U. Chowdry, A.W. Sleight: Phys. Rev. B 38, 225 (1988)
5.25 A.K. Cheetham, A.M. Chippindale, S.J. Hibble: Nature 333, 21 (1988)
5.26 R. Retoux, F. Studer, C. Michel, B. Raveau, A. Fontaine, E. Dartyge: Phys. Rev. B 41, 193 (1990)
5.27 Z.Z. Sheng, A.M. Hermann, E. El. Ali, C. Almasan, J. Estrada, T. Datta, R.J. Matson: Phys. Rev. Lett. 60, 937 (1988)
5.28 Z.Z.Sheng, A.M. Hermann: Nature 332, 55 (1988)
5.29 Z.Z. Sheng, A.M. Hermann: Nature 332, 138 (1988)
5.30 Z.Z. Sheng, D. Kiel, J. Bennett, A. El Ali, D. Marsh, G.D. Mooney, F. Arammash, J. Smith, D. Viar, A.M. Hermann: Appl. Phys. Lett. 52, 1738 (1988)
5.31 R.M. Hazen, L.W. Finger, R.J. Angel, C.T. Prewitt, N.L. Ross, C.G. Hadidiacos, P.J. Heaney, D.R. Veblen, Z. Sheng, A. El Ali, A.M. Hermann: Phys. Rev. Lett. 60, 1657 (1988)

5.32 S. Parkin, V.Y. Lee, E.M. Engler, A.I. Nazzal, T.C. Huang, G. Gorman, R. Savoy, R. Beyers: Phys. Rev. Lett. **60**, 2539 (1988)

5.33 C. Politis, H. Luo: Mod. Phys. Lett. B **20**, 793 (1988)

5.34 A. Maignan, C. Michel, M. Hervieu, C. Martin, D. Groult, B. Raveau: Mod. Phys. Lett. B **2**, 681 (1988)

5.35 M.A. Subramanian, J.C. Calabrese, C.C. Torardi, J. Gopalakrisnan, T.R. Askew, R.B. Flippen, K.J. Morrissey, O. Chowdry, A.W. Sleight: Nature **332**, 420 (1988)

5.36 M. Hervieu, C. Michel, A. Maignan, C. Martin, B. Raveau: J. Solid State Chem. **74**, 428 (1988)

5.37 C. Martin, C. Michel, A. Maignan, M. Hervieu, B. Raveau: C. R. Acad. Sci. **307**, Ser.II, 27 (1988)

5.38 S. Parkin, V.Y. Lee, A.I. Nazzal, R. Savoy, R. Beyers, S. Laplaca: Phys. Rev. Lett. **61**, 750 (1988)

5.39 A. Sulpice, B. Giordanengo, R. Tournier, M. Hervieu, A. Maignan, C. Martin, C. Michel, J. Provost: Physica C **156**, 243 (1988)

5.40 B. Domenges, M. Hervieu, B. Raveau: Solid State Commun. **69**, 1085 (1989)

5.41 M. Hervieu, A. Maignan, C. Martin, C. Michel, J. Provost, B. Raveau: J. Solid State Chem. **75**, 212 (1988)

5.42 S. Parkin, V.Y. Lee, A.I. Nazzal, R. Savoy, T.C. Huang, G. Gurman, R. Beyers: Phys. Rev. B **38**, 6531 (1988)

5.43 C. Martin, D. Bourgault, C. Michel, M. Hervieu, B. Raveau: Mod. Phys. Lett. B **3**, 93 (1989)

5.44 M. Hervieu, A. Maignan, C. Martin, C. Michel, J. Provost, B. Raveau: Mod. Phys. Lett. **2**, 1103 (1988)

5.45 J. Bok: Solid State Commun. **67**, 251 (1988)

5.46 H. Ihara, R. Sugise, M. Hirabayashi, N. Terada, M. Jo, K. Hayashi, A. Negishi, M. Tokumoto, Y. Kimura, T. Shimomura: Nature **334**, 511 (1988)

5.47 R. Sugise, M. Hirabayashi, N. Terada, M. Jo, T. Shimomura, H. Ihara: Physica C **157**, 131 (1989)

5.48 S. Nakajima, M. Kikuchi, Y. Syono, T. Oku, D. Shindo, K. Hiraga, N. Kobayashi, H. Iwasaki, Y. Muto: Physica C **158**, 471 (1989)

5.49 Z.Z. Sheng, A.M. Hermann, D.C. Vier, S. Schultz, S.B. Oseroff, D.J. George, R.M. Hazen: Phys. Rev. B **38**, 7074 (1988)

5.50 M.A. Subramanian, T.R. Askew, R.B. Flippen, A.W. Sleight: Science **242**, 249 (1988)

5.51 C. Martin, J. Provost, D. Bourgault, B. Domenges, C. Michel, M. Hervieu, B. Raveau: Physica C **157**, 460 (1989)

5.52 A.K. Ganuli, R. Nagarajan, K.S. Nanjun Daswamy, C.N.R. Rao: Mater. Res. Bull. **24**, 103 (1989)

5.53 S. Nakajima, M. Kikuchi, N. Kobayashi, H. Iwasaki, D. Shindo, Y. Syono, Y. Mato: 2nd Int'l Symp. on Superconductivity, Tskuba, Ibaraki, Japan, Nov. 1989

5.54 C.N.R. Rao, A.K. Ganguli, R. Vijayaraghavan: Phys. Rev. B **40**, 2565 (1989)

5.55 R. Vijayaraghavan, A.K. Ganguli, N.Y. Vasanthacharya, M.K. Rajumon, G.V. Kulkarni, G. Sankar, D. Sarma, A.K. Sood, N. Chandrabhas, C.N.R. Rao: Supercond. Sci. Technol. **2**, 195 (1989)

5.56 S. Li, M. Greenblatt: Physica C, **157**, 365 (1989)

5.57 Y.T. Huang, R.S. Liu, N.W. Wang, P.T. Wu: Jpn. J. Appl. Phys. **28**, L 1514 (1989)

5.58 E.A. Hayri, M. Greenblatt: Physica C **156**, 775 (1988)

5.59 D. Bourgault, C. Martin, C. Michel, M. Hervieu, I. Provost, B. Raveau: J. Solid State Chem. **78**, 326 (1989)

5.60 C. Martin, D. Bourgault, C. Michel, J. Provost, M. Hervieu, B. Raveau: Eur. J. Inorg. Solid State Chem. **26**, 1 (1989)

5.61 M. Greenblatt: Solid State Chem. Gordon Conf., Plymouth (July 1990)

5.62 D. Bourgault, C. Martin, C. Michel, M. Hervieu, B. Raveau: Physica C **158**, 511 (1989)

5.63 B. Morosin, R.J. Banghman, D.S. Ginley, J.E. Schirberand, E.L. Venturini: Physica C **161**, 115 (1990)

5.64 S. Kondoh, Y. Ando, M. Onoda, M. Sato, A Kimitsu: Solid State Commun. **65**, 1329 (1988)

5.65 Y. Kubo, Y. Shimikawa, T. Manako, T. Sato, S. Ljima, T. Ichihashi, H. Igarashi: Physica C **162-164**, 991 (1989)

5.66 K.V. Ramanujachary, S. Li, M. Greenblat: Physica C **165**, 377 (1990)

5.67 I.K. Gopalakrishnan, P. Sastry, H. Rajagpal, A. Saqweira, J.V. Yakhimi, R.M. Iyer: Physica C **159**, 811 (1989)

5.68 A. Schilling, H.R. Ott, F. Hulliger: Physica C **157**, 144 (1989)

5.69 C. Martin, A. Maignan, J. Provost, C. Michel, M. Hervieu, R. Tournier, B. Raveau: Physica C **168**, 8 (1990)

5.70 A. Maignan, C. Martin, J. Provost, M. Huvé, C. Michel, M. Hervieu, B. Raveau: Physica C **170**, 350 (1990)

5.71 Y. Shimikawa, Y. Kubo, T. Manako, T. Satoh, S. Ijima, T. Ichiashi, H. Igarashi: Physica C **157**, 279 (1989)

5.72 A. Sulpice, B. Giordanengo, R. Tournier, M. Hervieu, A. Maignan, C. Martin, C. Michel, J. Provost: Physica C **156**, 243 (1988)

5.73 R.J. Cava, B. Battlog, J.J. Krajewski, L.W. Rupp, L.F. Schneemeyer, T. Siegrist, R.B. Van Dover, P. Marsh, W.F. Peck, P.K. Callagher, S.H. Glarum, J.H. Marshall, R.C. Farrow, J.V. Waszczak, R. Hull, P. Trevor: Nature **336**, 211 (1988)

5.74 R. Retoux, C. Michel, M. Hervieu, B. Raveau: Mod. Phys. Lett. B **3**, 591 (1989)

5.75 H.W. Zandbergen, W.T. Fu, J.M. van Ruitenbeek, L.J. de Jongh, G. van Tendeloo, S. Amelinckx: Physica C **159**, 81 (1989)

5.76 T. Rouillon, R. Retoux, D. Groult, C. Michel, M. Hervieu, J. Provost, B. Raveau: J. Solid State Chem. **78**, 323 (1989)
V. Caignaert: Private communication

5.77 A. Tokiwa, M. Nagoshi, T. Oku, N. Kobayashi, M. Kikuchi, K. Hiraga, Y. Syono: Physica C **168**, 285 (1990)

5.78 T. Rouillon, J. Provost, M. Hervieu, D. Groult, C. Michel, B. Raveau: Physica C **159**, 201 (1989)

5.79 T. Rouillon, J. Provost, M. Hervieu, D. Groult, C. Michel, B. Raveau: J. Solid State Chem. **84**, 375 (1990)

5.80 T. Rouillon, A. Maignan, M. Hervieu, D. Groult, C. Michel, B. Raveau: Physica C **171**, 7 (1990)

5.81 C. Martin, D. Bourgault, M. Hervieu, C. Michel, J. Provost, B. Raveau: Mod. Phys. Lett. B **3**, 993 (1989)

5.82 J. Akimitsu, S. Suzuki, M. Watanabe, H. Sawa: Jpn. J. Appl. Phys. 27, L 1859 (1988)

5.83 H. Sawa, S. Suzuki, M. Watanabe, J. Akimitsu, H. Matsubara, H. Watabe, S. Uchida, K. Kokusho, H. Asano, F. Izumi, E. Takayama-Muromachi: Nature, **337**, 347 (1989)

5.84 Y. Tokura, T. Arima, H. Takagi, S. Uchida, T. Ishigaki, H. Asano, R. Beyers, A.I. Nazzal, P. Lacorre, J.B. Torrance: Nature **342**, 890 (1989)

5.85 T. Rouillon, D. Groult, M. Hervieu, C. Michel, B. Raveau: Physica C **167**, 107 (1990)

5.86 H. Sawa, K. Obara, J. Akimitsu, Y. Matsui, S. Horiuchi: J. Phys. Soc. Jpn. **58**, 2252 (1989)

5.87 T. Wada, A. Ichinose, Y. Yaegashi, H. Yamauchi, S. Tanaka: Phys. Rev. B **41**, 1984 (1990)

5.88 T. Siegrist, S.M. Zahurak, D.W. Murphy, R.S. Roth: Nature **334**, 231 (1988)

5.89 L. Er-Rakho, C. Michel, Ph. Lacorre, B. Raveau: J. Solid State Chem. 73, 531 (1988)

Chapter 6

6.1 J.C.M. Spence: *Experimental High Resolution Electron Microscopy* (Oxford Univ. Press, New York 1980)
6.2 J.J. Capponi, C. Chaillout, A.W. Hewat, P. Lejay, M. Marezio, N. Nguyen, B. Raveau, J.L. Soubeyroux, J.L. Tholence, R. Tournier: Europhys. Lett. 12, 1301 (1987)
6.3 M. Hervieu, B. Domenges, C. Michel, J. Provost, B. Raveau: J. Solid State Chem. 71, 263 (1987)
6.4 E.A. Hewat, M. Dupuy, A. Bourret, J.J. Capponi, M. Marezio: Nature 327, 4 (1987)
6.5 A. Ourmazd, J.C.H. Spence: Nature 329, 425 (1987)
6.6 Y. Muto, N. Kobayashi, Y. Syono: In *Novel Superconductivity*, ed. by A. Wolf, V. Kresin (Plenum, New York 1988) p.787
6.7 N.P. Huxford, D.J. Eaglesham, C.J. Humphreys: Nature 329, 812 (1987)
6.8 D. Tang, W. Zhou, J.M. Thomas: Angew. Chem. 26, 1048 (1987)
6.9 M. Hervieu, B. Domenges, C. Michel, G. Heger, J. Provost, B. Raveau: Phys. Rev. Lett. B 36, 3920 (1987)
6.10 P. Bordet, J.J. Capponi, C. Chaillout, J.L. Hodeau, M. Marezio: In a special issue of *Progress in High T_c Superconductivity*, ed. by C.N.R. Rao (World Scientific, Singapore 1988)
6.11 H.W. Zandbergen, C.J.D. Hetherington, R. Gronsky: J. Supercond. 1, 21 (1988)
6.12 G. Van Tendeloo, H.W. Zandbergen, S. Amelinckx: Solid State Commun. 63, 603 (1987)
6.13 S. Iigima, T. Ichihashi, Y. Kubo, J. Tabuchi: Jpn. J. Appl. Phys. 26, 1790 (1988)
6.14 S. Iijima, T. Ichihashi, Y. Kubo, J. Tabuchi: Jpn. J. Appl. Phys. 26, 1478 (1988)
6.15 H.W. Zandbergen, G. Van Tendeloo, T. Okabe, S. Amelinckx: Phys. Status Solidi (A) 103, 45 (1987)
6.16 B. Raveau, M. Hervieu, C. Michel, J. Provost: Mod. Phys. B 1, 733 (1987)
6.17 G. Roth, D. Ewert, G. Heger, M. Hervieu, C. Michel, B. Raveau, F. D'Yvoire, A. Revcolevschi: Z. Phys. B 69, 21 (1987)
6.18 B. Raveau, C. Michel, M. Hervieu: In *Chemistry of High Temperature Superconductors*, ed. by L. Nelson, ACS Symposium Series, Vol.351 (American Chemical Society, Washington, DC 1987)
6.19 J.Y. Henry, P. Burlet, A. Bourret, G. Roult, P. Bacher, M.J.G.M. Jurgens, J. Rossat-Mignod: Solid State Commun. 64, 1037 (1987)
6.20 E.A. Hewat, M. Dupuy, A. Bourret, J.J. Capponi, M. Marezio: Solid State Commun. 64, 517 (1987)
6.21 G. Van Tendeloo, H.W. Zandbergen, S. Amelinckx: Solid State Commun. 63, 389 (1987)
6.22 C.J. Jou, J. Washburn: J. Mater. Res. 4, 795 (1989)
6.23 G. Van Tendeloo, D. Broddin, M.W. Zandbergen, S. Amelinckx: Physica C 167, 627 (1990)
6.24 V. Caignaert, M. Hervieu, J. Wang, G. Desgardin, B. Raveau, F. Boterel, J.M. Haussonne: Physica C 170, 1, 139 (1990)
6.25 M. Hervieu, B. Domenges, C. Michel, B. Raveau: Europhys. Lett. 4, 205 (1987)
6.26 B. Raveau, C. Michel, M. Hervieu: In Proc. Int'l Discussion Meeting, Mauterndorf, Austria, February 1988 (Plenum, New York 1988)
6.27 B. Domenges, M. Hervieu, C. Michel, B. Raveau: Europhys. Lett. 4, 211 (1987)
6.28 G. Van Tendeloo, H.W. Zandbergen, T. Okabe, S. Amelinckx: Solid State Commun. 63, 969 (1987)

6.29 H.W. Zandbergen, R. Gronsky, G. Thomas: Phys. Status Solidi A **105**, 207 (1988)

6.30 P. Bordet, Ch. Chaillout, J. Chenavas, J.L. Hodeau, M. Marezio, J. Karpinski, E. Kaldis: Nature **334**, 596 (1988)

6.31 C. Chaillout, P. Bordet, J. Chenavas, J.L. Hodeau, M. Marezio: Solid State Commun. **70**, 275 (1988)

6.32 M. Hervieu, B. Domenges, B. Raveau, J.M. Tarascon, M. Post, W.R. McKinnon: Mater. Chem. Phys. **21**, 181 (1989)

6.33 E. Bouteloup, M. Hervieu, B. Mercey, H. Murray, G. Poullain, B. Raveau, T. Rouillon: J. Cryst. Growth **91**, 418 (1988)

6.34 C. Michel, B. Raveau, M. Hervieu: "High T_c superconductive ceramics" SEE (September 1988)

6.35 M. Hervieu, B. Domenges, B. Raveau, M. Post, M.W. McKinnon, J.M. Tarascon: Mater. Lett. **8**, 73 (1989)

6.36 G. Roth, B. Renker, G. Heger, M. Hervieu, B. Domenges, B. Raveau: Z. Phys. B **69**, 53 (1987)

6.37 B. Domenges, M. Hervieu, V. Caignaert, B. Raveau: J. Microsc. Spectrosc. Electron. **13**, 75 (1988)

6.38 M.W. Zandbergen, R. Gronsky, M.Y. Chy, L.C. Dejonghe, G.F. Holland, A. Stacy: In Proc. of the MRS Fall Meeting, Boston, MA (1987)

6.39 R.A. Camps, J.E. Evelts, B.A. Glowacki, S.B. Newcomb, R.E. Somekh, W.M. Stobbs: High T_c Superconductors and Potential Applications (1987) p.325

6.40 T. Iichihashi, S. Iijima, Y. Kubo, Y. Shimakawa, J. Tabuchi: Jpn. J. Appl. Phys. **27**, L594 (1988)

6.41 C. Chaillout, M.A. Alario Franco, J.J. Capponi, J. Chenavas, J.L. Hodeau, M. Marezio: Phys. Rev. B **36**, 7118 (1987)

6.42 M.A. Alario Franco, C. Chaillout, J.J. Capponi, J. Chenavas: Mater. Res. Bull. **22**, 1685 (1988)

6.43 C. Chaillout, M.A. Alario Franco, J.J. Capponi, J. Chenavas, P. Strobel, M. Marezio: Solid State Commun. **65**, 283 (1988)

6.44 M. Hervieu, B. Domenges, B. Raveau: Chem. Scr. **25**, 361 (1985)

6.45 J. Reyes-Gasga, T. Krekels, G. Van Tendeloo, J. Van Landuyt, W.H.M. Bruggink, H. Verweij, S. Amelinckx: Solid State Commun. **70**. 269 (1989)

6.46 T. Krekels, T.S. Shi, J. Reyes-Gasga, G. Van Tendeloo, J. Van Landuyt, S. Amelinckx: Physica C **167**, 677 (1990)

6.47 M. Hervieu, B. Domenges, J. Provost, F. Deslandes, B. Raveau, B. Chevalier, J. Etourneau: Angew. Chem. **27**, 440 (1988)

6.48 M. Hervieu, B. Domenges, F. Deslandes, C. Michel, B. Raveau: C.R. Acad. Sci. **307**, Ser. II, 1441 (1988)

6.49 W.I.F. David, W.T.A. Harrisson, M.R. Ibberson, M.T. Weller, J.R. Grosmeder, P. Lanchester: Nature **328**, 328 (1987)

6.50 B. Domenges, M. Hervieu, C. Michel, A. Maignan, B. Raveau: Phys. Status Solidi A **107**, 73 (1988)

6.51 C. Michel, L. Er Rakho, M. Hervieu, J. Pannetier, B. Raveau: J. Solid State Chem. **68**, 143 (1987)

6.52 V. Caignaert, M. Hervieu, N. Nguyen, B. Raveau: J. Solid State. Chem. **62**, 281 (1988)

6.53 V. Caignaert, M. Hervieu, B. Raveau: Mater. Res. Bull. **21**, 1147 (1986)

6.54 A. Hussain, L. Kihlborg: Acta Crystallogr. A **32**, 551 (1976)

6.55 A. Ramanan, J. Gopalakrishnan, M.K. Uppal, D.A. Jefferson, C.N.R. Rao: Proc. R. Soc. London A **395**, 127 (1984)

6.56 D.A. Jefferson, M.K. Uppal, D.J. Smith: J. Solid. State Chem. **53**, 101, (1984)

6.57 J.L. Hutchinson, G.R. Anstis, R.J.D. Tilley: Int'l Cong. on E.M., Hamburg (1982)
6.58 M. Hervieu, B. Raveau: J. Solid State Chem. **32**, 299 (1982)
6.59 M. Hervieu, B. Raveau: Chem. Scr. **22**, 117 (1983)
6.60 B. Domenges, M. Hervieu, R.J.D. Tilley, B. Raveau: J. Solid State Chem. **54**, 10 (1984)
6.61 M. Hervieu, B. Domonges, B. Raveau: J. Solid State Chem. **58**, 233 (1985)
6.62 J.L. Hutchison, A.J. Jacobson: Acta Crystallogr. B **31**, 1442 (1975)
6.63 J.L. Hutchison, A.J. Jacobson: J. Solid State Chem. **20**, 417 (1977)
6.64 V. Caignaert, M. Hervieu, B. Domenges, N. Nguyen, J. Pannetier, B. Raveau: J. Solid State Chem. **73**, 107 (1988)
6.65 C. Michel, M. Hervieu, M.M. Borel, A. Grandin, F. Deslandes, J. Provost, B. Raveau: Z. Phys. B **68**, 421 (1987)
6.66 G. Van Tendeloo, H.W. Zandbergen, S.A. Amelinckx: Solid State Commun. **66**, 927 (1988)
6.67 M. Hervieu, C. Michel, B. Domenges, Y. Laligant, A. Lebail, G. Ferey, B. Raveau: Mod. Phys. Lett. **2**, 491 (1988)
6.68 M. Hervieu, B. Domenges, C. Michel, B. Raveau: Mod. Phys. Lett. **2**, 835 (1988)
6.69 G. Van Tendeloo, H.W. Zandbergen, J. Van Landuy, S. Amelinckx: Appl. Phys. A **46**, 233 (1988)
6.70 E. Kakayama, E. Muromachi, Y. Uchida, A. Ono, F. Izumi, M. Onoda: Jpn. J. Appl. Phys. **27**, L365 (1988)
6.71 T. Kijima, J. Tanaka, Y. Bando, M. Onoda, F. Izumi: Jpn. J. Appl. Phys. **27**, L369 (1988)
6.72 P.L. Gai, P. Day: Physica C **152**, 335 (1988)
6.73 B. Raveau, C. Michel, M. Hervieu, J. Provost: Physica C **153–155**, 3 (1988)
6.74 E.A. Hewat, P. Bordet, J.J. Capponi, C. Chaillout, J.L. Hodeau, M. Marezio: Physica C **153–155**, 619 (1988)
6.75 E. Hewat: Physica C **152**, 413 (1988)
6.76 M. Hervieu, C. Michel, N. Nguyen, R. Retoux, B. Raveau: Eur. J. Solid State Inorg. Chem. **25**, 1 (1988)
6.77 Y. Matsui, H. Maeda, Y. Tanaka, S. Horiuchi: Jpn. J. Appl. Phys. **27**, L372 (1988)
6.78 R. Retoux, C. Michel, M. Hervieu, N. Nguyen, B. Raveau: Solid State Commun. **69**, 599 (1989)
6.79 H.W. Zandbergen, W.A. Groen, G. Van Tendeloo, S. Amelinckx: Appl. Phys. A **48**, 305 (1989)
6.80 S. Olivier, W.A. Groen, C. Van Der Beek, H.W. Zandbergen: Physica C **157**, 531 (1989)
6.81 C.C. Torardi, J.P. Parise, M.A. Subramanian, J. Gopalakrishnan, A.W. Sleight: Physica C **157**, 115 (1989)
6.82 R. Retoux, V. Caignaert, J. Provost, C. Michel, M. Hervieu, B. Raveau: J. Solid State Chem. **79**, 157 (1989)
6.83 J.M. Tarascon, W.R. McKinnon, P. Barboux, D.M. Hwang, B.G. Bagley, L.H. Greene, G. Hull, Y. Lepage, N. Stoffel, M. Giroud: Phys. Rev. B **37**, 9382 (1988)
6.84 H.W. Zandbergen, P. Groen, G. Van Tendeloo, J. Van Landuyt, S. Amelinckx: Solid. State Commun. **66**, 397 (1988)
6.85 S. Pekker, J. Sasvari, Gy. Hutiray, L. Mihaly: J. Less-Common Met. **150**, 277 (1989)
6.86 Y. Oka, N. Yamamoto, H. Kitaguchi, K. Oda, J. Takada: Jpn. J. Appl. Phys., Lett. **28**, L 213 (1989)
6.87 M. Nobumasa, K. Shimizu, Y. Kitano, T. Kawai: Jpn. J. Appl. Phys., Lett. **27**, L 846 (1989)

6.88 N. Murayama, T. Sudo, M. Awano, K. Kani, Y. Torü: Jpn. J. Appl. Phys. Lett. 27, L 1269 (1989)
6.89 J.S. Luo, F. Faudot, J.P. Chevalier, R. Portier, D. Michel: J. Solid State Chem., in press
6.90 U. Endo, S. Koyama, T. Kawai: Jpn. J. Appl. Phys. 27, 1476 (1988)
6.91 A. Maignan, C. Michel, M. Hervieu, C. Martin, D. Groult, B. Raveau: Mod. Phys. Lett. B 2, 681 (1988)
6.92 M. Hervieu, C. Michel, A. Maignan, C. Martin, B. Raveau: J. Solid State Chem. 74, 428 (1988)
6.93 M. Hervieu, C. Martin, J. Provost, B. Raveau: J. Solid State Chem. 75, 212 (1988)
6.94 M. Hervieu, B. Domenges, B. Raveau: J. Microsc. Spectrosc. Electron. 13, 279 (1988)
6.95 M. Hervieu, A. Maignan, C. Martin, C. Michel, J. Provost, B. Raveau: Mod. Phys. Lett. 2, 1103 (1988)
6.96 H.W. Zandbergen, G. Van Tendeloo, J. Van Landuyt, S. Amelinckx: Physica C 156, 325 (1988)
6.97 S. Iijima, T. Ichichashi, Y. Skimakawa, T. Manako, Y. Kubo: Jpn. J. Appl. Phys. 27, L1054 (1988)
6.98 L. Kihlborg: Chem. Scr. 14, 187 (1978)
6.99 B. Domenges, M. Hervieu, C. Martin, D. Bourgault, C. Michel, B. Raveau: Phase Trans. 19, 231 (1989)
6.100 B. Domenges, M. Hervieu, B. Raveau: Solid State Commun. 69, 1085 (1989)
6.101 B. Morosin, D.S. Ginley, P.F. Lava, M.J. Karr, R.J. Baughmann, J.E. Shirber, E.L. Venturini, J.F. Kwak: Physica C 152, 413 (1988)
6.102 L. Er-Rakho, C. Michel, Ph. Lacorre, B. Raveau: J. Solid State Chem. 73, 531 (1988)
6.103 H.K. Müller-Buschbaum, W. Wollschlager: Z. Anorg. Allgem. Chem. 414, 76 (1975)
6.104 C. Martin, D. Bourgault, M. Hervieu, C. Michel, J. Provost, B. Raveau: Mod. Phys. Lett. 3, 93 (1989)
6.105 T. Rouillon, J. Provost, M. Hervieu, D. Groult, C. Michel, B. Raveau: Physica C 159, 201 (1989)
6.106 T. Rouillon, J. Provost, M. Hervieu, D. Groult, C. Michel, B. Raveau: J. Solid State Chem. 84, 375 (1990)
6.107 T. Rouillon, A. Maignan, M. Hervieu, C. Michel, D. Groult, B. Raveau: Physica C 171, 7 (1990)
6.108 C. Martin, J. Provost, D. Bourgault, B. Domenges, C. Michel, M. Hervieu, B. Raveau: Physica C 157, 460 (1989)
6.109 M.A. Subramanian, C.C. Torardi, J. Gopalakrishnan, P.L. Gai, J.C. Calabrese, T.R. Askew, R.B. Flipper, A.W. Sleight: Science 242, 249 (1988)
6.110 R.J. Cava, B. Batlogg, J.J. Krajewski, L.W. Rupp, L.F. Schneemeyer, T. Siegrist, R.B. van Dover, P. Marsh, W.F. Peck, P.K. Gallager, S.H. Glarum, J.M. Marshall, R.C. Farrow, J.V. Waszczak, R. Hull, P. Trevor: Nature 336, 211 (1988)
6.111 M. Hervieu, V. Caignaert, R. Retoux, B. Raveau: Mater. Sci. Eng. 6, 211 (1990)
6.112 T. Rouillon, R. Retoux, D. Groult, C. Michel, M. Hervieu, J. Provost, B. Raveau: J. Solid State Chem. 78, 322 (1989)
6.113 R. Retoux, C. Michel, M. Hervieu, B. Raveau: Mod. Phys. Lett. B 3, 7, 591 (1989)
6.114 P. Goodman, D.G. Jensen, J.J. White: Physica C 158, 173 (1989)
6.115 R. Hull, J.M. Bonar, L.F. Schneemeyer, R.J. Cava, J.J. Krajewski, J.V. Waszczak: Phys. Rev. B 39, 9685 (1989)
6.116 E.A. Hewat, J.J. Capponi, J.C. Cava, C. Chailloux, M. Marezio, J.L. Tholence: Physica C 157, 509 (1989)

6.117 H.W. Zandbergen, K. Kadokawi, M.J.V. Menken, A.A. Menovsky, G. Van Tendeloo, S. Amelinckx: Physica C **158**, 155 (1989)
6.118 V. Caignaert, R. Retoux, C. Michel, M. Hervieu, B. Raveau: Physica C **167**, 483 (1990)
6.119 M. Hervieu, V. Caignaert, C. Michel, R. Retoux, B. Raveau: Microscopy, Microanal. microstruct. 1, 2 (1990)
6.120 O. Eibl: Physica C **168**, 239, 249 (1990)
6.121 J. Darriet, A. Le Lirzin, E. Marquestaut, B. Lepine, B. Chevalier, J. Etourneau: Solid State. Commun. **69**, 739 (1989)
6.122 A.K. Ganguli, R. Nagarajan, K.S. Nanjrindaswamy, C.N.R. Rao: Mater. Res. Bull. **24**, 103 (1989)
6.123 J.M. Tarascon, P. Barboux, G.W. Hull, R. Ramesh, L.H. Greene, M. Giroud, M.S. Hedge, W.R. McKinnon: Phys. Rev. B **39**, 4316 (1989)
6.124 S. Olivier, W.A. Groen, C. Van der Beek, H.W. Zandbergen: Physica C **157**, 531 (1989)
6.125 E. Hewat: J. Microsc. Spectrosc. Electron. **13**, 297 (1988)
6.126 J.L. Hodeau, P. Bordet, J.J. Capponi, C. Chaillout, J. Chenevas, M. Godinho, A.W. Hewat, E.A. Hewat, H. Renevier, A.M. Spreser, P. Strobel, J.L. Tholence, M. Marezio: *Proc. 1st AS Pacific Conference on HT$_C$S*, 1988 (World Scientific, Singapore 1989)
6.127 R. Retoux: Thesis, University of Caen (1990)
6.128 J.M. Tarascon, Y. Lepage, W.R. McKinnon, R. Ramesh, M. Eibschutz, E. Tselepis, E. Wang, G.W. Hull: Physica C **167**, 20 (1990)
6.129 H.W. Zandbergen, G. Van Tendeloo, J. Van Landuyt, S. Amelinckx: Appl. Phys. A **46**, 233 (1988)
6.130 H.W. Zandbergen, W.A. Groen, F.C. Mijlhoff, G. Van Tendeloo, S. Amelinckx: Physica. C **156**, 325 (1988)
6.131 D. Bourgault, C. Martin, C. Michel, M. Hervieu, B. Raveau: J. Solid State Chem. **78**, 326 (1989)
6.132 J.P. Zhang, D.J. Li, H. Shibahara, L.D. Marks: Supercond. Sci. Technol. 1, 132 (1988)
6.133 C.C. Torardi, M.A. Subramanian, J.C. Calabrese, J. Gopalakrishnan, K.J. Morrissey, T.R. Askew, R.B. Flippen, V. Chowdhry, A.W. Sleight: Nature **240**, 631 (1988)
6.134 M. Hervieu, C. Michel, B. Raveau: J. Less-Common Met. **150**, 59 (1989)
6.135 D.M. de Leeuw, W.A. Groen, J.C. Jol, M.B. Brom, H.W. Zandbergen: Physica C **166**, 349 (1990)
6.136 Y. Kubo, Y. Shimakawa, T. Manako, T. Satoh, S. Iijima, T. Ichihashi, M. Igarashi; Physica C **162**, 991 (1990)
6.137 T.C. Huang, V.Y. Lee, R. Karimi, R. Beyers, S.S.P. Parkin: Mater. Res. Bull. **23**, 1307 (1988)
6.138 J.B. Parise, C.C. Torardi, M.A. Subramanian, J. Gopalakrishnan, A.W. Sleight, E. Prince: Physica C **159**, 239 (1989)
6.139 K.V. Ramanuyachary, S. Li, M. Greenblatt: Physica C **165**, 377 (1990)
6.140 D. Bourgault, C. Martin, C. Michel, M. Hervieu, B. Raveau: Physica C **158**, 511 (1989)
6.141 C. Michel, C. Martin, M. Hervieu, D. Groult, D. Bourgault, J. Provost, B. Raveau: Proc. Taiwan Int'l Symp. on Superconductivity (April 1989)
6.142 H.W. Zandbergen, R. Gronsky, G. Thomas: Physica C **153–155**, 1002 (1988)

Chapter 7

7.1 F. Rullier-Albenque: Ann. de Chim. 12, 573 (1987)
Ø. Fischer, M.B. Maple (eds.): *Superconductivity in Ternary Compounds I and II*, Topics Current Phys., Vols.32 and 34 (Springer, Berlin, Heidelberg 1982)
7.2 A.M. Campbell, J. Evetts: Adv. Phys. 21, 199 (1972)
7.3 C. Lehmann: *Interaction of Radiation with Solids*, Defects in Crystalline Solids Series, Vol.10 (North-Holland, Amsterdam 1977)
7.4 H.A. Bethe: Ann. Physik 5, 325 (1930)
7.5 J. Lindhard, M. Scharff: Phys. Rev. 124, 128 (1961)
7.6 J. Lindhard, M. Scharff, H.E. Schiott: Mat.-Fys. Medd. K. Dan. Vidensk. Selsk. 33, no.14 (1963)
7.7 J.P. Biersack: Z. Phys. 211, 495 (1968)
7.8 F. Hubert, A. Fleury, R. Bimbot, D. Gardes: Ann. de Phys. 5, 214 (1980)
7.9 L.C. Northcliffe, R.F. Schilling: Nucl. Data Tables 7, 233 (1970)
7.10 J.F. Ziegler: *Stopping and Ranges of Ions in Matter*, Vol.5 (Pergamon, New York 1980)
7.11 A. Audouard, E. Balanzat, G. Fuchs, J.C. Jousset, D. Lesueur, L. Thome: Europhys. Lett. 3, 327 (1987); ibid. 5, 241 (1988)
7.12 R.L. Fleischer, P.B. Price, R.M. Walker: *Nuclear Tracks in Solids: Principles and Applications* (Univ. California Press, Berkeley 1975)
7.13 E. Dartyge, J.P. Duraud, Y. Langevin, M. Maurette: Phys. Rev. B 23, 5213 (1981)
7.14 E. Dartyge, P. Sigmund: Phys. Rev. B 32, 5249 (1985)
7.15 C. Houpert, M. Hervieu, D. Groult, F. Studer, M. Toulemonde: Nucl. Instrum. Methods B 32, 393 (1988)
7.16 G. Fuchs, F. Studer, E. Balanzat, D. Groult, M. Toulemonde, J.C. Jousset: Europhys. Lett. 3, 321 (1987)
7.17 C. Houpert, F. Studer, D. Groult, M. Toulemonde: Nucl. Instrum. Methods B 39, 720 (1989)
7.18 G.H. Kinchin, R.S. Pease: Rep. Prog. Phys. 18, 1 (1955)
7.19 N.J. Norgett, M.T. Robinson, I.M. Torrens: Nucl. Eng. Des. 33, 50 (1975)
7.20 F. Rullier-Albenque, J. Provost, F. Studer, D. Groult, B. Raveau: Solid State Commun. 66, 417 (1988)
7.21 B. Stritzker, W. Zander, F. Dworschak, U. Poppe, K. Fischer: Mater. Res. Soc. Symp. Proc. 99, 491 (1988)
7.22 N. Moser, A. Hofmann, P. Schüle, R. Henes, H. Kronmüller: Z. Phys. B 71, 37 (1988)
7.23 A. Hofmann, P. Schüle, N. Moser, H. Kronmüller, H. Jager, F. Dworschak: Physica C 153-155, 341 (1988)
7.24 M.A. Kirk, M.C. Baker, J.Z. Lin, D.J. Lam, H.W. Weber: Mater. Res. Soc. Symp. Proc. 99, 209 (1988)
7.25 F. Rullier-Albenque, S. Paidassi, Y. Quéré: J. de Phys. 41, 515 (1980)
7.26 M. Gurvitch, A.K. Gosh, C.L. Snead, M. Strongin: Phys. Rev. Lett. 39, 1102 (1977)
7.27 M. Tokumoto, H. Thara, T. Matsubara, M. Hirabayashi, N. Tereda, A. Oyanagi, K. Murata, Y. Kimura: Jpn. J. Appl. Phys. 26, L1565 (1987)
7.28 K. Kishio, J. Shimoyama, T. Hasegawa, K. Kitazawa, K. Fucki: Jpn. J. Appl. Phys. 26, L1228 (1987)
7.29 P. Monod, M. Ribault, F. d'Yvoire, J. Jegoudez, G. Collin, A. Revcolevschi: J. Physique 48, 1369 (1987)
7.30 J.M. Tarascon, L.H. Greene, B.G. Bagley, W.R. McKinnon, P. Barboux, G.W. Hull: In *Chemistry of High Temperature Superconductors*, ed. by L. Nelson, ACS Symposium Series, Vol.351 (American Chemical Society, Washington, DC 1987) p.198

7.31 R.J. Cava, B. Batlogg, A.P. Ranurez, D. Werder, C.H. Chen, E.A. Rietman, S.M. Zahurak: Mater. Res. Soc. Symp. Proc. **99**, 19 (1988)

7.32 V.F. Zelenskij, I.M. Neklyudov, Yu.T. Petrusenko, A.N. Sleptsov, V.A. Finkel: Physica C **153-155**, 850 (1988)

7.33 S.T. Sekula, D.K. Christen, H.R. Kerchner, J.R. Thompson, L.A. Boatner, B.C. Sales: Jpn. J. Appl. Phys. **26**, 1185 (1987)

7.34 H. Küpfer, L. Apfelstedt, W. Schauer, R. Flükiger, R. Meier-Hirmer, H. Wuhl, H. Scheurer: Z. Phys. B **69**, 167 (1987)

7.35 A. Umezawa, G.W. Crabtree, J.Z. Lin, H.W. Weber, W.K. Kwok: Phys. Rev. B **36**, 7151 (1987)

7.36 J.R. Cost, J.O. Willis, J.D. Thompson, D.E. Peterson: Phys. Rev. B **37**, 1563 (1988)

7.37 J.O. Willis, J.R. Cost, R.D. Brown, J.D. Thompson, D.E. Peterson: Mater. Res. Soc. Symp. Proc. **99**, 391 (1988)

7.38 P. Müller, H. Gerstenberg, M. Fisher, W. Schindler, J. Strobel, G. Saemann-Ischenko, H. Kammermeier: Solid State Commun. **65**, 223 (1988)

7.39 A.R. Sweedler, D.E. Cox, S. Moehlecke: J. Nucl. Mater. **72**, 50 (1978)

7.40 J.B. Gibson, A.N. Goland, M. Milgram, G.H. Vineyard: Phys. Rev. **120**, 1229 (1960)

7.41 H. Küpfer, L. Apfelstedt, W. Schauer, R. Flükiger, R. Meier-Hirmer, H. Wühl: Z. Phys. B **69**, 159 (1987)

7.42 C.P. Bean: Phys. Rev. Lett. **8**, 250 (1962)

7.43 T.R. Dinger, T.K. Worthington, W.J. Gallagher, R.L. Sandstrom: Phys. Rev. Lett. **58**, 2687 (1987)

7.44 P. Ganguly, C.N.R. Rao: Mater. Res. Bull. **8**, 405 (1973)

7.45 J.D. Jorgensen, H.B. Schutter, D.G. Hinks, D.W. Capone, K. Zhang, M.B. Brodsky, D.J. Scalapino: Phys. Rev. Lett. **58**, 1024 (1987)

7.46 W. Kang, G. Collin, M. Ribault, J. Friedel, D. Jerome, J.M. Bassat, J.P. Coutures, Ph. Odier: J. Physique **48**, 1181 (1987)

7.47 J. Beille, B. Chevalier, G. Demazeau, F. Deslandes, J. Etourneau, O. Laborde, C. Michel, P. Lejay, J. Provost, B. Raveau, A. Sulpice, J.L. Tholence, R. Tournier: Physica B **146**, 307 (1987)

7.48 J. Beille, R. Cabanel, C. Chaillout, B. Chevalier, G. Demazeau, F. Deslandes, J. Etourneau, P. Lejay, C. Michel, J. Provost, B. Raveau, A. Sulpice, J.L. Tholence, R. Tournier: C.R. Acad. Sci. **304**, 1097 (1987)

7.49 P.M. Grant, S.S.P. Parkin, V.Y. Lee, E.M. Engler, M.L. Ramirez, J.E. Vazquez, G. Lim, R.D. Jacowitz, R.L. Greene: Phys. Rev. Lett. **58**, 2482 (1987)

7.50 D. Vaknin, S.K. Sinha, D.E. Moneton, D.G. Johnston, J. Newsam, C.R. Safinya, H. King: Phys. Rev. Lett. **58**, 2802 (1987)

7.51 T. Freltoft, J.E. Fischer, G. Shirane, D.E. Moneton, S.K. Sinha, D. Vaknin, J.P. Remeika, A.S. Cooper, D. Harshman: Phys. Rev. B **36**, 826 (1987)

7.52 J.I. Budnick, A. Golnik, Ch. Niedermayer, E. Recknagel, M. Rossmanith, A. Weidinger, B. Chamberland, M. Filipkowski, D.P. Yang: Phys. Lett. **124**, 103 (1987)

7.53 D. Groult, J. Provost, B. Raveau, F. Studer, S. Bouffard, J.C. Jousset, S.J. Lewandowski, M. Toulemonde, F. Rullier-Albenque: Europhys. Lett. **6**, 151 (1988)

7.54 L. Forro, F. Beuneu: Solid State Commun. B **44**, 623 (1982)

7.55 H. Mutka: Phys. Rev. B **28**, 2855 (1983)

7.56 M. Toulemonde, G. Fuchs, N. Nguyen, F. Studer, D. Groult: Phys. Rev. B **35**, 6560 (1987)

7.57 D. Bourgault, D. Groult, S. Bouffard, J. Provost, F. Studer, N. Nguyen, B. Raveau, M. Toulemonde: Phys. Rev. B **39**, 6549 (1989); Rev. Phys. Appl. **24**, 507 (1989)

7.58 R.J. Cava, B. Batlogg, P. Ramirez, D. Werder, C.H. Chen, E.A. Rietmann, S.M. Sahurak: Mater. Res. Soc. Symp. Proc. 99, 19 (1988)

7.59 M. Hervieu, B. Domengès, B. Raveau, M. Post, W.R. McKinnon, J.M. Tarascon: Mater. Lett. 8, 73 (1989)

7.60 M. Hervieu, B. Domengès, B. Raveau, J.M. Tarascon, M. Post, W.R. McKinnon: Mater. Chem. Phys. 21, 181 (1989)

7.61 D. Bourgault, M. Hervieu, S. Bouffard, D. Groult, B. Raveau: Nucl. Instrum. Methods B 42, 61 (1989)

7.62 D. Groult, M. Hervieu, N. Nguyen, F. Studer, M. Toulemonde: Defects and Diffusion Forum 57-58, 391 (1988)

7.63 D.A. Thompson: Rad. Effects 56, 105 (1981)

7.64 G.J. Clark, F.K. Legoues, A.D. Marwick, R.B. Laibowitz, R. Koch: Appl. Phys. Lett. 51, 1462 (1987)

7.65 C. Houpert, M. Hervieu, D. Groult, F. Studer, M. Toulemonde: Nucl. Instrum. Methods B 32, 393 (1988)

7.66 G.J. Clark, A.D. Marwick, R.H. Koch, R.B. Laibowitz: Appl. Phys. Lett. 51, 139 (1987)

7.67 R.H. Koch, C.P. Umback, G.J. Clark, P. Chaudhari, R.B. Laibowitz: Appl. Phys. Lett. 51, 200 (1987)

7.68 G.J. Clark, A.D. Marwick, F. Legoues, R.B. Laibowitz, R. Koch, P. Madakson: Nucl. Instrum. Methods B 32, 405 (1988)

7.69 O. Meyer, B. Egner, G.C. Xiong, X.X. Xi, G. Linker, J. Geerk: Nucl. Instrum. Methods B 39, 628 (1989)

7.70 G.C. Xiong, H.C. Li, G. Linker, O. Meyer: Phys. Rev. B 38, 240 (1988)

7.71 A.E. White, K.T. Short, D.C. Jacobson, J.M. Poate, R.C. Dynes, P.M. Mankiewitch, W.J. Skocpol, R.E. Howard, M. Anzlowar, K.W. Baldwin, A.F.J. Levi, J.R. Kwo, T. Hsieh, M. Hong: Phys. Rev. B 33, 3755 (1988); Mater. Res. Soc. Symp. Proc. 99, 531 (1988)

7.72 B. Egner, J. Geerk, H.C. Li, G. Linker, O. Meyer, B. Strehlau: Jpn. J. Appl. Phys. 26-3, 2141 (1987)

7.73 M.O. Ruault, H. Bernas, J. Lesueur, L. Dumoulin, M. Nicolas, J.P. Burger, M. Gasgnier, H. Noel, P. Gougeon, H. Potel, J.C. Levet: Europhys. Lett. 7, 435 (1988)

7.74 W.C. Mosley: J. Am. Ceram. Soc. 54, 475 (1971)

7.75 O. Hauser, M. Schenk: Phys. Status Solidi 18, 547 (1966)

7.76 J. Geerk, G. Linker, O. Meyer, C. Politis, F. Ratzel, R. Smithey, B. Strehlau, G.C. Xiong: Z. Phys. B 67, 507 (1987)

7.77 J.M. Valles Jr., A.E. White, K.T. Short, R.C. Dynes, J.P. Garno, A.F.J. Levi, M. Anzlowar, K. Baldwin: Phys. Rev. B 39, 11599 (1989)

7.78 G.P. Summers, E.A. Burke, D.B. Chrisey, M. Nastasi, J.R. Tesmer: Appl. Phys. Lett. 55, 1469 (1989)

7.79 B. Roas, B. Hensel, S. Henke, G. Saemann-Ischenko, L. Schultz, S. Klaumünzer, B. Kabius, W. Watanabe, K. Urban: Europhys. Lett. 11, 669 (1990)

7.80 B. Hensel, B. Roas, S. Henke, R. Hopfengärtner, M. Lippert, J.P. Ströbel, M. Vildič, G. Saemann-Ischenko, S. Klaumünzer: Phys. Rev. B 42, 4135 (1990)

7.81 V. Hardy, D. Groult, M. Hervieu, J. Provost, B. Raveau, S. Bouffard: Nucl. Instrum. Methods B 54, 472 (1991)

Chapter 8

8.1 J. Labbé, J. Bok: Europhys. Lett. 3, 1225 (1987)

8.2 J. Friedel: J. Phys 48, 1787 (1987)

8.3   C. Martin, D. Bourgault, C. Michel, M. Hervieu, B. Raveau: Mod. Phys. Lett **3**, 93 (1989)
8.4   T. Rouillon, R. Retoux, D. Groult, C. Michel, M. Hervieu, J. Provost, B. Raveau: J. Solid State Chem. **78**, 322 (1989)
8.5   F. Baudelet, G. Collin, E. Dartyge, A. Fontaine, J.P. Kappler, G. Krill, J.P. Itie, J. Jegoudez, M. Maurer, Ph. Monod, A. Revcolevschi, H.T. Tolentino, G. Tourillon, M. Verdaguer: Z. Phys. B **69**, 141 (1987)
8.6   D.O. Sarma, C.N.R. Rao: Solid State Commun. **65**, 47 (1988)
8.7   F. Studer, D. Bourgault, C. Martin, R. Retoux, C. Michel, B. Raveau, E. Dartyge, A. Fontaine, G. Tourillon: Physica C **159**, 609 (1989); and Mod. Phys. Lett. B **3**, 1085 (1989)
8.8   R. Retoux, F. Studer, C. Michel, B. Raveau, A. Fontaine, E. Dartyge: Phys. Rev. B **41**, 193 (1990)
8.9   B. Raveau, C. Michel, J. Provost, M. Hervieu: Solid State Ionics **40/41**, 785 (1990) (Proc. 7th Int'l Conf. on Solid State Ionics, Hakone, Jpn, Nov. 1989)
8.10 P. De Rango, B. Giordanengo, R. Tournier, A. Sulpice, J. Chaussy, G. Deutscher, J.L. Genicon, P. Lejay, R. Retoux, B. Raveau: J. Physique **50**, 2857 (1989) (special issue dedicated to J. Friedel)
8.11 B. Raveau, C. Michel, M. Hervieu, J. Provost: Physica C **153-155**, 3-8 (1988)
8.12 R.J. Cava, B. Batlogg, J.J. Krajewski, L.W. Rupp, L.F. Schneemeyer, T. Siegrist, R.B. Van Dover, P. Marsh, W.F. Peck, P.K. Gallagher, S.H. Glarum, J.H. Marshall, R.C. Farrow, J.V. Waszczak, R. Hull, P. Trevor: Nature **336**, 211 (1988)
8.13 M. Hervieu, C. Michel, N. Nguyen, R. Retoux, B. Raveau: Eur. J. Solid State Inorg. Chem. **25**, 1 (1988)
8.14 R. Retoux, C. Michel, M. Hervieu, N. Nguyen, B. Raveau: Solid State Commun. **69**, 599 (1989)
8.15 R. Retoux, C. Michel, M. Hervieu, B. Raveau: Mod. Phys. Lett. B **3**, 591 (1989)

Subject Index